T0413707

The Internet of Things

The Internet of Things

Foundation for Smart Cities, eHealth, and Ubiquitous Computing

Edited by
Ricardo Armentano
Robin Singh Bhadoria
Parag Chatterjee
Ganesh Chandra Deka

CRC Press
Taylor & Francis Group
Boca Raton London New York

CRC Press is an imprint of the
Taylor & Francis Group, an **informa** business

A CHAPMAN & HALL BOOK

CRC Press
Taylor & Francis Group
6000 Broken Sound Parkway NW, Suite 300
Boca Raton, FL 33487-2742

International Standard Book Number-13: 978-1-4987-8902-8 (Hardback)

Library of Congress Cataloging-in-Publication Data

Names: Armentano, Ricardo, editor.
Title: The Internet of things : foundation for smart cities, eHealth and ubiquitous computing / editors, Ricardo Armentano [and 3 others].
Other titles: Internet of things (Armentano)
Description: Boca Raton : CRC Press, 2018. | Includes bibliographical references and index.
Identifiers: LCCN 2017016990| ISBN 9781498789028 (hardback : alk. paper) | ISBN 9781315156026 (ebook)
Subjects: LCSH: Internet of things. | Ubiquitous computing. | Medical informatics.
Classification: LCC TK5105.8857 I56 2018 | DDC 004.67/8--dc23
LC record available at https://lccn.loc.gov/2017016990

Visit the Taylor & Francis Web site at
http://www.taylorandfrancis.com

and the CRC Press Web site at
http://www.crcpress.com

Contents

Section I Introduction to the Internet of Things: Definition and Basic Foundation

Section II Frameworks for the Internet of Things: An Architectural Perspective

Section III Interdisciplinary Aspects of the Internet of Things

Section IV Future Research, Scope, and Case Studies of the Internet of Things and Ubiquitous Computing

Preface

The Internet of Things (IoT) has emerged as a massive technology, which discovers new fields that touch life especially in building protocols for smart cities and health care. It poses a new trend of IoT, which thrives to proffer personalized services to the users by developing a ubiquitous pool of services. Identifying these game-changing approaches of IoT has led us to write this book, which would showcase a concrete state-of-the-art research in these fields.

Research in the field of IoT cannot be confined to a specific area; it is enabled by the handshaking of several domains of research such as sensors, networking, cloud computing, edge computing, big data, machine learning intelligence, security, and privacy. This makes the domain of this book quite interdisciplinary where researchers from more than 15 countries in the world have shared their leading research in the field of IoT. All of the 21 chapters showcase the best research outcomes of the authors, having paramount significance in the field of IoT.

Smart cities hold one of the key sectors of development led by IoT. Ranging from smart transports to efficient management of citizens' data and highway safety measures, IoT has revolutionized every aspect for the development of a city. With a two-fold approach, this book discusses the core concepts of IoT as well as different case studies related to application for building smart cities with health care. We hope that this book would be a perfect companion for all the researchers in IoT. The chapters on eHealth showcase interesting research on developing comprehensive smart health care models.

With the advent of IoT, there has been a humongous increase in the amount of data generated by the devices. However, this gives more power to the systems, as the more these data are analyzed, deeper insights into the users are obtained. Some chapters have also discussed these aspects of data in IoT.

This book has the following four sections:

- Introduction to the Internet of Things: Definition and Basic Foundation
- Frameworks for the Internet of Things: An Architectural Perspective
- Interdisciplinary Aspects of the Internet of Things
- Future Research, Scope, and Case Studies of the Internet of Things and Ubiquitous Computing

Ricardo Armentano

Robin Singh Bhadoria

Parag Chatterjee

Ganesh Chandra Deka

Editors

Ricardo Armentano works as a distinguished professor at the Favaloro University, Buenos Aires, Argentina. In addition, he is the director of the PhD program (signal and image processing) at the National Technological University, Buenos Aires, Argentina and a Grade 5 Investigator of the URU/84/002 United Nations Program for the Development of Basic Sciences. He received his engineering degree in 1984 and by the end of 1994 he earned his first PhD in physiological sciences from the University of Buenos Aires, Argentina. In 1999, he earned his second PhD degree from Université de Paris VII Denis Diderot, Paris, France. He has acquired international recognition in the field of cardiovascular hemodynamics and arterial hypertension and has profound research and teaching experience in the fields of cardiovascular dynamics applied to an extensive area of engineering in medicine and biology. He has more than 200 publications, including books, book chapters, and peer-reviewed articles. Armentano is the vice-chair of the Global Citizen Safety and Security Working Group of the International Federation of Medical and Biological Engineering (IFMBE). He is a member of the EMBS IEEE since 1985 and chair of the Argentinean Chapter of the Engineering in Medicine and Biology Society (EMBS), IEEE (2004). He was the chair of the 32nd International Conference of the EMBS IEEE Buenos Aires 2010 and was the Latin-American IEEE EMB representative for 2015.

Robin Singh Bhadoria earned his PhD from Indian Institute of Technology Indore, India. He has worked in different fields such as data mining, frequent pattern mining, cloud computing, service-oriented architecture, and wireless sensor networks. He has published more than 50 research articles in several international and national conferences and journals of high repute such as IEEE, Elsevier, and Springer that also includes book chapters. Presently, he is serving as the associate editor of *International Journal of Computing, Communications and Networking* (*IJCCN*) ISSN 2319-2720. In addition, he is serving as an editorial board member for different journals of the world. Presently, he is a professional member of different professional research bodies such as IEEE (U.S.), International Association of Engineers (IAENG) (Hong-Kong), Internet Society, Virginia (U.S.), and International Association of Computer Science and Information Technology (IACSIT) (Singapore).

Parag Chatterjee works as a researcher at the UTN–GIBIO (Research and Development Group in Bioengineering) and as a professor at the postgraduate Department of Information Systems Engineering in the National Technological University (*Universidad Tecnológica Nacional*), Buenos Aires, Argentina. In addition, he is a visiting scientist to the University of the Republic (*Universidad de la República*), Montevideo, Uruguay. In 2015, Chatterjee received his MSc degree with a first class in computer science from the St. Xavier's College, Kolkata (University of Calcutta), India. His research has been published in several international journals and was presented at IEEE International Conferences and sessions of the Indian Science Congress. In addition, he delivered keynote lectures and invited talks at highly reputed institutions such as the Indian Institute of Science (IISc Bangalore, India), Jain University (India), Favaloro University (Argentina), University of the Republic (Uruguay), and many international conferences. He has cochaired sessions at international conferences such as SICC-2017 in Rome, and delivered talks and moderated workshops during the IoT Week 2017, Geneva. Besides serving as an editorial board member and reviewer of several international journals, Chatterjee is a member of international scientific bodies such as Indian Science Congress Association, Machine Intelligence Research Labs, and IEEE. His current research is focused on the Internet of Things applied to health care.

Ganesh Chandra Deka works as the deputy director of training, under the Directorate General of Training and Employment, Ministry of Skill Development and Entrepreneurship, Government of India, New Delhi. His research interests include big data, cloud computing, data mining, NoSQL databases, and vocational education and training. He has published more than 50 research papers in various conferences, workshops, and international journals of repute, including IEEE and Elsevier. He is editor-in-chief of the *International Journal of Computing, Communications, and Networking* (ISSN 2319-2720). So far, he has organized eight *IEEE International Conferences* in India as the technical chair. He is the coauthor for four textbooks on fundamentals of computer science. He has edited five books (3 by IGI Global, U.S. and 2 by CRC Press/ Taylor & Francis Group) in the field of cloud computing, big data and NoSQL.

Contributors

Jemal Abawajy
Parallel and Distributed Computing
  Laboratory
School of Information Technology
Deakin University
Burwood, Victoria, Australia

Nor Fadzilah Abdullah
Faculty of Engineering and Built
  Environment
Universiti Kebangsaan Malaysia
Bangi, Malaysia

Matts Ahlsen
CNet Svenska AB
Stockholm, Sweden

Raphael Ahrens
User-Centered Ubiquitous Computing
  Department
Fraunhofer FIT
Sankt Augustin, Germany

Charilaos Akasiadis
Institute of Informatics and
  Telecommunications
National Center for Scientific Research
  "Demokritos"
Athens, Greece

Mimonah Al Qathrady
Computer Engineering
University of Florida
Gainesville, Florida

Alexandre Alapetite
Data Science and Engineering Lab
Alexandra Institute
Copenhagen, Denmark

Kelechi Anabi
Faculty of Engineering and Built
  Environment
Universiti Kebangsaan Malaysia
Bangi, Malaysia

K.V. Arya
ABV-Indian Institute of Information
  Technology and Management, Gwalior
Gwalior, India

Mathias Axling
CNet Svenska AB
Stockholm, Sweden

Bojan Bakmaz
Faculty of Transport and Traffic Engineering
University of Belgrade
Belgrade, Serbia

Miodrag Bakmaz
Faculty of Transport and Traffic
  Engineering
University of Belgrade
Belgrade, Serbia

Carlo Bellecci
Department of Industrial Engineering
University of Rome Tor Vergata
Rome, Italy

Bidyut K. Bhattacharyya
Department of Electrical Engineering
National Institute of Technology Agartala
Agartala, India

Zoran Bojkovic
Department of Electrical Engineering
University of Belgrade
Belgrade, Serbia

Dario Bonino
Pervasive Technologies
Istituto Superiore Mario Boella (ISMB)
Turin, Italy

Jose Angel Carvajal Soto
User-Centered Ubiquitous Computing
    Department
Fraunhofer FIT
Sankt Augustin, Germany

Leandro Cymberknop
National University of Technology
Buenos Aires, Argentina

Upena Dalal
Department of Electronics and
    Communication Engineering
Sardar Vallabhbhai National Institute
    of Technology, Surat
Surat, India

Maria Teresa Delgado
Pervasive Technologies
Istituto Superiore Mario Boella (ISMB)
Turin, Italy

Meera Dhabu
Department of Computer Science and
    Engineering
Visvesvaraya National Institute of
    Technology
Nagpur, India

Aradea Dipaloka
Department of Informatics Engineering
Faculty of Engineering
Siliwangi University
Tasikmalaya, Indonesia

Diego Dujovne
School of Informatics and
    Telecommunications
Diego Portales University
Santiago, Chile

Nitesh Funde
Department of Computer Science and
    Engineering
Visvesvaraya National Institute of
    Technology
Nagpur, India

Pasqualino Gaudio
Department of Industrial
    Engineering
University of Rome Tor Vergata
Rome, Italy

Sara Ghanavati
Parallel and Distributed Computing
    Laboratory
School of Information Technology
Deakin University
Burwood, Victoria, Australia

Francesco Gilardi
Department of Biomedicine and
    Prevention
University of Rome Tor Vergata
Rome, Italy

Thomas Gilbert
Data Science and Engineering Lab
Alexandra Institute
Aarhus, Denmark

Sadik Kamel Gharghan
College of Electrical and Electronic
    Engineering Techniques
Middle Technical University
Baghdad, Iraq

Shirsha Ghosh
Department of Computer Science
    and Engineering
National Institute of Technology
    Arunachal Pradesh
Yupia, India

Joyeeta Goswami
Department of Computer Science and
  Engineering
National Institute of Technology
  Arunachal Pradesh
Yupia, India

Ahmed Helmy
Computer and Information Science and
  Engineering (CISE) Department
University of Florida
Gainesville, Florida

Davood Izadi
Parallel and Distributed Computing
  Laboratory
School of Information Technology
Deakin University
Burwood, Victoria, Australia

Marco Jahn
User-Centered Ubiquitous Computing
  Department
Fraunhofer FIT
Sankt Augustin, Germany

Rahul Jichkar
Persistent Systems Ltd.
Nagpur, India

Ayesha Khan
Infocepts Technologies Pvt. Ltd.
Nagpur, India

Awanish Kumar
Department of Biotechnology
National Institute of Technology Raipur
Raipur, India

Jacopo Maria Legramante
Department of Medicina dei Sistemi
University of Rome Tor Vergata
Rome, Italy

Diogo Ferreira Lima Filho
Department of Electronic Systems
  Engineering
University of São Paulo
São Paulo, Brazil

Khalid Lmuzaini
King Abdulaziz City for Science and
  Technology
Riyadh, Saudi Arabia

Alak Majumder
Department of Electronics and
  Communication Engineering
National Institute of Technology
  Arunachal Pradesh
Yupia, India

Andrea Malizia
Department of Industrial Engineering
and
Department of Biomedicine and
  Prevention
University of Rome Tor Vergata
Rome, Italy

Sandro Mancinelli
Department of Biomedicine and
  Prevention
University of Rome Tor Vergata
Rome, Italy

Arka Prokash Mazumdar
Department of Computer Science and
  Engineering
Malaviya National Institute of Technology,
  Jaipur
Jaipur, India

Mohit Mittal
Department of Computer Science
Gurukul Kangri University
Haridwar, India

Laura Morciano
Department of Biomedicine and
    Prevention
University of Rome Tor Vergata
Rome, Italy

Rosdiadee Nordin
Faculty of Engineering and Built
    Environment
Universiti Kebangsaan Malaysia
Bangi, Malaysia

Leonardo Palombi
Department of Biomedicine and Prevention
University of Rome Tor Vergata
Rome, Italy

Claudio Pastrone
Pervasive Technologies
Istituto Superiore Mario Boella (ISMB)
Turin, Italy

Jigisha Patel
Electronics and Communication
    Engineering Department
Sardar Vallabhbhai National Institute of
    Technology, Surat
Surat, India

Vishnu Prabhakaran
ABV-Indian Institute of Information
    Technology and Management, Gwalior
Gwalior, India

Siddharth S. Prasad
Discipline of Computer Science and
    Engineering
Indian Institute of Technology Indore
Indore, India

Edvin Ramadhan
Department of Informatics Engineering
Faculty of Computer Science
Sriwijaya University
Palembang, Indonesia

Miguel Arjona Ramírez
University of São Paulo
São Paulo, Brazil

A. Pravin Renold
Department of Electrical and Electronics
    Engineering
Velammal Engineering College
Chennai, India

Luis Romero
National University of San Juan
San Juan, Argentina

Peter Rosengren
CNet Svenska AB
Stockholm, Sweden

Shweta Shah
Electronics and Communication
    Engineering Department
Sardar Vallabhbhai National Institute of
    Technology, Surat
Surat, India

Avani Sharma
Department of Computer Science and
    Engineering
Malaviya National Institute of Technology,
    Jaipur
Jaipur, India

Pilli Emmanuel Shubhakar
Department of Computer Science and
    Engineering
Malaviya National Institute of Technology,
    Jaipur
Jaipur, India

Maurizio Spirito
Emerging Trends and Opportunities
Istituto Superiore Mario Boella (ISMB)
Turin, Italy

Evaggelos Spyrou
Institute of Informatics and
    Telecommunications
National Center for Scientific Research
    "Demokritos"
Athens, Greece

S.A. Srinivasa Moorthy
Andhra Pradesh Electronics and IT Agency
Visakhapatnam, India

Iping Supriana
School of Electrical Engineering and
  Informatics
Bandung Institute of Technology
Bandung, Indonesia

Kridanto Surendro
School of Electrical Engineering and
  Informatics
Bandung Institute of Technology
Bandung, Indonesia

Sándor Szénási
Neumann János Faculty of Informatics
Óbuda University
Budapest, Hungary

and

Faculty of Economics
Selye János University
Komárno, Slovakia

Grigorios Tzortzis
Institute of Informatics and
  Telecommunications
National Center for Scientific Research
  "Demokritos"
Athens, Greece

B. Venkatalakshmi
Department of Electronics and
  Communication Engineering
Velammal Engineering College
Chennai, India

S. Vijayakumar
Department of Electrical and Electronics
  Engineering
Velammal Engineering College
Chennai, India

Archana Vimal
Department of Biotechnology
National Institute of Technology
  Raipur
Raipur, India

Marcel Stefan Wagner
University of São Paulo
São Paulo, Brazil

Otilia Werner-Kytölä
User-Centered Ubiquitous Computing
  Department
Fraunhofer FIT
Sankt Augustin, Germany

Janusz Wielki
Department of E-Business and Electronic
  Economy
Opole University of Technology
Opole, Poland

Wagner Luiz Zucchi
University of São Paulo
São Paulo, Brazil

Section I

Introduction to the Internet of Things
Definition and Basic Foundation

1

The IoT Vision from an Opportunistic Networking Perspective

Zoran Bojkovic, Bojan Bakmaz, and Miodrag Bakmaz

CONTENTS

1.1 Introduction

Currently, the Internet of Things (IoT) is still in the stages of development and deployment. However, IoT will have an important impact on everyday life, just like Internet has today. IoT allows us to interact and control systems as well as environments by creating digital representation of the physical world. It is expected that IoT will have a huge influence to the development not only on industry but our society as a whole. Differences in traffic characteristics along with the energy constraints and specific features of the IoT communication environment are the motivation for research activities.

Inspired by the vision of Nikola Tesla "when wireless is perfectly applied the whole earth will be converted into a huge brain, which in fact it is, all things being particles of a real and rhythmic whole," this chapter deals with IoT from the opportunistic networking perspective. Starting from the summation of the technologies that are giving fundamentals for understanding IoT conception, this chapter continues with software-defined IoT (SD-IoT), as a new solution. In software-defined networking, the control intelligence is moved from data plane devices (switches, routers) and implemented in a logically centralized controller that interacts with data plane through standard interfaces. In accordance with this, SD-IoT decouples the control logic from functions of the physical devices through a logically centralized controller that manages the devices via standard interface. Questions that often arise are how to eliminate bottlenecks, with the idea to process the IoT data and on the other hand, not to place a strain on the network. Fortunately, there exists a solution in Internet traffic routing. Next, we deal with interaction of cloud computing and IoT, which can provide intelligent participation and connection not only from rational, but also machine-type communications. Special requirements go to interactions with sensors. In connection with this are storage big data, huge computation power, and real-time audio and video streaming. Big data analytics, as one of the trending research topics in science and technology communities when dealing with IoT vision from opportunistic networking perspective, is presented. One of the important challenges in IoT vision is how to move to real-time control for smart and connected communities (SCC). Some opportunities of IoT in SCC are identified as mobile crowdsensing and cyber-physical cloud computing. In the final part of this chapter, information-centric networking (ICN) will be invoked. Improved data dissemination efficiency and robustness in communication scenarios indicate the high potential of ICN as a model that is different from the traditional IP-centric one. This approach consists of the retrieval of content by names, regardless of origin server location, application, and distribution channel, thus enabling in-network caching/replication and content-based security.

1.2 IoT-Enabling Communications Technologies

The main task of the IoT is to enable ubiquitous objects to coordinate together. In this way, the traditional objects become smart by exploiting pervasive computing, embedded devices, and underlying communications technologies and protocols. Thus, world's economy and quality of everyday life will be significantly improved. On the other hand, emerging technologies and different applications have to match the market demands as well as customers' needs. Devices are offered to fit the requirements of anywhere and anytime service availability. IoT-enabling technologies and protocols contain the Internet and IP-based standards that will be fundamental in providing a common, well-paved road for the development and deployment of new IoT applications (Al-Fuqaha et al., 2015).

IoT devices can be either stationary (e.g., home smart meters, thermostats, traffic signalization, etc.) or mobile (e.g., fleet management devices, eHealth sensors, etc.). They can be connected to the core Internet using either wired or wireless access links. Although the wired solutions can provide higher data rates, reliability, security, and lower latency, they may not be suitable for many IoT applications due to its cost, lack of scalability, and mobility support. Wireless access networks can be either capillary/short range (e.g., ZigBee, Bluetooth, Wi-Fi, etc.) or cellular (e.g., high speed packet access [HSPA], long term evolution [LTE], etc.). Capillary solutions, mainly used for shared short-range links, are rather cheaper to roll out and generally scalable. However, small coverage, weak security, severe interference, and lack of universal infrastructure pose restriction on their applications. On the other hand, mobile systems offer wide coverage, seamless connectivity, acceptable security, and ready-to-use infrastructure, making cellular networking a promising solution for IoT applications. After decades of conceptual inception of the IoT, a variety of communication technologies has gradually emerged, reflecting a large divrsity of application domains and communication requirements.

1.2.1 Short-Range Communications Technologies

IEEE 802.15.4 is generally characterized by low data rate, high message throughput, low power consumption, and low cost. Also, it provides a reliable and secure communication, but without quality of service (QoS) guarantees. This standard is the base for the well-known ZigBee protocol, as they both focus on offering low data rate services over power-constrained devices, and they build a complete network protocol stack for wireless sensor networks. The IEEE 802.15.4 Task Group has defined star and mesh network topologies to suit different application requirements. These topologies consist of different devices that can be classified as full function device (FFD) and reduced function device (RFD). The FFD includes all features of the IEEE 802.15.4 standard and can serve as a regular device or as a personal area network coordinator/controller. Also, the FFD can serve as a relay in a mesh topology or as an end node with limited functions. On the other hand, RFD implements the minimum required functionalities by the standard and can only function as an end node such as home automation sensor.

In 2010, the Bluetooth Special Interest Group published their Core Specification 4.0, for Bluetooth Low Energy (BLE) technology. BLE utilizes a short-range links with a minimal amount of power (0.01 to 10 mW) and lower energy consumption compared with its previous versions. Almost all the latest versions of mobile platforms have been working to provide native support for BLE and are already everywhere in the market. Its coverage is 10 times that of the classic Bluetooth, whereas its latency is 15 times shorter. Compared with ZigBee, BLE is more efficient in terms of energy efficiency, and with these characteristics it is a good candidate for certain IoT applications (Chang, 2014). Implementation of BLE nodes in the IoT environment requires invoking IPv6, which was not originally supported by this standard. Usage of BLE for Internet applications poses challenges beyond IPv6 packet transport, including gateway operation, as well as application protocol efficiency and security. A holistic solution for enabling and optimizing IPv6 over BLE is proposed by Nieminen et al. (2014). Recently, the Bluetooth SIG announced release 5.0, which will provide significantly increased speed, data-broadcasting capacity, and operation range. Extending range will provide robust, reliable IoT connections that make full-home and building and outdoor use cases a reality.

The IEEE 802.11ah Task Group was formed in 2010 to define an amendment to the 802.11 standard that operates in the sub-1-GHz license-exempt frequency spectrum to support the following three use cases: sensors and meters, backhaul sensor and meter data, and extended range Wi-Fi. The sensors and meters use case includes sub-use cases such as smart grid, environmental monitoring, eHealth, home/building automation, and so on. They all fit into IoT use cases that typically support a much larger number of devices per access point for both indoor and outdoor environments with a longer transmission range but at lower data rates than the use cases and requirements of conventional IEEE 802.11 (Park, 2015). IEEE 802.11ah (a.k.a. HaLow) is designed to support outdoor applications that need a transmission range up to 1 km at 150 kb/s. For indoor applications, the long-range capability can be used for energy-efficient sensor communications by using a very low transmit power. IEEE 802.11ah provides energy-efficient medium access control protocols and frame formats that are optimized for IoT applications, which need to support a large number of nodes and infrequent small packet transmissions. The basic characteristics of prominent short-range wireless technologies for IoT use cases are summarized in Table 1.1.

6LoWPAN (IPv6 over Low-Power Wireless Personal Area Networks) protocol was originally designed as an adaptation layer between IPv6 and 802.15.4 that provides a set of mechanisms including fragmentation (RFC 4944), header compression (RFC 6282), and

TABLE 1.1

Characteristics of Short-Range Wireless Technologies for IoT

Standard	IEEE 802.15.4	Bluetooth LE	IEEE 802.11ah
Frequency band (MHz)	868/915/2,400	2,400	900
Data rates (Mb/s)	0.25	1	347
Coverage (m)	30	100	1,000
Number of nodes	65,000	6,000	6,000

optimized IPv6 neighbor discovery (RFC 6775). Recently, its application was expanded over a variety of other radio networking technologies including BLE and IEEE 802.11ah. 6LoWPAN offers all the advantages from the IPv6 such as ubiquity, scalability, flexibility, and end-to-end connectivity to the IoT environment.

1.2.2 Cellular Technologies and IoT

Although today all cellular technologies can support more or less some of the IoT applications; native support for massive communications is expected with development of future generation of mobile systems, (i.e., fifth generation [5G]). Practically, in order to realize the vision of IoT, emerging mobile networks need to provide connectivity to thousands of devices, whereas, for example, standard LTE systems were originally designed to support up to 600 connected users per cell. It is obvious that each connected thing will transmit small data blocks sporadically. Current cellular architectures are not designed to simultaneously serve the aggregated traffic generated from a large number of devices. For instance, current systems could easily serve five devices at 2 Mb/s each, but not 10,000 devices each requiring 1 kb/s (Boccardi et al., 2014).

Apart from higher capacity, low latency is another challenging issue for future cellular systems. Generally, IoT comprises a vast amount of real-time applications with extremely low-latency requirements. Motivated by the tactile sense of the humans, which can distinguish latencies on the order of 1 ms, 5G can be applied to managing and control scenarios, implying a disruptive change from today's content-driven communications. As stated by Fettweis (2012), a 1-ms round-trip time for a typical tactile interaction requires a budget of 100 ms on the physical layer. This is far shorter than current cellular systems allow, missing the target by nearly two orders of magnitude.

1.3 Software-Defined IoT

It is well known that Internet has got great success, but it still shows some lacks. For example, control is implemented by various routing and management protocols and is embedded in every router/switch being hard to change. In that way, Internet infrastructure is developing slowly, whereas vendor-dependent interfaces make the networking management very complex and unreliable. In order to support sustainable evolution, efficient management, and heterogeneous application requirements, IoT architecture should avoid these problems. It is clear that a change is needed in how networks are designed and operated. Software-defined networks (SDNs) offer the ability to address the abovementioned problems (Vaughan-Nichols, 2011). The main idea of SDNs integration in the IoT infrastructure is to extend software-defined approach from network devices to sensor platforms and the cloud. This solution will provide well-defined service application programming interfaces (APIs) in terms of data acquisition, transmission, and processing.

1.3.1 Benefits of Software-Defined Networking

SDN approach supports the changing nature of future networks while taking advantage of savings in equipment investments and operating costs. This will give operators the agility to create or program highly flexible and dynamic networks capable of integrating and monitoring terminals, intelligent machines, and smart devices while providing new services.

In an attempt to overcome the inconvenience of vertical integration in today's networks, which means that the control and data planes are bundled together; SDN control logic is separated from the underlying routers and switches (Zilberman et al., 2015). In this way, the centralization of network control is promoted while increasing flexibility. SDN can be considered to utilize a common and defined interface between a control plane and date plane. The control plane is used for overall coordination, for example, routing and error recovery, whereas the data plane usually manages packet-by-packet operations. Comparison between functionality of the legacy router in classical network and the equivalent SDN architecture is shown in Figure 1.1.

The separation of control plane from data plane would be realized through an API between SDN switch agent and controller. Several processing functions will be carried out such as buffering while headers are processed, header examination to identify switching action, and finally queuing packet for transmission. On the other hand, the control plane manages the mechanisms by which the data-plane forwarding tables are created. This includes running one or more routing protocols as well as exchanging local routing information with other routers, together with deriving a local forwarding table. SDNs hold great potential for application in optical networks (Channegowda et al., 2013), 5G mobile networks (Cho et al., 2014), direct communications (Zhou et al., 2016), and in IoT environment (Liu et al., 2015).

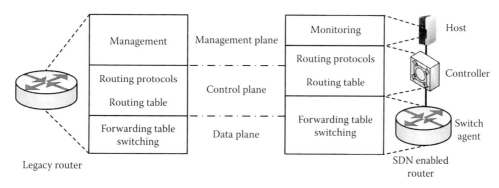

FIGURE 1.1
Functionality of legacy router and equivalent SDN architecture.

1.3.2 Software-Defined IoT Architecture and System Design

SD-IoT architecture proposed by Liu et al. (2015) consists of physical infrastructure layer, control layer, and application layer, as presented in Figure 1.2.

Physical infrastructure layer includes various physical devices, that is, sensors, actuators, gateways, access points, routers, and servers. The physical devices transmit data from one node to another and process them to extract required information. They interact with control layer through standard interfaces, that is, a southbound interface. The control layer acts between the infrastructure layer and application layer, managing the physical devices with various functions through different southbound interfaces. Also, this layer provides services to the application layer through APIs known as northbound interfaces. The control layer provides data acquisition, transmission, and processing service. The data acquisition service provides APIs for applications to specify their data requirements. The controller obtains the required data from sensor platform. Data specification contains general attributes: data type, targeted geographical areas, and time duration. In SD-IoT, sensor platform is equipped with more than one sensor shared by many applications. The sensor controller has a global overview of the underlying sensor platform. The network is used to transmit data from the sensor platforms to servers in the cloud. To realize data transmission, the network uses an SDN architecture. The controller steers packets to different destinations and schedules traffic to satisfy application requirements for network quality and optimize the usage of network resource. Data-processing APIs allow applications to specify the required resources that include running submitted programs on specific platform and deploying corresponding software entities. Finally, in application

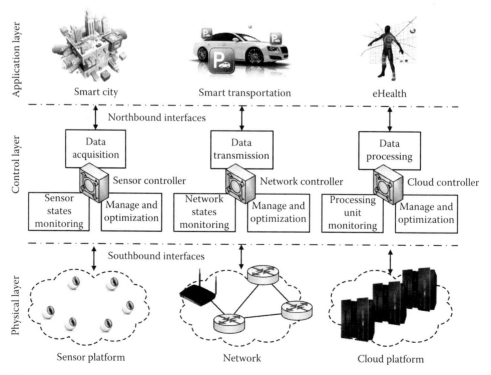

FIGURE 1.2
SD-IoT architecture.

layer, developers build sensing applications using provided APIs. They can customize data acquisition, transmission, and processing. As the infrastructure is shared by multiple applications, the overall costs are reduced.

In SD-IoT system design, some open problems related to the southbound interface, control layer, and so on often arise. By providing processing and transmission procedure in the sensor platform, the sensor controller is decoupled from sensor platforms. To reduce energy consumption at the sensor platform, middleware software can be implemented in the controller. In order to decrease the frequency of interactions with sensor platforms when they are not active, the controller is used. High scalability, performance, and robustness can be achieved with a design of a logically centralized control layer. In order to achieve these objectives, deploying multiple controllers is an obligation. The network controller in SD-IoT provides end-to-end QoS-guaranteed data transmission. Anyhow, some challenges need to be emphasized. Namely, the forwarding devices have a limited number of queues for QoS provisioning. Thus, it is difficult to support a great number of QoS requirements. Next, it is a challenging issue to design efficient traffic-scheduling algorithms to satisfy QoS requirements in a large-scale network environment. Ahmed and Boutaba (2014) proposed a vertical approach to organize multiple controllers for wide-area SDNs, which can also be extended to the control of sensor platforms and clouds.

1.3.3 Transition to Cloud Computing

It is well known that most objects in physical world are tiny not only in size but also in capacity. In order to monitor these types of objects, a great number of sensors with the goal to generate a huge amount of data to be stored and processed successfully are required. IoT itself is not capable of storing, computing, or analyzing data collected from millions and millions of sensors. Fortunately, this weakness is the potency of cloud computing in which the storage and computation can be regarded as infinite. Thus, the IoT–cloud integration is welcome. By providing infinite storage and computation, a shared pool of resources becomes dynamically allocated and obtained when using any of IoT applications.

As an example, consider the importance of the cloud controller. In SD-IoT system, this entity needs to decide how to map application service requests to the physical devices. The constraints include server and storage capacity, as well as the type of software or platform. If the cloud allows applications to store data and after that to rent virtual resources to process them, the cloud controller needs to decide where to store the data and which servers should be used to host virtual resources for post-data processing.

1.4 Cloud Computing and IoT

As a paradigm that is envisioned to involve all things in the world, IoT can never be observed and treated as an isolated island. Its potential can never be fully explored before the complete formation of cyber-physical space, where humans, computers, and smart objects are pervasively interconnected. As a result of this process, the integration of IoT with the existing communications systems and the corresponding enterprise networks and cloud computing can be achieved. Social and mobile networks are included, too. Recently, the integration of heterogeneous communications systems has become a main source of networking innovation and has stimulated the proposition of novel interdisciplinary concepts such as the Web of Things (Christophe et al., 2011), Social IoT (Atzori et al., 2011), and Cloud of Things (CoT) (Distefano et al., 2012).

Starting from the fact that cloud computing can have virtually unlimited capabilities in terms of storage and processing power, it represents a model for accessing a set of shared and configurable resources that include networks, servers, storage facilities, and so on (Mell and Grance, 2011). Therefore, IoT can be abstracted of its limitations regarding heterogeneity, connectivity, and security by cloud computing integration. Although conceptual analyses are abundant in the open literature, cloud computing and IoT integration are still a challenging issue. Considering that objects are huge in number is a challenging task because they are largely dynamic, their life spans vary, and the provided applications require different QoS supports.

Taking into account relations between IoT and cloud computing, two different approaches exist: bringing the cloud to the things and vice versa (Cavalcante et al., 2016). The possibility of bringing the cloud to the things understands taking advantages offered by cloud services. In that way, constraints of IoT such as storage, processing, energy consumption are taken into account. This is of importance due to the energy limitation for the IoT devices. To overcome this limitation, such devices could play the simple role of data providers and send data to be processed and stored directly on the cloud. In the second case, bringing the things to the cloud, it becomes a layer between the things and applications. In that way, CoT can be referred to everything as a service (XaaS), which represents a flexible and scalable model, offering anything that could be consumed as cloud services. An example of this approach is the concept of Sensing as a Service (SenaaS), which can virtualize, share, and reuse sensing data to be ubiquitously consumed from the cloud.

1.4.1 Cloud of Things

The concept of IoT, with underplaying physical objects abstracted according to think-like semantics, seems a valid starting point for orchestration of the various resources. In this context, the cloud could play the central role to connect IoT with the Internet of People through the Internet of Services. This new concept of pervasive and ubiquitous computing can be addressed as the CoT, going beyond the interconnection and hyperlink of things (Petrolo et al., 2015). It is based on the SenaaS model of IoT architecture, which involves four conceptual layers (Perera et al., 2014): sensor and sensor owners' layer, sensor publishers, extended service providers, and sensor data consumers. First layer consists of sensors, whereas the owners manage them and allow or not the publication in cloud. Sensor publishers detect available sensors and communicate with the sensor owners. They also get permission to publish sensors in cloud. In order to select required sensors, extended service providers communicate with multiple sensor publishers. Finally, for consuming sensed data, sensor data consumers need to be registered. The benefits of invoking this model are sharing and reusing of sensed data, reduction of data acquisition cost, and collect data previously unavailable.

However, there are some challenges that CoT needs to deal with such as heterogeneity in sensor types, protocols, and communication technologies, as well as interoperability among different hardware and cloud solutions, and so on. In order to bridge the gap between the disparate technologies, defining the abstraction layer is an obligation. The technologies promote interoperability between IoT resources, information models, data providers, and consumers, as well as simplify effective data access and integration.

In IoT, processing, storing, and representing big data are provided using the CoT conception. Generally, some questions such as verifying, filtering, and analyzing data still remain to be solved. One of the reasons is not only the lack of corresponding standards, but also the heterogeneity of involved technologies. Second, the amount of generated data requires the improvement as well as optimization of integration issues.

1.4.2 Big Data Analysis

Big data is the term that describes the large volume of data, both structured and unstructured, that inundates our life on a day-to-day basis. These data can be analyzed for insights that lead to better decisions and strategic business moves. Big data and the trending research in communications technology with great applications such as economy, climate, health, social science, data mining, machine learning, and so on are continuing to grow everyday using managing software tools. This will be possible owing to supporting the networking and computers complexity when real-time services are required.

General architecture of big data analysis is shown in Figure 1.3. Volume, velocity, and variety are common characteristics of big data (Lu et al., 2014). As it is not recommended to move collected data for centralized storing, distributed big data is proposed, whereas big data should be processed in parallel. In this way, new knowledge and innovation can be mined in a reasonable amount of time. Big data processing is divided into two groups: intraprocessing and interprocessing. In intra-big data processing, all data belong to the same organization. On the other hand, if big data are part of different organizations, it will be inter-big data processing. It should be emphasized that inter-big data processing will be more challenging because big data sharing will first be executed before processing. Also, many new security and privacy issues will arise during the big data sharing. The following problems such as storage of high-volume data, process streaming, and analysis of real-time data are identified.

Big data and IoT have been combined to work together. Namely, big data is one of the promising solutions for IoT data management. Recently, a next-generation operational data historian system is proposed by Huang et al. (2014). The main characteristic of that system is being designed for long-time storage and analyze data, as well as process real-time queries. Here, the system combines the advantages of a rational model and a time series data model to support throughput and fast queries. First of all, it is of importance for sampled data to be packaged forming regular time series, which are compressed. Also, irregular time series, as well as hybrid grouping structures according to data characteristics, have to be taken into

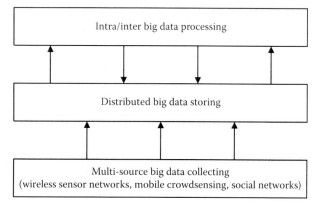

FIGURE 1.3
General architecture of big data analysis.

consideration. In that way, compared with traditional databases, data-processing performance in this system becomes improved by one to two orders of magnitude.

Two main challenges for big data analytics are data heterogeneity and decision-making. Big data analytics enable to move from IoT to real-time control, important from many SCC applications.

First of all, two applications of IoT in SCC are identified: mobile crowdsensing and cyber-physical cloud computing. Mobile crowdsensing serves as a relay for data collection from large number of mobile sensing devices (Ganti et al., 2011). Based on the type of shape being measured or mapped, mobile crowdsensing applications are categorized into three groups: environmental, infrastructural, and social. Compared with the classical wireless sensor networks, mobile crowdsensing advantages include significantly more storage, computation, and networking resources equipped with multimodality sensing capabilities. Also, mobile devices avoid the cost and time of deploying large-scale sensor networks. Next, IoT represents a networking infrastructure for cyber-physical systems (CPSs). These are smart networked systems with embedded sensors and actuators designed to sense and interact with physical world including the human users. Moreover, CPSs support real-time, guaranteed performance in safety-critical applications. A new computing paradigm that can modify and rapidly build these systems is cloud computing. Cyber-physical cloud computing is characterized by efficient use of resources, modular composition, rapid development, and scalability. Cyber-physical cloud computing is of great significance for numerous SCC applications such as smart grid, smart health care, and smart disaster management.

The data relevant to SCC are heterogeneous by nature. Thus, it is a challenge how to transform physical, biological, or social variables into an electrical signal. Taking into account, massy data and multisource data streams are also needed. In order to successfully deal with data heterogeneity, the questions that often arise are related to quality and accuracy improvement, unified data representation and processing, interpretation and interoperability improvement, knowledge creation, and short-term and long-term storage conduction (Sun et al., 2016).

As for decision-making, there are many sources of uncertainty that have to be considered in SCC (facing cities is incomplete, lack the data needed to specify the boundary conditions with sufficient accuracy, etc.). This process can be improved under uncertainty by understanding assessment, representation and propagation of uncertainty, developing robust optimization methods, and designing optimal sequential approach.

1.5 Information-Centric Networking toward Efficient IoT Environment

The large-scale deployment of IP-based IoT solutions still provides challenging issues, regardless of the great efforts and valuable achievements of standardization bodies, (e.g., Internet Engineering Task Force [IETF], Institute of Electrical and Electronics Engineers [IEEE], etc.). The general constraints and challenges are related to the limited expressiveness of addressing, complex mobility and multicast support, and opportunistic massive access under the strict reliability and security requirements.

Simultaneously with ongoing development of IoT environment, research community is currently promoting a new communication paradigm that is fundamentally different from traditional IP-centric networking. Having in mind the fact that the Internet is increasingly used for information dissemination, rather than for peer-to-peer communication, this revolutionary paradigm aims to reflect future needs better than the traditional architecture (Ahlgren et al., 2012). Significant approach of several future Internet research activities is known as ICN. In this concept, instead of accessing and manipulating information bypassing of servers hosting them, appealing of the named information object in the center of networking, from the viewpoint of information flow and storage, is applied.

Defining network service in terms of named information objects, independently of where and how they are stored or distributed, caching information storage becomes the fundamental part of network infrastructure. As a result, an efficient and application-independent large-scale information distribution is enabled (Ahlgren et al., 2011). Here, a user request can be sent directly to the network, independently of considering the original content location.

During the last decade, several prospective ICN approaches have been proposed in the open literature. Although characterized by different design issues, they are sharing common core principles, objectives, and architectural properties that can be summarized as follows (Rao et al., 2014): content-based naming, name-based routing (NBR) and name resolution, and in-network caching.

Considering the main principles of ICN (i.e., unique location-independent naming, content-based security, in-network cashing), this revolutionary paradigm can match a wide set of IoT applications that are information-centric in nature (Amadeo et al., 2016). It is obvious that ICN names can directly address heterogeneous IoT contents and services independent from the location, especially in the presence of mobility. Moreover, by caching data closer to the user, ICN can significantly reduce traffic load, delay, and massive access to resource-constrained devices. Consequently, scientific community is considering IoT integration with ICN within the Internet Research Task Force (Pentikousis et al., 2015) and through recently published research works (Baccelli et al., 2014; Amadeo et al., 2015). However, ICN deployment in IoT environment is still in discussion phase.

Generally, ICN can be deployed as overlay over the existing infrastructure or as clean-slate solution directly over physical layer instead of IP. Bearing in mind the fact, that most of IoT devices are resource-constrained, overlay solution seems to be unsustainable due to its complexity and overhead as a result of overlay management and encapsulation inside traditional protocols. On the other hand, a clean-slate solution can easily be deployed if there is no need to maintain backward compatibility or to provide global connectivity. A most likely short-term solution would allow coexistence with IP-based technologies. Similarities between ICN hierarchical names and uniform resource identifiers of web resources could facilitate such coexistence. The translation between them may be implemented easily in the node (e.g., gateway) interconnecting ICN islands and the rest of the Internet. In what follows, some challenging issues regarding ICN and IoT integration will be analyzed.

1.5.1 Content-Based Naming

In ICN, instead of specifying a host pair for communication, a piece of information itself is named, addressed, and matched independently of its location, being located anywhere in the network. In that way, moving from the host-naming model to the information-naming model contributes that the information retrieval becomes receiver driven as an interfering implication and benefit. In ICN, there is no possibility for data to be received unless it is explicitly requested by the receiver. As for the network, it is responsible for locating the best source that will provide the desired information after sanded request. Both hierarchical and flat names are possible, with the former appearing as uniform resource identifier-like identifiers with variable lengths, whereas the latter comprises unique fixed-length identifiers without semantic structure. As routing entries for contents might be aggregated, hierarchical nature helps mitigate the routing scalability. On the other hand, flat naming aggravates the routing scalability problem due to the absence of aggregation.

Moreover, security and privacy are closely related to the naming scheme. In the case of hierarchical naming, security-related information is embedded into a specific field of the content unit, thus requiring a public key infrastructure for integrity checks. Flat name principle instead enables the self-certifying namespaces, allowing integrity checks without the need for a public key infrastructure.

Naming IoT content enables information to be structured into scopes and allows users to specifically request the desired content. This flexibility exploits the higher addressing potential of ICN, allowing a name in the IoT context to identify not only content but also a service or a device function. By offering name resolution at the network layer and forwarding content by its name, ICN also has the potential to reduce the signaling overhead in IoT environment. ICN nodes have the ability to identify requests for the same named information, avoiding the need to forward them differently on the same path. In addition, content becomes cached in traversing nodes, allowing requests to be satisfied by the first available copy, preventing source overquerying, and supporting connectionless scenarios. Moreover, in ICN, data can be transmitted to multiple consumers by supporting native multisource dissemination.

An ICN-naming scheme for IoT should be highly expressive and scalable. As stated before, flat names are typically obtained through hash algorithms applied to existing contents and can hardly be assigned to dynamic and unpublished contents. On the other hand, hierarchical names facilitate dynamic on demand contents (e.g., a parameter measured by a sensor), provided that naming conventions have been specified during the system configuration. On account of that, hierarchical naming has been mainly considered in the literature in order to support such properties (Baccelli et al., 2014; Shang et al., 2014). The basic idea is to define a hierarchy of name segments that identifies the IoT application (e.g., building automation systems, smart city) and the attributes that describe the related contents/services (Figure 1.4).

However, hierarchical naming entails length constraints, for instance, to fit the maximum payload size of some protocols. In parallel, variable-length names make line-speed name lookup extremely challenging. Especially under large-scale scenarios, naming schemes should be designed together with processing techniques (e.g., name component encoding) that accelerate name lookup (Zhang et al., 2015). This is of practical importance for

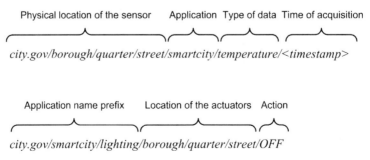

FIGURE 1.4
Examples of hierarchical naming for smart city application.

safety-critical applications (e.g., smart transport and eHealth) in which content access latency reduction is crucial. By sharing a common name prefix for multiple contents/ services, hierarchical names scale better than flat names, as they facilitate the definition of name aggregation rules, which is of importance for big data analytics. In IoT, the name prefix can identify application types, physical locations, or other macro-categories that broadly identify groups of data and services (Amadeo et al., 2016).

1.5.2 Information Distribution

Efficient information distribution should make use of any available data source to reduce network load and latency and increase information availability. ICN architectures have to solve the problem of how to retrieve data based on a location-independent identifier. There are two major solutions to this problem: NBR and name resolution system (NRS). The main difference between these two solutions is in the fact that the first one involves a path from the provider to the requesting host, whereas the second is going to involve matching an information name to a provider or any source that can supply that information. These two functions can be integrated or independent. In the case of integrated functions, the information request is routed to the provider, which subsequently sends the information to the requesting host by utilizing the reverse path over which the request was forwarded. The independent approach is characterized by the name resolution function, which does not determine or restrict the path that is used on the provider–subscriber route. Whether it is a NBR or a NRS approach, a list of common desirable properties are identified by Bari et al. (2012) as follows:

- ICN routing mechanism should provide low-latency network level primitive operations for content (original, replica, or cached) registration, metadata update, and deletion. It should be able to route a content request to the closest copy, based on some network metric.
- Message propagation for name resolution and retrieval should not leave the network domain that contains both the source and the content.
- The routing mechanism should provide guarantees on discovery of any existing content, regardless of the content's popularity and replication level.
- As the number of contents is in the order of trillions, any NBR/NRS scheme needs to scale to at least this many contents and possibly beyond to accommodate future growth. The trade-off between routing stretch (routing path length and minimum length path ratio) and routing table size needs to be analyzed, while keeping in mind the huge number of names and physical limitations imposed by memory technologies.
- Ideally, the content retrieval process should be a one-step process, either by combining name resolution and routing in a single step or by completely eliminating the name resolution part.

Offered by ICN for content discovery, NBR and NRS schemes may suit specific IoT scenarios, mainly depending on the content characteristics (e.g., popularity, dynamic generation) and network features (e.g., infrastructureless vs. infrastructured). NBR, coupled with data delivery performed by maintaining some soft-state at each Interest forwarder, is suitable to access popular contents in infrastructured scenarios, whereas it is the only viable solution in isolated networks with an alternately available infrastructure or ad-hoc environment. Its inherent benefits are as follows:

- Robust and resilient retrieval through adaptive forwarding coupled with in-network caching proposed by Amadeo et al. (2014).
- Easy resource discovery in infrastructureless networks through direct interdevice communication by means of Interest packet broadcasting analyzed by Baccelli et al. (2014).

The downside of deploying NBR is mainly related to the growth of the forwarding table size and routing updates in the case of a huge number of names, and the overhead of maintaining the soft state. However, name-prefix aggregation can successfully cope with such challenges, together with adaptive forwarding. NRS is useful in infrastructured scenarios with unpopular and popular content. It can also be beneficial with off-path caching, that is, an alternately connected IoT source can push data in a predefined always-on location (e.g., the cloud) accessible by consumers. In these cases, deploying a global name resolution service based on a hierarchically organized distributed hash tables is desirable, whereas additional scalability properties for name lookup can be obtained by using data center capacities in the cloud, as suggested by Xylomenos et al. (2014). Moreover, NBR and NRS solutions may complement each other, and together application of cloud computing, multilevel distributed hash tables, name-prefix aggregation, and adaptive forwarding, an effective discovery and delivery platform can be provided with the potential to scale even for a huge number of IoT resources.

1.5.3 In-Network Caching

In order to improve the content distribution and network utilization, an important approach of ICN is to provide ubiquitous and transparent in-network caching process (Zhang et al., 2013). Although caching is not a wholly new technique, the lack of a unique identification of identical objects makes it incompletely utilized in the current Internet architecture. As mentioned earlier, ICN names content in a unified, consistent, and network-aware way. This characteristic feature makes caching a general and transparent service, independent of applications.

In the context of in-network caching, the crucial problems that arise are related to the selection of the storage locations and of the specific content items to store in the different caches. Generally speaking, there exist two opposite approaches that can be distinguished for addressing such problems: coordinated and uncoordinated. The main property of coordinated case is that routers exchange information to achieve a better estimation of the contents popularity, avoiding at the same time the storage of too many content copies. In the uncoordinated case, each cache operates autonomously using transparent route-caching policy. Here, each router in the end-to-end path decides whether to cache transiting pieces of content and which cache-replacing policy without interacting with other routers to perform. In the cases where the popularity of objects does not change frequently, the least frequently used approach would achieve the highest performance, whereas the most proposed solutions adopt the least recently used replacing policy (Kim and Yeom, 2013).

It is of interest to point out that the effectiveness of in-network caching is higher when coordinated solution is applied. Also, it should be noted that the overhead required to manage the coordination between caches may become extremely high. Consequently, time uncoordinated solutions are now adopted more frequently. As a perspective solution,

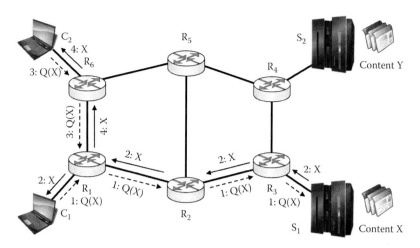

FIGURE 1.5
An example of in-network caching concept.

some trade-offs between the coordinated and uncoordinated approaches are proposed by Sourlas et al. (2012) and Chai et al. (2013).

As an illustrative example of in-network caching concept, the network shown in Figure 1.5 is considered. Suppose that consumer C_1 sends an interest message requesting a piece of content X published by server S_1. Consequently, the request message Q(X) traverses routers R_1, R_2, and R_3. Server S_1 receives the interest message issued by consumer C_1 and forward the desired piece of content X. Content data will traverse nodes R_3, R_2, and R_1 back, which will cache local copies. Now, assuming that consumer C_2 generates an interest message for the same piece of content, the corresponding request message Q(X) will traverse routers R_6 and R_1. The router R_1 can perceive that a copy of X existing in its cache and there is no need to forward the interest message to the next hop toward S_1. Instead R_1 sends the content X to consumer C_2 only via router R_6.

In-network caching acquires special significance in IoT environment. Generally, caching and related operations can be quite ineffective in terms of processing and energy consumption. However, in-network caching is beneficial because it speeds up data retrieval and increases its availability. It can reduce the number of lossy hops toward the source of information by limiting the network load and the overall energy consumption. In addition, caching is viable as data generated by IoT devices typically have a small size and a short lifetime.

Overall, IoT data can be cached in network routers and resource-constrained devices by implementing caching decision and replacement policies that account for the peculiarities of IoT traffic, for example, compatibly with freshness requirements and device capabilities. Specifically, the indiscriminate *cache everything everywhere* approach is inefficient due to the high level of content redundancy and poor utilization of the available cache resources. Alternative caching decision policies (e.g., probabilistic) have been proposed by Vural et al. (2014) for more efficient usage of the available caching resources, alleviating the load of nodes and bandwidth consumption. Some of these policies address the space usage issue by also considering the content popularity for topology-related centrality metrics. These could also be feasible approaches for IoT, when the same content is requested by different applications. However, they are suitable for static content, which is not entirely the case for IoT, where content is usually transient.

1.6 Concluding Remarks

This chapter focuses on some most relevant networking solutions for IoT realization that will play a vital role in the development of future communications systems. Reaching a level of maturity that allows for practical realization of IoT, the enabling technologies will be useful in many applications. This will help us to clear uncertainty that prevents a massive adoption. Application innovation will be accelerated when using software-defined architecture. In that way, not only control and physical infrastructure management but also cloud computing will be invoked as sustainable solution to overcome or alleviate several challenges faced by the IoT paradigm. Moreover, information-centric approach is advocated as inspiring networking solution for future Internet design, as well as IoT environment as its indispensable part. Proposed opportunistic networking vision is a good start, although many open challenges still remain. Also, designing better infrastructures and services for the future smart communities remain open questions for the future work.

References

Ahlgren, B. et al., 2011. Content, connectivity, and cloud: Ingredients for the network of the future. *IEEE Communications Magazine* 49 (7): 62–70.

Ahlgren, B. et al., 2012. A survey of information-centric networking. *IEEE Communications Magazine* 50 (7): 26–36.

Ahmed, R. and Boutaba, R., (2014). Design considerations for managing wide area software defined networks. *IEEE Communications Magazine* 52 (7): 116–123.

Al-Fuqaha, A. et al., 2015. Internet of things: A survey on enabling technologies, protocols, and applications. *IEEE Communications Surveys & Tutorials* 17 (4): 2347–2376.

Amadeo, M. et al., 2014. Named data networking for IoT: An architectural perspective. *Proceedings of the EuCNC*. Bologna, Italy, pp. 1–5.

Amadeo, M. et al., 2015. Information centric networking in IoT scenarios: The case of a smart home. *Proceedings of the IEEE International Conference on Communications*. London, GB: 648–653.

Amadeo, M. et al., 2016. Information-centric networking for the internet of things: Challenges and opportunities. *IEEE Networks* 30 (2): 92–100.

Atzori, L., Iera, A., and Morabito, G., 2011. SIoT: Giving a social structure to the internet of things. *IEEE Communications Letters* 15 (11): 1193–1195.

Baccelli, E. et al., 2014. Information centric networking in the IoT: Experiments with NDN in the wild. *Proceedings of The ACM ICN'14*. Paris, France, pp. 77–86.

Bari, M.F. et al., 2012. A survey of naming and routing in information-centric networks. *IEEE Communications Magazine* 50 (12): 44–53.

Boccardi, F. et al., 2014. Five disruptive technology directions for 5G. *IEEE Communications Magazine* 52 (2): 74–80.

Cavalcante, E. et al., 2016. On the interplay of internet of things and cloud computing: A systematic mapping study. *Computer Communications* 89–90: 17–33.

Chai, W.K. et al., 2013. Cache "Less for More" in Information-centric networks (extended version). *Computer Communications* 36 (7): 758–770.

Chang, K.-H., 2014. Bluetooth: A viable solution for IoT? [Industry Perspectives]. *IEEE Wireless Communications* 21 (6): 6–7.

Channegowda, M., Nejabati, R., and Simeonidou, D., 2013. Software-defined optical networks technology and infrastructure: Enabling software-defined optical network operations. *IEEE/OSA Journal of Optical Communications and Networking* 5 (10): A274–A282.

Cho, H.H. et al., 2014. Integration of SDR and SDN for 5G. *IEEE Access* 2: 1196–1204.

Christophe, B. et al., 2011. The web of things vision: Things as a service and interaction patterns. *Bell Labs Technical Journal* 16 (1): 55–61.

Distefano, S., Merlino, G., and Puliafito, A., 2012. Enabling the cloud of things. *Proceedings of the 6th (IMIS)*. Palermo, Italy, pp. 858–863.

Fettweis, G., 2012. A 5G wireless communications vision. *Microwave Journal* 55 (12): 24–36.

Ganti, R.K., Ye, F., and Lei, H., 2011. Mobile crowdsensing: Current state and future challenges. *IEEE Communications Magazine* 49 (11): 32–39.

Huang, S. et al., 2014. The next generation operational data historian for IoT based on informix. *Proceedings of the ACM SIGMOD*. Snowbird, UT: 169–176.

Kim, Y. and Yeom, I., 2013. Performance analysis of in-network caching for content-centric networking. *Computer Networks* 57 (13): 2465–2482.

Liu, J. et al., 2015. Software-defined internet of things for smart urban sensing. *IEEE Communications Magazine* 53 (9): 55–63.

Lu, R. et al., 2014. Toward efficient and privacy-preserving computing in big data era. *IEEE Network* 28 (4): 46–50.

Mell, P. and Grance, T., 2011. The NIST definition of cloud computing. *Recommendations of the National Institute of Standards and Technology*. Gaithersburg, MD.

Nieminen, J. et al., 2014. Networking solutions for connecting bluetooth low energy enabled machines to the internet of things. *IEEE Network* 28 (6): 83–90.

Park, M., 2015. IEEE 802.11ah: Sub-1-GHz license-exempt operation for the internet of things. *IEEE Communications Magazine* 53 (9): 145–151.

Pentikousis, K. et al., 2015. Information-centric networking: Baseline scenarios. *IRTF Information-Centric Networking Research Group (ICNRG) RFC 7476*.

Perera, C. et al., 2014. Sensing as a service model for smart cities supported by internet of things. *Transactions on Emerging Telecommunications Technologies* 25 (1): 81–93.

Petrolo, R., Loscri, V., and Mitton, N., 2015. Towards a smart city based on cloud of things, a survey on the smart city vision and paradigms. *Transactions on Emerging Telecommunications Technologies* (early view).

Rao, K.R., Bojkovic, Z.S., and Bakmaz, B.M., 2014. *Wireless Multimedia Communication Systems: Design, Analysis, and Implementation*. Boca Raton, FL: CRC Press.

Shang, W. et al., 2014. Securing building management systems using named data networking. *IEEE Network* 28 (3): 50–56.

Sourlas, V. et al., 2012. Autonomic cache management in information-centric networks. *Proceedings of the IEEE/IFIP NOMS 2012*. Maui, HI, pp. 121–129.

Sun, Y. et al., 2016. Internet of things and big data analytics for smart and connected communities. *IEEE Access* 4: 766–773.

Vaughan-Nichols, S.J., 2011. OpenFlow: The next generation of the network? *Computer* 44 (8): 13–15.

Vural, P. et al., 2014. In-network caching of internet-of-things data. *Proceedings of the IEEE ICC 2014*. Sydney, NSW, pp. 3185–3190.

Xylomenos, G. et al., 2014. A survey of information-centric networking research. *IEEE Communications Surveys & Tutorials* 16 (2): 1024–1049.

Zhang, G., Li, Y., and Lin, T., 2013. Caching in information centric networking: A survey. *Computer Networks* 57 (16): 3128–3141.

Zhang, Y. et al., 2015. ICN based architecture for IoT-requirements and challenges. *IETF Internet Draft*.

Zhou, M. et al., 2016. Design and implementation of device-to-device software-defined networks. *Proceedings of the IEEE ICC*. Kuala Lumpur, Malaysia, pp. 1–6.

Zilberman, N. et al., 2015. Reconfigurable network systems and software-defined networking. *Proceedings of the IEEE* 103 (7): 1102–1124.

2

DEFINE—An Architecture Framework for Designing IoT Systems

S.A. Srinivasa Moorthy

CONTENTS

2.1 Introduction

The easiest way to describe an embedded system is to compare it with a computer. A typical computer includes subsystems such as CPU, hard drives, display, and keyboard. It can be graphically described as in Figure 2.1.

An embedded system is similar to a computer, but it is not so big. The CPU and all the peripherals can be accommodated in a single printed circuit board as illustrated in Figure 2.2.

Embedded systems have started pervading in our daily life, and their dominance is increasing with the passage of time. The embedded systems that are used by us in day-to-day life are interconnected and can send data from the activities they monitor or can

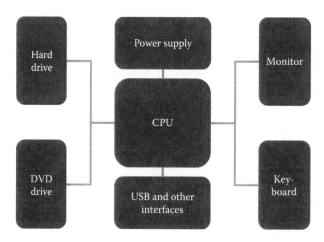

FIGURE 2.1
Block diagram of a typical computer.

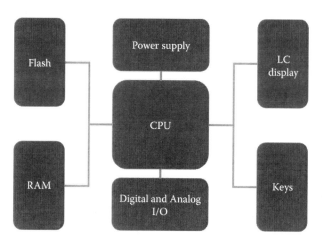

FIGURE 2.2
Block diagram of typical embedded system.

communicate to a central system that is called the *Internet of Things* (*IoT*). This central system could be a server or a cloud platform.

One of the best examples of IoT in our daily life is the set-top box that we use to watch cable TV. A set-top box starts sending information to the central sever about the channels that we watch and the programs we frequently watch. This enables broadcasters to tailor the programs for the individual users (an option that is available today only in IP TV!).

IoT is a collection of *Things* (which are basically embedded systems of various sizes and capability) that interact with a real environment with an infrastructure mechanism to connect them to the central repository through communication links, and a central storage and process mechanism (Cloud with applications to process the data collected).

As the applications of IoT start expanding, one of the biggest challenges that the designers face is the development of *Things* (embedded systems) to work in different environments, to be used by different users, and to work with different communication infrastructures. Designing systems with so many combinations is a bit difficult. Major challenge faced by most designers is designing a Thing (a.k.a. embedded system) that can work unattended 24 × 7 without any human intervention and also send data to the central location. The ability to design a system like this calls for skills that spread across every aspect of product design covering electrical, software, mechanical, manufacturing, and compliance. Although expertise for individual skills may be available for implementation, their availability during the early stage of product design is always a challenge for small companies and startups. Second, access to these skills at the early stage is also an expensive proposition. So, a tool or a framework to help in the early stage design will be really beneficial to the designers in the long course. This chapter describes one such framework called *DEFINE* that enables the designers to develop the system architecture in all aspects by clearly understanding the challenges involved in the Thing (System) design.

Functionally, all the embedded systems are very similar in architecture with the main CPU, associated peripherals, and the connected sensors. These are controlled by dedicated software called *Firmware* running on them to do the necessary functions. However, for an embedded system to work successfully and reliably in different applications scenario, many factors need to be addressed.

Normally, new product ideas come from needs or problems and many a time, the user having the need will not be a technical person. These needy users reach out to a designer to solve their problem. So the solution to the problem will be quite abstract in the beginning, and starting the design with such abstract inputs will definitely be very challenging. In addition, most startups and small companies will not be having all the skills required for a complete product design, especially skills such as enclosure design, manufacturing, test engineering, and so on. This forces them to reach out to external experts for help at the very beginning of the project. In such situations, the designers will be only validating their concepts, and a firm decision to design a product would not have been taken. Involving experts at this stage will be expensive. This reiterates the need for a tool or a framework for the designers to do the initial design inexpensively.

This chapter will discuss a framework that will help the designers to develop the initial design in a robust way. Before discussing the framework, it is important that designers understand the concept of *Product Lifecycle* (*PLC*). A brief introduction of a typical PLC is described to help the users in understanding the activities involved in each phase or stage in the development of a product. A typical PLC starts with the *Concept* stage and progresses toward the final *Disposal* stage. Figure 2.3 illustrated the typical life cycle of an electronic product. The PLC has six distinct phases/stages starting from concept, design, engineering, new product introduction, manufacture, and finally to the disposal stage. Figure 2.3 clearly lists out the distinct activities in each stage. The key aspect of the PLC is the interdependence of the different activities as well as concurrent (parallel) activities in some stages.

With this understanding, let us see how the *DEFINE* framework will help the developers in designing a robust and successful product.

Dictionary meaning of framework is *the basic structure of something: a set of ideas or facts that provide support for something*. The *DEFINE tool* does exactly the same. It allows the developer to design a product from abstract inputs so that it can meet all needs of the user successfully. DEFINE framework essentially consists of elements and each element has characteristics. Every characteristic addresses a need of the product.

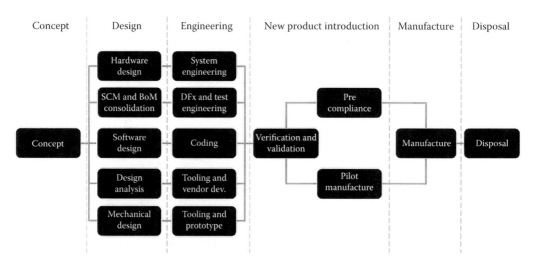

FIGURE 2.3
Modern product life cycle.

When all the elements and the characteristics are covered, the basic design requirement of a product can be easily derived.

As the basic requirements are captured completely through the *DEFINE* framework, the designers are able to have a solid design for the product.

2.2 DEFINE Framework—An Outline

Based on the years of experience and study as a designer and a product manager, the *DEFINE* framework was developed to help the designers to get their design (especially the architecture) right when they design embedded products or IoT—*Things*. With explosive growth of IoT, this framework will enable designers to quickly arrive at a cost-effective, reliable, and robust design that is able to meet the Time to Market target easily.

Basic foundation of this *DEFINE* framework is the elements that are critical for the working of *Things*. *Things* in a IoT network have to work in all kinds of environment and communication systems, with any type of user and usage (sometimes misuse) unlike the computers that typically work in a controlled and comfortable environment (most of the time, computers work in the same space where humans work, which are comfortable). This essentially forces the designers to develop a system that is robust, reliable, and easily usable. This is feasible only when the systems working from various aspects are well understood by the designers.

- *DEFINE* is a framework that enables the designers to understand the various aspects to develop a robust product in the easiest way.
- *DEFINE* is an intuitive framework and stands for six key elements that impact the performance of the systems.

  D   Domain
  E   Environment
  F   Functionalities
  I   Implementation
  N   Necessities
  E   Enablers

These six key elements when understood properly will enable the designer to design a robust, functional embedded system or a *Thing*.

Note: From now, Embedded System and Thing are interchangeably used for easy understanding.

This chapter would cover the details pertaining to each element and how they contribute to the success of the product being designed.

2.2.1 Domain

The first element in the framework is *Domain*. Classical meaning of Domain is—*A territory over which a rule or control is exercised* or *A sphere of activity, concern or function*. Domain defines the working environment of a product or a Thing. Based on the type of the domain, the product's characteristics will change the way it needs to be designed and manufactured.

Most electronic products can be classified under the following domains:

- Avionics
- Automotive
- Consumer electronics
- Industrial systems
- Medical devices
- Military
- Networking and telecom

There are six common characteristics that differentiate these domains from each other. Understanding and using these six characteristics will ensure that the product meets specific needs of any domain.

- First, product requirements will be captured in the right way by selecting the domain.
- Second, these six characteristics will form the basis for rest of the other requirements of the product.

Later in this chapter, we will discuss the interrelationship between the requirements and how these six characteristics impact them. Now, let us see the six characteristics of the domains and how they can be used to develop the product requirements in the following section.

2.2.2 Characteristics of Domains

The following six characteristics typically define the product usage, cost of the product, working environment, manufacturing process, and so on.

Six characteristics of the domains are as follows:

1. Environment in which the product is expected to work—*Operating environment.*
2. Who uses the product and how it is used?—*Usage of the product.*
3. What is product sensitivity to cost?—*Cost of the product.*
4. Process used for developing the product?—*Product development process.*
5. How long the customer will use the product?—*Product's life.*
6. What is the support needed for the product?—*Product support needed during its useful life.*

2.2.2.1 Operating Environment

Operating environment describes the actual working environment of a product. It includes parameters such as temperature, dust, humidity, and electrical power. However, these tend to be specific to the domains in which the products are to be used. For example, in the automotive domain, temperature of 120°C and power fluctuation from 2 to 16 V DC are very normal.

TABLE 2.1

Domain and Operating Temperature

Domain	Avionics	Automotive	Consumer	Industrial	Medical	Military	Telecom and Networking
Temperature variation	Extremely varying	Moderately varying	Regular	Moderately varying	Regular	Extremely varying	Moderately varying

Products need to be designed to withstand these types of variations. Defining the domain allows the designer to understand the characteristics of the environment in which the product has to work and factor them into the design.

Table 2.1 gives the domain and the temperature encountered.

2.2.2.2 Usage of the Product

Second characteristic is the usage of the product. This characteristic defines how the product will be used and who will be using the product. The focus is more on the user and how it has to be easy for a user to use the product. For example, let us consider the avionics domain. Typical aircraft controls and displays are almost consistent across different aircrafts. In addition, in most cases, other electronics subsystems also are common. This is essentially done to minimize and standardize the training needs for operations (for both pilots and maintenance engineers). Another example could be a basic health parameter measuring devices (devices for measuring temperature, BP, sugar, etc.). Their user interface has to be standard so that the users who have minimal exposure to different types of equipment can be trained very quickly.

Table 2.2 gives the list of all the usage parameters for each domain.

2.2.2.3 Product Cost

Although cost of the product looks as an easy parameter, it needs to be analyzed with the customer's perspective. Customers view the product's entire life and the money spent on operation and maintenance. Let us take an example of an Inkjet Printer the cost of which is generally based on the manufacturer's strategy to reach the market (they sell the printer cheap and sell the printer toner for a high cost). However, customers view these products from the point of recurring costs such as the consumption of paper and the cost of toner during the usage. Low initial cost and higher operating cost will result in product failure. Most designers tend to take the short-term view of a product but not its entire life, which results in a good product failing due to high operation and maintenance cost. At the same time, you have products in the medical field in which some of them are disposable, and the cost of disposable part will decide the success of the product. So, when the designer

TABLE 2.2

Domain and User Training Needs

Domain	Avionics	Automotive	Consumer	Industrial	Medical	Military	Telecom and Networking
Usage	Training needed	No training	No training	Partial training	Training needed	Training needed	Partial training

TABLE 2.3

Domain and the Cost of Products

Domain	Avionics	Automotive	Consumer	Industrial	Medical	Military	Telecom and Networking
Cost	Very high	Moderate	Low	Moderate	High	Very high	Moderate

chooses the domain, this characteristic will allow them to understand the cost implication on the product at a very early stage.

Table 2.3 gives the cost of different domain-based products for reference.

2.2.2.4 Product Development Process

For most of the readers, this will sound a little puzzling as to why development process is being discussed in the domain characteristics. Products are typically developed on the basis of the competence and the experience of the team. However, products that are used in the Medical and Avionics domain need to mandatorily follow standard processes as defined by the certifying agencies such as FAA (Federal Aviation Administration) and FDA (Food and Drug Administration). These two agencies regulate the products that are used in their respective markets. In addition, these products have to be approved by the agencies before they are released to the users. This type of approval is required even in industrial and automotive domains when safety of public life is involved. If the products are developed for any of these domains, the designer needs to factor the compliance to the process and the documentation needs that have to be submitted for the approval. The effort to develop these types of products is large especially the documentation alone, which can take up substantial time. Designers need to consider this when they develop the products.

2.2.2.5 Product Life

This is a parameter that defines the life of the product used by the customer. Once the product's life is over, customer disposes it and changes to a new one. Typically, life of a product is adversely impacted by the obsolescence of the components used in the products. With the trend of shorter life of electronic components, product that has to be used for a long period needs careful design. Today, the average life of a typical electronics component is in the range of 7–8 years. So, when the designer is developing a product, one of the key elements to be factored is the life expectancy of the overall product. Products used in aviation, medical, automotive, and industrial domains have a product life at any point between 12 and 18 years. Designers need to be careful in selecting the parts so that component obsolescence does not impact the product life. Without a structured components engineering process, selection of components can bring grief to the designers.

Table 2.4 gives the domain and the expected life with remarks.

TABLE 2.4

Domain and the Product's Life

Domain	Avionics	Automotive	Consumer	Industrial	Medical	Military	Telecom and Networking
Life	Very long	Long	Short	Long	Long	Very long	Long

Note: Very long—greater than 20 years, Long—between 10 and 15 years, Short—less than 5 years.

2.2.2.6 Product Support Requirement

Product support has direct correlation to the product life. A typical case is a product the life of which is going to be 15 years, but the components used in the product may be obsolete within 8–9 years of the first release. However, customers expect the manufacturer to support the product throughout its life. This essentially mandates the designers to factor the support for the product to its full life, into its design. This implies that the design has to address component obsolescence. This forces the designers to restrict the use components of very few suppliers who have the policy of offering drop in compatible upgrades for obsolete parts (Free scale was one vendor who practices this). Typically, development teams do not have the expertise of component engineering, and this results in the usage of wrong parts or shorter lifespan parts. One way to address this is to identify the contract manufacturer (CM) in the initial stage of design as most CMs have components engineering teams. Another approach is to replace the entire module that has obsolete parts with new parts that have longer life spans. However, this approach is expensive.

2.2.3 Environment

Second element in the *DEFINE* framework is the environment. This is the actual environment in which the product is actually used. In the domain section, we briefly discussed about environment associated with domain, but that was restricted with respect to the domain-specific parameters. Once we factor the domain, this section will help developers the ways to handle the impact of those parameters on the product in detail. There are six important characteristics that will impact the performance of any product by way of environmental issues. They are

1. Temperature
2. Humidity
3. Pressure
4. Dust
5. Fluid Ingress
6. Electrical (power, lightening, EMC/EMI, etc.)

2.2.3.1 Temperature

Temperature impacts any product in two ways. One is due to ambient (outside environment) temperature. The other is the temperature generated inside the product and its impact on the working of the product. Once the designer chooses the domain, outer contour of the temperature will be known and that will help the designer to understand the impact of ambient temperature on the product. However, the temperature inside the product which is due to the working of the components inside depends entirely on the design. It also has a bearing on the product enclosure design by way of the material used for packing of the product, efficiency of the circuit design, and thermal management. For example, let us assume that we are designing an automotive product in which the circuit will dissipate 30 W of power. Unless this heat is taken out of the system, product will have issue by way of heating. For example, if the developer up-front chooses a plastic enclosure, the decision can impact the performance of the product due to heating. So it is important that the designers understand the real working environment with respect to temperature and how it can be handled.

Table 2.1 helps in understanding the effect of temperature on products.

2.2.3.2 Humidity

Similar to temperature, humidity also plays a vital role in the design process. Humidity becomes critical if the *Thing or system* uses mechatronics parts or high voltage circuitry, in which presence of humidity may lead to arcing that may result in fire. Solution to humidity is a bit complicated, and it starts from the enclosure design, extending up to product manufacturing process. Humidity also mandates strict adherence of design rules (especially in the case of electrical interconnections) to ensure that there is no arcing due to presence of humidity. Sometimes, special processes such as Conformal Coating have to be used. Humidity is an unavoidable problem, and the solution is a bit complex and needs to be understood upfront in the design solution to be implemented.

2.2.3.3 Pressure

Third characteristic in the environment is the pressure. This is very critical in the case of IoT things as most of these will have sensors that depend on pressure for working (water, air, and other flow measurement). Atmospheric pressure also decides the heat dissipation from a product. As the altitude increases, atmospheric pressure decreases, which leads to less and inefficient heat transfer from the product. This gets aggravated further if the system uses fans for cooling, as performance of the fan deteriorates with the increase in altitude. So, it becomes critical to know the pressure of the locations in which the product has to work for the successful working of the product. IoT systems when deployed over a wide area have to work in different altitudes and locations, which make the pressure a critical parameter in development. It becomes even more critical if the sensors used are also dependent on pressure.

2.2.3.4 Dust Ingress

Dust is another factor that impacts all electronics systems. Dust impacts systems more with moving parts and the mechanical movement. If the *Thing* that is being designed has motors and other moving devices, then dust becomes the biggest threat to the product. Designing a dust proof product is highly challenging. It becomes critical when the system generates lots of heat. A system-generating heat needs ventilation by way of holes in the enclosure but detrimental to dust prevention. A good design should satisfy both the dust prevention and heat ventilation requirements, and the need has to be identified right in the beginning stage of the design process.

2.2.3.5 Fluid Ingress

Fourth characteristic in the environment is the fluid ingress. Fluid means all kinds of fluids from water to oil. If the products have components that are mechanical and optical (like cameras and lenses), fluid ingress becomes a critical issue. Printed circuit materials are sensitive fluids and can be damaged by fluids. For example, if designer is developing a *WaterFlow Meter* to measure the flow of water in a reservoir, the *Thing/System* will need to work in an environment that has lots of water. In such environment, the water vapor will impact the product adversely. Another situation could be when the system is installed in the open air and exposed to rain. So, the designers have to factor the fluid ingress in the system as a parameter, which will impact the design of the system enclosure.

2.2.3.6 Electrical Power

The final environment characteristic to be considered is the electrical power that powers the system (Thing). Conventional systems operate either on AC mains electrical power or DC power that is derived from AC mains and batteries (charged through mains and provides power to the system when there is no mains power). However, in the case of IoT Things, they may be installed in an area where there is no power, and they may need to work on solar or other nonconventional power supply. This criterion has to be identified upfront while designing the system, as finding a power solution later to the designed system is not an easy option. One of the key challenges in this scenario is considering the use of nonconventional energy. The system software has to play a vital role in managing the power, and adding power is never an easy job once the design is frozen.

Table 2.5 gives an interdependence matrix for the six parameters against the domains.

2.2.4 Functionality

Third key element in the *DEFINE* framework is the functionality. This element defines the core functions to be performed by the system (Thing). Six key characteristics that are to be captured or identified at the beginning of a successful product design are listed below:

1. Usability and ergonomics
2. Manufacturability
3. Product interconnect and interfaces
4. Product performance
5. Product safety
6. Product sustenance

TABLE 2.5

Domain and the Six Key Parameters

Domain	Avionics	Automotive	Consumer	Industrial	Medical	Military	Telecom and Networking
Temperature	High	High	Low	High	Low	High	High
Humidity	High	Moderate	Low	High	High	High	High
Pressure	High	Low	Low	High	Moderate	High	Moderate
Dust ingress	High	High	Low	High	High	High	Moderate
Fluid ingress	High	High	Low	Moderate	Moderate	High	High
Electric power	High	High	Low	Low	Low	High	Moderate

2.2.4.1 Usability and Ergonomics

This characteristic essentially covers two key interrelated elements of the product: the user and the usage. Usability defines how the user can learn and use the system. Ergonomics ensures that the product is convenient and safe to use. Usability describes the flow of actions while using the device and how the system interacts with the user. Ergonomics details the shape of the product, colors used, audio and display used by the system, and location of the system in the operating environment. Both these are key elements in ensuring that user gets the best comfortable experience while using the product. One of the best examples is the Point of Sale Terminal (which is a Thing) in which the buttons are used and the display plays a very crucial role when user interacts with the system.

2.2.4.2 Manufacturability

Biggest success factor for a product is the volume sale of the product. This depends mainly on how easily and cost effectively the product can be produced. Most designers think about manufacturability in the end of the design cycle. This invariably leads to a heavy price for the product's manufacturing, resulting in an increased product cost. Manufacturability becomes critical with increased use of CM. Designers have to plan for the product's manufacturing early on in the development. This is because most CMs have design rules for optimal cost and maximum yield in manufacturing. These have to be implemented in the design right from the beginning to meet the design objectives. Delaying the identification of manufacturing partner till the end of design will lead to reduced yield and escalated cost. Identification of the CM and using their design for x process (DFx—x stands for assembly, manufacture, test, cost, support, etc.) is critical for the successful manufacturing of the product. As very few established CMs tend to have a well-defined DFx process, evaluation of CM becomes a key aspect to be addressed as part of manufacturability.

2.2.4.3 Product Interconnect and Interface

Interconnection to the outside world is the key to success of any IoT product. The interconnect mechanism may vary from pure software protocols to power and data communication to IoT Thing. When a product uses subsystems such as power supply, communication modules, and so on, their interface is critical for the product to succeed. Second, if the product contains mechatronics items such as Solenoids, Print Heads, Motors, and so on, the power needs are different, and selection of right power supply (which is again a bought out subsystem) becomes critical. Most designers fail to address these types of issues and face problems later in the field when they install the products. Sometimes, the subsystems may be fine, but the interconnect assemblies consisting of the cable harness and connectors may be inadequate for the need and they have to be designed properly. In the wireless systems, software driving the communication port may have interconnect and interoperability issues at protocol level and need to be addressed in the design. Hence, it is important that the product interconnect issues are addressed properly at the start of the design.

2.2.4.4 Product Performance

This characteristic may initially seem to be bit abstract and generic. However, most electronic products tend to slow down in performance as their usage starts increasing. Most designers tend to look at the current needs of a product as the requirement instead of the longer life span of the product's whole life, while designing and developing the system. They later realize that the performance of the system starts failing as the usage of the system increases. Poor product performance may be because of low CPU speed, insufficient memory (both FLASH and RAM), bad power management, poorly written, and tested software. Even when all these aspects are well designed, poor installation can ruin the system's performance. In extreme cases, poor communication infrastructure could be the root cause for bad performance. This essentially means that the designers need to have a holistic view of the solution and ensure that performance parameters are defined and captured properly. For example, best performance measured for a small home router is the number of packets it can process per second. In a typical router design, the performance depends on the CPU memory, CPU speed, and the CPU and I/O Bus width (32 or 64 bits). By defining the performance figure, rest of the components can be derived.

2.2.4.5 Product Safety

Another key element is the safety of the product. When the system is installed and working, safety of the user and the environment is critical. Product safety could be right from safety of the user in terms of shape and size to electrical safety (by way of no electrical shock to the users). When the system uses chemicals for its functionality (which is common in medical and analytical instruments), it is important that the chemicals are safe to use, and any accidental emission or spillage will not cause any problems to the user or environment. Systems that use LASER and X-RAY need to be carefully designed so that they are safe to the user and the environment. When a system (Thing) is developed, designer needs to understand the safety aspects and also the type of users and design the system with adequate safety measures. System producing low-frequency sound is also a health hazard if exposed for long duration. Designers need to understand issues similar to these when designing the system. Most electronic packaging design rules cover safety aspect in great detail.

2.2.4.6 Product Sustenance

Most designers tend to think that once the product goes to the manufacturing line, their responsibility is over and completely overlooks the lifespan of the product and the user support. To accommodate product sustenance, designers have to factor the product's life and how it can be sustained even in the case of nonavailability of the original design team, components used as also the technology obsolescence. This is very critical in Avionics and Military equipment as their life extends up to 25 years and more. During the procurement, one of the critical conditions they need to evaluate is how the product will be supported by the manufacturer during the life of the product. This effectively depends on six factors that will ensure the product's long life without any support issues.

1. Component engineering—Ensuring parts used in the product have support for availability of upgrade.
2. Stable manufacturing technology—Product uses stable manufacturing technology that will stay and support the product manufacturing life.
3. Well-established Test and Calibration process that is well documented so that even if the original design team is no longer available, it can still be used for the support of product.

4. Systematic supply chain process that ensures all components obsolescence is well tracked. Availability of alternate strategy for component procurement that may include strategic stocking of parts in the case of obsolete parts through Time Buy mode.

5. Proper version controlled product documentation from the original design team and maintained using Product Life Cycle Management tools.

6. Established Vendor Development program for alternate vendors and distributed supplier base for continuous availability of parts.

2.2.5 Implementation

Fourth element in the *DEFINE* Framework is the Implementation. It covers the complete process sequence from the Design Phase to Support Phase of a product. As this element is the most critical element that decides the success or failure of the product, it should be given highest priority by the designers for a successful product.

There are six processes that describe Implementation and they are as follows:

1. Development process
2. Manufacturing process
3. Verification and validation process
4. Calibration process
5. Product support process
6. Documentation and control process

2.2.5.1 Development Process

Most designers tend to treat this step as a burden and ignore it in the product development effort. However, as the development matures and product goes into testing, they realize that not following Product Development Process leads to extra costs and also the failure of the product in many cases. Most developers get confused between Product Development Process with Certification Processes such as ISO9000 that demand a large amount of documentation. In reality, basic development process has a simple method for controlling the different versions of artifacts developed and how they are released to the field and manufacturing. If the product developed belongs to either medical or avionics (what is known as Regulated Industry product), it is mandatory that the development processes have to comply to FDA and FAA guide lines and certifications such as AS9100C and ISO135485 are mandatory. Product Development Process needs some amount of effort, and designers need to factor them in the development cycle to avoid future problems. They also need to stick to a simple development process for product success.

2.2.5.2 Product Manufacturing Process

With the increased trend toward CM also known as Electronic Manufacturing Services, Product Manufacturing Process has transformed and is different from conventional in-house manufacturing. When the products belong to either Avionics or Medical domain, rigor of the FDA/FAA mandated process becomes a necessity even in manufacturing. Most CMs follow a very strict process in their manufacturing line. Understanding them is important for the designers to deliver an easy-to-manufacture product. As most CMs operate on very thin margins, any additional work that they need to do include to make the product manufacture worthy will cost a lot. Essentially it means if the same job is done as part of the design will cost a lot less. In addition, regulated industry product manufacturing also mandates corresponding certifications for manufacturing, thereby increasing the cost. In the recent past, most automotive vendors do mandate their CM to follow ISO16949 certification for manufacturing, to ensure the quality of the parts produced. All these compliances add additional cost, and designers need to understand their implication well during the design cycle to ensure that the product is within the cost and performance norm set. In the manufacturing, the focus is based on tracking the components used, and tracking of the products produced so that their performance in the field can be monitored and any fault can be tracked back to the root cause of failure up to components level. Use of CM thus becomes rigid, which is different from the in-house manufacturing. Strict adherence to the process laid out by the CM/Electronic Manufacturing Services vendors is a must for the success of the product. This essentially forces the product designers to choose the CM early in the design cycle.

2.2.5.3 *Product Verification and Validation Process*

This is the most crucial part of product development step, and failure to do a proper verification and validation (V&V) leads to failure of the product in the field. Most designers think this process is the job of the manufacturing team by confusing it with testing. However, V&V is a process integrated into the product development. Regulated industry products need to have documented V&V process. The V&V process should include the test cases against every product requirement with both success and failures recoded for submission as formal document to the regulatory agencies. If this practice is followed, it will help the designers to develop a robust product, even if it is not mandatory. This will also result in having a product manufactured with very few problems in the field, leading to customer satisfaction and product success. Although the process sounds complex, in practice it is a very simple to-do-list that needs to be religiously followed, and all the test cases have to be executed with success in all the cases. Simplest technique for a comprehensive V&V is to create a test case for every product requirement. This step ensures that the product fairs well with very few problems when installed in the field and thereby reducing field repair expenses.

2.2.5.4 *Product Support Process*

Key to success of any product is the support that product gets after it is deployed in the field (bought by customers). Product Support process is not just the logistics associated with the product support, but it basically depends on how the product is designed to support quick and timely correction of the problem.
Some of the aspects that help in product support are as follows:

- Ability to upgrade the product software easily without a service engineer attending to it.
- If the product is connected through wireless network, how the Over The Air (OTA) update can be implemented.
- Resources, such as FLASH memory size and RAM, that are additionally needed for the remote product update.
- Ability to diagnose the product remotely and isolate the problem so that the support engineers can carry the right spares avoiding delay in solving the problem.

- When the product uses consumables or needs periodic replacements of sensors and other devices, how the support will be implemented in the system to forewarn the user as well as manufacturer about the impending replacement?

These issues have to be addressed right in the beginning of the design, so that product is supportable, and the resources needed for this are accounted in the product costing.

2.2.5.5 Product Documentation and Control

Although this may be covered under the product support, primarily it is kept as a separate element due to the following reasons:

- Regulated industry products and products with long life mandate detailed documentation.
- When the product is manufactured through CMs, documentation is the only key for success as the product may have multiple versions in the field, and they have to be maintained.

These are seen as luxuries by the designers, and they tend to avoid them but later when the product hits the market they struggle to meet the field needs and find it extremely difficult to correct the problems. It is important that minimal set of documents that need to be created has to be planned during the design phase and the time for doing that has to be factored in.

2.2.6 Necessities

Fifth element in the *DEFINE* framework is the *Necessities*. The components that fall under *Necessities* are the *must be there* elements in the product and are not negotiable. Every product should have Necessities which consists of the following six elements:

1. Compliance to global standards
2. Product reliability
3. Accuracy (meeting the specification throughout the life)
4. Design for support (entire life cycle of the product)
5. Interoperability
6. Compliance to environment standards

2.2.6.1 Compliance to Global Standards

With the increased globalization, concept of designing a product for specific locations has become obsolete. With the market being wide, designing a product meeting a custom standard makes them uncompetitive. Second, local products can easily lose out to products complying with global standards. Unless the compliance to global standard is designed into a product early in the design cycle, implementing them at a later period will never be economical and may not be feasible in some cases. This compliance requirement also calls for the designers to understand the global standards so that the product can be designed to meet the standards. This typically needs in-depth knowledge of standards and domain that will come only with experience and does cost a lot of money to get. This activity has to be factored in to the design cost up front.

2.2.6.2 *Product Reliability*

One of the least understood aspects of an electronic product is the reliability of the product. Reliability is not something that happens by itself. It has to be designed into the reliability figure, and one of the main design goals should be the reliability figure. For most systems, reliability is typically described by a figure called Mean Time Between Failure. In simple terms, it is estimated as the duration for which the product will run continuously without any breakdown (Please note that Software also can contribute to failure!). Basically reliability is a predicted value that is derived by using the reliability figures collected over a period of time from the components used in different systems. However, designer will get pointers about the reliability of the parts by the way they are constructed. Parts with mechanical movements will have lower MTBF value (Relays, Solenoid, motors, etc.). In addition, working environment also contributes to the reliability. A dusty environment with parts that have moving elements is a good candidate for lower reliability. With the IoT, reliability becomes a critical element as entire system performance is dependent on the Things—one good example in which Reliability is the key in the case of wearable diagnostic devices. These devices actually monitor the health of the patients and have to be extremely robust and reliable. Unless a proper reliability calculation is done and right parts are chosen, the product may not meet its ultimate goal.

2.2.6.3 *Product Accuracy*

Accuracy of a product is defined as its ability to maintain the precision and tolerance for which it was originally designed throughout its life. Designers should be aware of the fact that the electronics components especially passive components (resisters, capacitors, etc.) tend to vary from their original value, as the product ages with time. This results in reduced system performance and errors creeping into the measurements. With the increased use of semiconductors, the accuracy can be maintained to a fair degree. However, analog circuits invariably use resisters and capacitors and end up impacting the accuracy. One way of mitigating is to periodically calibrate the system against a defined standard and ensure that the system's precision is maintained to the designed value. However, if calibration has to be facilitated it needs to be incorporated in the design up front. Although options for automatic calibration in systems are feasible, it will not be possible to implement them in all the products. Another challenge is that calibration is a special process, and knowledge of the system is needed. This mandates that there should be a separate process for calibration, and system has to be calibrated with special equipment. This also means that the manufacturer has to plan for infrastructure to meet the need for the calibration of equipment. Most designers miss this point and end up with a design that has to be discarded after the system goes out of tolerance measurement. One another key aspect is the need for a centralized system that keeps track of the system's age and ensures that every system is calibrated, a task most designers and companies tend to miss. So it is essential that a plan for the system calibration to maintain its accuracy has to be made while the product is designed.

2.2.6.4 *Design for Support*

Fourth element in the Necessities is the plan for designing the support for the product throughout its life. As we have seen, identifying the domain does give a fair idea of what the product life is and how the product will be used. With globalization, usage patterns change with countries and culture. Best example is the mobile phones. In North America, faulty mobile phone (especially low- and middle-priced ones) gets discarded, and user goes for a new phone. While in India, the users go in for fixing the problem. This essentially forces the phone manufacturer to have different strategy for supporting the product. One of the essential elements in the product design is to have a easy disassemble and assemble plan for a product for repair and rectification. This becomes very important when the product is water resistant. These water-resistant products need specialized tools as well as additional components to be repaired to function as robust as the original. This calls for designers to understand the support needs and implement them in the design so that it is easy for the product support team to open, fix, and close the products that need to be repaired. With the increased use of Personal Electronics as part of the IoT, *Design for Support* becomes very crucial.

2.2.6.5 Product Interoperability

This is another key aspect of any product especially with reference to IoT. Although regular standalone products may not have this issue of interoperability, all IoT products will have to support interoperability as their functionality depends on networking with other devices and servers. With IoT, interworking becomes very critical using the publically available communication and networking infrastructure. Interoperability has three critical elements to address. They are as follows:

1. Networking interface (wireline, wireless, etc.)
2. Networking protocols (TCP/IP, GPRS, LTE, and new [like Constrained Application Protocol]) or proprietary protocol defined by the manufacturers
3. System's operational power and associated aspects like Interference/Emission, and so on

These aspects vary from market to market, and designers have to be aware of the place where the product will be sold and supported. Despite higher level protocols being the same, minor location-specific variations can still stall the work. Designers have to ensure that these aspects are fully covered in their design so that later discovery does not create any problems. Typical problems that were seen in these types of situations are the lack of provision of additional resources (memories) in the system that prevents the designers to implement the required changes in the software. So, they need to plan for the product interoperability and also plan for additional resources that may be needed when they have to implement the new requirements. This essentially means that they have a full-scale structure of the product and its market at the initial stage and plan accordingly. The other hard aspect of interoperability is the product operating voltage and how it is derived. Main challenge is when the Thing has to interact with the mains frequency that varies from country to country. Also, some applications such as Avionics use 400 Hz that is much higher than the conventional mains frequency. This calls for a completely different power supply design for the product. Similarly, unattended installations need power derived from different sources such as solar, wind, and so on and the system need to be designed for these solutions too. Similarly, when the product is a medical device, the requirements are different and the focus is on safety. So designers have to plan for the interoperability requirement up front during the design phase so that changes in the later stage are not needed.

2.2.6.6 Environmental Compliance

This environment is different from the products working environment. Here, we are discussing the impact of the product on the global environment or compliance to green initiatives. With the increased awareness about the impact of electronic waste on our environment due to use of lead or associated materials in the manufacturing processes, most countries have now drafted compliance laws covering the following three aspects:

1. Reduction of e-waste by encouraging recycling as well as use of recyclable materials.
2. Elimination of lead and similar toxic material that are used in the manufacturing process of electronic products.
3. Compliance to using only approved chemicals and raw materials.

The three universal standards with respect to Environmental Compliance are as follows:

1. *RoHS*—Reduction of Hazardous Substance (this typically applies to components which are treated with alternate material reducing the usage of lead to bare minimum).
2. *WEEE Directive*—Waste Electronics and Electrical Equipment which that mandates that all the product design should be recyclable, and only a miniscule percentage of the total material used in a product can be nonrecyclable.
3. *REACH*—Registration, Evaluation, Authorization, and Restriction of Chemicals is a directive that very clearly specifies the process of what materials have to be used and how they should be recorded and submitted along with the product, as documentation in compliance with the REACH directives. Sole purpose of REACH is to address the production and use of chemical substances, and their potential impacts on both human health and the environment.

Environmental compliance is an activity that needs to done in the design phase. Although RoHS compliance can be met by selecting the process and the components to meet the standards, WEEE and REACH need proper selection and design of the system to meet these needs. This calls for specialized skills as well as knowledge of the compliance rules and how to design products to comply with the rules. Primary reason for the compliance is that any product that does not meet these directives will not be allowed for use. Figure 2.4 shows how the e-waste standards stack up against the compliance processes.

FIGURE 2.4
Green laws and the level of recyclability.

2.2.7 Enablers

Last element of the *DEFINE* framework are the Enablers. These are items that enable the designers to execute the design faster, better, and reliable. In addition, enablers also allow the designers to evaluate up front the options available, take decisions that ensure that the product meets the design objectives and makes it successful. Enablers are the external elements that could include Reference Design, Reference Software and Libraries, provided by the device vendors, tool vendors, open-source platforms, and so on.

There are six aspects which constitute the enablers and they are as follows:

1. Reference hardware platforms
2. Reference software platforms
3. Operating systems
4. OEM/ODM components
5. Intellectual properties and patents
6. Tools and libraries

2.2.7.1 Reference Hardware Platform

With the increased pressure on Time to Market, component vendors are finding multiple ways to help the designers. One of them is the Reference Hardware Platforms. The device vendor supplies the complete working circuit design with complete design files. This helps the designers to evaluate the capabilities of the device before committing to design. It has to be kept in mind that reference designs are not final product designs, but they form a foundation to the product design. However, most reference design does not undergo the rigorous design and engineering analysis that product design demand and caution have to be exercised. One good example will be the timing tolerance design that a product design need to undergo so that the circuit works properly even when the external parameters like Temperature, Humidity, and Power Supply change. Most reference designs are mainly catering to functional demonstration of the devices only. Although the availability of reference platform helps in the design process, using the reference design without any design analysis will bring grief to the designers. When an embedded architecture is selected, reference platform does help but designers have to clearly evaluate their needs against the adequacy of the reference platform before selecting the final design.

2.2.7.2 Reference Software Platform

Similar to reference hardware platform, vendors also provide Reference software. Most of the reference solutions will be based on the open-source software (Linux) and drivers for the peripherals that are in the processor. Unlike the hardware platform, Software vendor supplies only basic things. However, use of Reference Software will accelerate the development time. But the actual system software has to be developed only by the designers. Biggest caveat is that most reference software are not fully debugged as their purpose is only to demonstrate the functioning of the devices and not the actual software itself. Blindly using the reference software platform will lead to a system with bugs and will lead to a suboptimal system with low performance. Most young designers do not understand this aspect as they believe the vendor supplied software to be robust which may not be actually true. In the context of IoT, real-time performance is very critical and using the reference does not guarantee that. Designers can use the reference platforms for comparison, but they should estimate the effort for the main software development. This aspect gets complicated when the product is a Medical or Avionics or Safety critical industrial product in which the software need to be completely validated independently with the target hardware (the actual product hardware). With IoT taking a big lead in the industrial automation, software platform selection is very important and need to be carefully selected.

2.2.7.3 Operating System Used

Third important element in the Enablers is the Operating system used. Although more than 50% do not use any commercial operating systems but use home grown schedulers or monolithic software. However, with the designers having the tendency to use the open-source hardware platforms like Raspberry Pi or Arduino, they use the OS that comes with them. Primary reason for this is the ready availability of HW and SW platforms for development. However, a cost-effective product in large volume cannot be realized out of the open-source hardware platforms. Designers need to keep in mind that when a product has to be made in volumes, the best and economical way is to have dedicated hardware and the software to go with it. Second important issue is that the designers tend to ignore implications of use of open-source software. When an open-source license is used, it is important to understand what things are already covered and what needs to be covered, to avoid future litigations in terms of infringements. Most designers tend to miss this point and run into legal issues later. The issue gets complicated when the product belongs to regulated industries. In these cases, the OS has to be validated on the target hardware separately and submitted for compliance to agency (FDA/FAA) regulations. If the designers choose off the shelf operating system, then they have to ensure that the OS is available for their chosen processor. In effect, elements in the Enabler segment are interrelated and have impact on each other. In addition, the designers need to look for five key elements in an operating system. They are:

1. Performance parameter of the OS
2. Resources needed by the OS
3. Availability of Drivers and networking protocol stack support for the OS
4. Security of the OS from hacking
5. Licensing cost of the OS and the cost of maintenance for the entire life of the product

It can be seen that OS plays a big part in the software development stage of the system and also on the performance of the system. Due care has to be taken to select the right OS and its support elements for the success of the product.

2.2.7.4 OEM/ODM Components

OEM/ODM (Original Equipment Manufacturer/Original Device Manufacturer) parts are basically subsystems (products) which are bought as a subsystem and integrated into the products. Best examples in the case of IoT products are Power Supplies and Wireless modules. These are basically finished products manufactured by OEM/ODM vendors that can be used in designs. Although use of OEM/ODM parts reduces the design cycle, we need to remember that cost of OEM part will be more than the same designed part of the circuit. Other than parts like Power Supplies that need specialized skills, OEM/ODM parts make economic sense when the volumes are less, but expensive when the volumes are high.

Another challenge that designers normally miss is the need for understanding the OEM/ODM part completely so that it can be successfully used the products. Most of the times designers miss the critical elements of the EOM/ODM part and product runs into trouble.
   Selection of OEM/ODM has to be based the following parameters;

1. Cost
2. Performance specification
3. Flexibility
4. Compliance to standards
5. Lead time availability

2.2.7.5 Intellectual Property and Patents

With the liberal patenting regime, most companies and individuals patent their ideas as soon as they get them. Although majority of the patents will not be useful to create a product but they ensure others from using the idea. These patents are called *Defensive Patents*. Second technique is to start the litigation against a company that uses one of their ideas only after the company brings the product to the market. Basic idea is to cause huge economic loss forcing the company to discontinue or pay a hefty royalty and in both the cases company stands to incur losses in the business. This essentially means designers have to validate their ideas against the existing patent to ensure that they do not fall prey to the IP-related issues. It is also imperative that if the idea being used by the company is unique it has to be patented so that the product is protected from copycats. Many designers are not aware of this and end up losing their idea to copies and incur losses. Challenge is that the IP and Patent search and filing is expensive but this step is essential for success and survival of the product.

2.2.7.6 Tools and Libraries

Last element in the DEFINE framework is the Tool and SW libraries that are used in the product development. Tools enable the designer to develop and debug the product or design quickly in short time and bring out the product to the market. Tools are either supplied by the device and OEM vendors or by third party suppliers. Best example for third-party suppliers of tools are Compiler Vendors who supply for different processors that are far more compact and optimized compared with the compilers supplied by the device vendors. This is just one example. However, what is important is the selection of the right tool. Also with the increased use of user interfaces, creating video and graphics-related utilities have became expensive. Most designers prefer to buy the libraries that are proven in performance and use them in the design. Tools also decide the programming languages that are used. Normally IoT and embedded products use C language for higher performance and reliability. In essence, choice of the tools makes the product development process simple and easy.

2.3 Putting All Together

How this framework can be used by the designers is the next step. Although actual framework has lot of good guideline and check list, templates explaining them need spate book!

However, to put it in a nut shell, Tables 2.6 through 2.8 are given. The table lists the product development life cycle and the six elements of DEFINE framework. Easiest way to use the table is to look at each element and by looking at the density of the starts in the activities, development effort can be estimated.

TABLE 2.6

DEFINE Parameters and Life Cycle Activities to Power Supply and Printed Circuit Boards

Elements		Material	Product Packaging	Power Supply	Components	Printed Circuit Board
			Product Life Cycle Activities			
Domain	Operating environment	*	*			
	Usage		*			
	Cost	*		*	*	
	Process	*			*	
	Product life	*		*	*	
	Product support	*		*	*	
Environment	Temperature	*	*	*	*	
	Humidity	*	*		*	*
	Pressure				*	
	Dust	*	*			*
	Fluid ingress	*	*			*
	Electrical power		*	*	*	*
Functionality	Usability and ergonomics	*	*			*
	Manufacturability	*	*		*	*
	Interconnect and interface		*			*
	Performance		*	*	*	*
	Safety	*	*	*	*	
	Product sustenance			*	*	
Implementation	Development process				*	*
	Manufacturing process	*	*		*	*
	Verification and validation					
	Calibration					*
	Support			*	*	*
	Document and control					*
Necessities	Compliance	*	*	*	*	*
	Reliability	*	*	*	*	*
	Accuracy			*	*	*
	Support			*	*	*
	Interoperability		*			*
	Environmental compliance	*	*	*	*	*
Enablers	Hardware platform				*	*
	Software platform					
	Operating systems					
	OEM/ODM parts	*	*	*		*
	Intellectual properties		*			
	Tools		*		*	*

TABLE 2.7

DEFINE Parameters and Life Cycle Activities to System Software and Manufacturing Process

Elements		Product Life Cycle Activities				
		Wires and Cables	Thermal Design	User Interface	System Software	Manufacturing Process
Domain	Operating environment		*		*	
	Usage	*	*	*		
	Cost				*	
	Process	*	*	*	*	
	Product life	*		*		
	Product support	*	*	*		
Environment	Temperature	*		*		
	Humidity				*	*
	Pressure	*				
	Dust	*			*	*
	Fluid ingress	*	*	*		*
	Electrical power					*
Functionality	Usability and ergonomics		*	*		*
	Manufacturability	*			*	*
	Interconnect and interface			*		*
	Performance	*	*	*	*	*
	Safety	*		*		
	Product sustenance			*		
Implementation	Development process	*	*	*		*
	Manufacturing process				*	*
	Verification and validation		*	*	*	
	Calibration		*	*	*	*
	Support			*		*
	Document and control	*	*	*	*	*
Necessities	Compliance		*	*	*	*
	Reliability	*			*	*
	Accuracy	*			*	*
	Support	*	*	*		*
	Interoperability		*	*		*
	Environmental compliance	*			*	*
Enablers	Hardware platform				*	*
	Software platform		*	*		
	Operating systems		*	*		
	OEM/ODM parts	*		*	*	*
	Intellectual properties	*	*	*		
	Tools	*	*	*	*	*

TABLE 2.8

DEFINE Parameters and Life Cycle Activities to Product Repair, Support, and Disposal

Elements		Product Life Cycle Activities				
		Testing Process	Product Compliance	Product Repair	Product Support	Product Disposal
Domain	Operating environment	*			*	
	Usage	*	*	*		
	Cost	*	*	*	*	
	Process	*	*	*		
	Product life	*	*	*		
	Product support	*	*	*	*	
Environment	Temperature					
	Humidity	*				*
	Pressure					
	Dust	*	*			*
	Fluid ingress	*				*
	Electrical power	*		*		*
Functionality	Usability and ergonomics		*	*		*
	Manufacturability	*	*	*	*	*
	Interconnect and interface	*	*	*	*	*
	Performance	*				*
	Safety	*	*			
	Product sustenance	*		*		
Implementation	Development process	*			*	*
	Manufacturing process	*	*		*	*
	Verification and validation	*				
	Calibration					*
	Support			*	*	*
	Document and control	*	*	*	*	*
Necessities	Compliance	*	*	*		*
	Reliability	*	*	*		*
	Accuracy		*	*		*
	Support	*	*	*		*
	Interoperability	*		*		*
	Environmental compliance	*				*
Enablers	Hardware platform	*	*	*	*	*
	Software platform	*	*	*		
	Operating systems				*	
	OEM/ODM parts	*	*	*	*	*
	Intellectual properties					
	Tools	*	*	*	*	*

TABLE 2.9

DEFINE Parameters and Life Cycle Activities to Consumer, Industry, and Medicine

Elements		Product Life Cycle Activities				
		Avionics	Auto	Consumer	Industrial	Medical
Domain	Operating environment	*			*	
	Usage	*	*			
	Cost	*	*	*	*	
	Process	*	*			
	Product life	*	*			
	Product support	*	*	*	*	
Environment	Temperature					
	Humidity	*				*
	Pressure					
	Dust	*	*			*
	Fluid ingress	*				*
	Electrical power	*		*		*
Functionality	Usability and ergonomics		*	*		*
	Manufacturability	*	*	*	*	*
	Interconnect and interface	*	*	*	*	*
	Performance	*		*		*
	Safety	*	*	*		
	Product sustenance	*				
Implementation	Development process	*			*	*
	Manufacturing process	*	*		*	*
	Verification and validation	*				
	Calibration					*
	Product support process	*	*	*	*	*
	Document and control	*	*		*	*
Necessities	Compliance	*	*	*		*
	Reliability	*	*	*		*
	Accuracy		*			*
	Support	*	*			*
	Interoperability	*		*		*
	Environmental compliance	*				*
Enablers	Hardware platform	*	*	*	*	*
	Software platform	*	*	*		
	Operating systems				*	
	OEM/ODM parts	*	*	*	*	*
	Intellectual properties					
	Tools	*	*	*	*	*

2.4 Summary

In this chapter, we have seen how the DEFINE framework helps in designing an electronic product including IoT. Although this approach can also be used for generic electronic products, this is very useful for IoT as the *Things* are expected to work in any kind of environment that needs careful design right from the beginning. To help the designers better, the author has provided a series of four tables that help the developers to estimate the effort and cost needed for the product development. Tables 2.6 through 2.8 indicate the impact of DEFINE elements on the product development. They are very important and allow the user to know the dependency. Table 2.9 helps one to understand the dependence of DEFINE elements on the domains that discussed. These two table sets will help the developers to quickly understand the complexity of the effort involved and also to ensure that the product meets the expectation of the end users.

Bibliography

Boothroyd, G. (2005) *Assembly Automation & Product Design.* Boca Raton, FL: CRC Press/Taylor & Francis Group.

Cundy, D. R. and R. S. Brown. (1997) *Introduction to Avionics.* London, UK: Pearson Education as Prentice Hall.

Fries, R. C. (2001) *Handbook of Medical Device Design.* New York: Marcel Dekker.

Haik, Y. (2005) *Engineering Design Process.* Boston, MA: Brooks/Cole, a Part of Cengage Learning.

Harper, C. A. (1997) *Electronic Packaging & Interconnection Handbook.* New York: McGraw-Hill.

Kalpakjian, S. and S. R. Schmid. (2001) *Manufacturing Engineering and Technology.* London, UK: Pearson Education.

Noergaard, T. (2005) *Embedded System Architecture: A Comprehensive Guide to Engineers & Programmers.* Burlington, MA: Newnes: An Imprint of Elsevier.

O'Connor, P. D. T. with D. Newton and R. Bromley. (2006) *Practical Reliability Engineering.* Hoboken, NJ: John Wiley & Sons.

Pecht, M. G. (Ed.). (2004) *Parts Selection and Management.* Hoboken, NJ: John Wiley & Sons.

Ulrich, K. T. and S. D. Eppinger. (2005) *Product Design and Development.* New York: McGraw-Hill.

3

MAC and Network Layer Issues and Challenges for IoT

Upena Dalal, Shweta Shah, and Jigisha Patel

CONTENTS

3.1 Introduction

Internet of Things (IoT) and its protocols are the emerging areas of research. The fast growth of the mobile Internet, micro-gazettes, ultra-high-speed computing, and machine-to-machine (M2M) communication systems has enabled the IoT technologies. Any communication system utilizes a set of rules and standards for networking to control the data exchange with reliable reception. Till now, the networking was among the traditional computers or smart mobiles, whereas IoT technologies allow Internet protocol (IP)-based connectivity of *things*. Things are the devices or machines and not strictly computers. Those act smartly and make collaborative decisions that are beneficial to specific applications. IoT allows things to sense, process, or act by allowing them to communicate in coordination with others in order to make decisions.

IoT is nothing but convergence among a number of disciplines and forms the heterogeneous system. The underlying technologies, such as wireless technologies for small and large networks, ubiquitous computing, embedded sensors, light communication, cellular technologies, and IPs, allow IoT to provide its significance; however, they impose lots of challenges and introduce the need for either modifications in the existing standards/ specifications or development of specialized standards and communication protocols. Especially, the critical design of protocols is required at the IP level because the medium access control (MAC) is normally the underlying technology specific, that is, Wi-Fi has its own specific MAC frame format and handling mechanisms at data link layer. However, few new MAC protocols are introduced, or the olders are optimized as per IoT challenges and requirements. Up to a certain extent, the major challenges of IoT are handled by the lower layers—especially MAC and Network implemented over specific wireless technology. The chapter deals with challenges as well as IoT protocols that are suitable at the data link layer and the network layer (IEEE 802.11ah—sub-GHz Wi-Fi) (Home of RF and Wireless Vendors and Resources).

3.2 IoT and TCP/IP Stack

IoT follows TCP/IP protocol architecture that becomes customized as per the application requirements. The challenges of IoT in general are mobility, reliability, scalability, management, availability, interoperability, security, and privacy. These challenges are because of a large number of different platforms/technologies used in IoT systems.

One of the disadvantages of the TCP/IP stack is that it is fairly complex. Conventional IP protocol as per the standard results in larger size data packets and therefore requires a significant amount of processing power, sufficient memory, more processing time, and expensive devices. What we require from IoT is sometimes fast and near real-time responses. For these reasons, for many well-established technology-based networks, such as the ZigBee/Bluetooth, the protocols must be simple, fast, and energy efficient; hence, new protocols for adaptation of IP are the requirement of time (IEEE 802.11ah— sub-GHz Wi-Fi) (Home of RF and Wireless Vendors and Resources). Protocol stack for web-based application and typical equivalent protocol stack for IoT is given in Table 3.1.

TABLE 3.1

Comparison of Web-Based and Typical IoT-Based Protocol Requirements and Solutions

	For Web Application (Hundreds/thousands of bytes to handle) • Inefficient content encoding • Huge overhead and difficult parsing • Requires full Internet devices	For IoT Application (Tens of bytes to handle) • Efficient objects • Efficient Web • Optimized IP access	Hybrid Scenario or Wireless Embedded Internet • Both the Web and IoT features are combined
Application	XML	Web Objects	Web Objects
	HTTP	CoAP	CoAP
	TLS	DTLS	DTLS
Transport	TCP (and/or UDP)	UDP	UDP
Network • Routing activity • Encapsulation activity	IPv6	6LoWPAN	IPv6 [RPL] over 6LoWPAN (Acting as adaptation layer)[a]
Host to Network Interface including Data link and Physical	Existing Wireless technologies like IEEE 802.15.4, IEEE 802.11ah, Bluetooth low power or other MAC layer Technologies with physical layer specifications [Few wireless technologies use dedicated licensed spectrum while few use shared/unlicensed spectrum]		

[a] IPv6 is an Internet Layer protocol for packet-switched internetworking and provides end-to-end datagram transmission across multiple IP networks. 6LoWPAN is an acronym of *IPv6 over Low-Power Wireless Personal Area Networks*. It gains popularity day-by-day and hence selected here to give an example. It is an adaptation layer for things to be communicated via IPv6. IPv6 must be supported with specific routing protocols like RPL (described later in this chapter). All these protocols working together is a good example of *Embedded Wireless Internet*.

3.2.1 Importance of Adaptation Layer below IPv6

The adaptation to IP by wireless-embedded devices is challenging due to several reasons:

1. Battery-powered wireless devices/things make and break the connections frequently, whereas IP is based on always connected devices.

2. Sometimes, it is difficult to route the traffic of *things* in a large coverage area via multihop wireless mesh networks with cost efficiency.

3. Low-power wireless networks have low bandwidth (few kbps) and frame size. IPv6 requires that every link must have a maximum transmission unit of 1280 bytes or greater; otherwise, link-specific fragmentation and reassembly must be provided below IPv6.

4. Multicasting is not supported basically in IEEE 802.15.4 or such technologies, but it is essential in many IPv6 operations.

5. Standard protocols may not perform well with TCP/IP directly. For example, TCP is not able to distinguish between packet losses due to congestion and those due to channel error and, hence, required to define how to carry IP datagrams over existing links and also perform necessary configuration functions to form and maintain an IPv6 subnet.

Three working groups formed by The Internet Engineering Task Force (IETF) defined an adaptation layer for IoT. Widely considered protocols, IEEE 802.15.4 and 6LowPAN, are

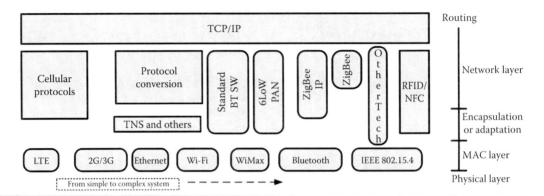

FIGURE 3.1
Representing major MAC and network layer protocols working in coordination along with a level of complexity.

explained in detail in Sections 3.4.3 and 3.4.6, respectively, to understand the requirement of adaptation layer. Figure 3.1 shows the various protocols suitable at the physical layer (PHY)/MAC and also the suitable protocols above specific MAC. 6LowPAN has different roles. It is a thin layer above IEEE 802.15.4 acting as an adaptation layer and creating embedded Internet, whereas it is whole and sole above Bluetooth.

In addition to the comments for MAC and network layer, a few comments are there for the transport layer. As TCP/IP laid the foundation for the Internet, IoT communication network mainly employs TCP and UDP protocols. Compared with UDP protocol, the TCP protocol is more complex, which makes it difficult to employ over the devices with constrained resources. Most of IoT uses UDP protocol, but as UDP is not stable, special techniques are used to incorporate it with application layer (Patil and Lahudkar, 2016).

3.2.2 IP and Non-IP IoT Devices

Heterogeneity of devices and applications due to different vendors can be adopted in modern IoT system due to open standards of TCP/IP protocol suite. The IoT is nothing but connecting things to the Internet and hence the things/devices need a logical connection, that is, an IP address and must use the IP suite to be able to exchange data with devices and servers over the Internet. Devices in a local network can use non-IP protocols to communicate within the local network. These non-IP devices can be communicated to Internet via an Internet gateway. The gateway communicates with both local devices using a non-IP protocol and other devices on the Internet using IP. The gateway here is an application layer gateway because it needs to restructure the data coming in from the local network with a TCP/IP stack to enable communication with an Internet service.

3.3 Challenges at MAC and Network Layers Due to Physical Layer Aspects

There are hundreds of protocols supported by IoT. Of these, wireless protocols play an important role in IoT development. Complexity of MAC and Network layers are dependent upon wireless technology adopted at the physical layer; thus, there are few challenges at the physical layer.

- *Adopting Present Techniques of Wireless Communication*—M2M devices are normally connected in a group and communicate on the basis of various wireless protocols such as ZigBee, Wi-Fi, and Bluetooth. For Internet access, they can have direct access as per facility or they can use Internet gateways or GPRS or EDGE as per their availability. That is why existing wireless protocols are adopted in IoT with required minor modifications.
- *Limitation in Current Technique*—Device makers are proving to get to do with existing communication protocols that do not satisfy the operational requirements of M2M. Having multiple set of rules also obstructs standardization, bandwidth requirements, power requirements, and the number of node limitations.
- *Mobility and Handovers to Manage*—It is a big challenge as both these tasks are related to location identification or tracking tasks and for that the efficient computing is also required. Near real-time systems are required in order to take quick decisions. To maintain seamless connectivity over large coverage areas, hard and soft handovers are to be managed by central control units. This is required for continuous communication as well as IP management also.
- *Spectrum/Resource Management*—Applications have to be operative within the limited bandwidth satisfying as many numbers of users or nodes intended to add. Also, channel assignment in near real time also requires fast computing.

Various organizations started to find out the compatibility of the various wireless technologies for IoT or M2M applications with the goal of embedding M2M communications in the upcoming 5G systems. Similarly, they started enhancing the MAC and network layer requirements. Few widely considered protocols of our focal domains are described in the consecutive sections (Reiter, 2012).

3.4 MAC Layer Protocols with Key Features and Elimination of Challenges

MAC layer has the specific challenges to deal with are as follows:

- *Data throughput*: Can be optimized by managing collisions in contention-based protocols, by managing the access of users to the available channel, by plan-based frameworks, the control overhead, and so on.
- *Scalability*: Node density is expected to increase or decrease, and it should be handled without disturbing the operation of the network and with an energy-efficient solution.
- *Latency*: Near real-time solutions are expected that can be achieved by reducing the channel-access latency or other fast decision-making strategies.
- *Coexistence*: Coexistence of licensed and unlicensed systems, interworking aspects, adaptation, and so on are to be managed.
- *Cost-effective solution*: Cost is always a trade-off with bandwidth, quality of service (QoS), power consumption, and speedy response of the system. Optimized solution is required finally.

Various wireless technologies with their key features are highlighted in this section. Sometimes, it is difficult to write only MAC layer issues without mentioning the key physical layer aspects, and hence as per requirement, the PHY is also described in the section (Rajandekar and Sikdar, 2015).

3.4.1 IEEE 802.11ah—Sub-GHz Wi-Fi

The two primary limitations of IEEE 802.11 technologies have opened the door for 802.11ah technology to be applied for M2M communication:

- *The absence of power saving mechanisms*: The energy constraints of wireless sensor networks are not considered in the IEEE 802.11 standard.
- *Unsuitable bands*: Existing Wi-Fi bands require the use of intermediate nodes due to their short wireless range and high obstruction losses, and thus the complex scenario.

The IEEE 802.11ah standardization task group has developed a global Wireless LAN standard that allows wireless access using carrier frequencies below 1 GHz in the industrial, scientific, and medical (ISM) band. It is an unlicensed, shared spectrum range that will help Wi-Fi-enabled devices to get guaranteed access for short-burst data transmissions. The channels of IEEE 802.11ah are defined on the basis of spectrum available in a given country. The basic channel width is 1 MHz, but it is possible to bond two adjacent channels together to form a 2 MHz channel for higher data throughput. IEEE 802.11ah is interoperable with 1 MHz and 2 MHz bandwidth modes globally, which results in new application and also compatible with 4, 8, and 16 MHz bandwidths for advanced data rate applications.

Key features of this standard are as follows:

- Wide coverage, that is, extended range of its access point (AP) device (up to 1 km) because of the enhanced propagation and diffusion of 900 MHz radio waves through strong obstacles such as walls and ceiling. Please refer to Figure 3.2.
- Less power consumption due to below 1 GHz, that is, power efficient.
- Unlicensed band, that is, no regulatory issues.
- Almost clear coexistence issues.
- Compatible with existing wireless standard versions, for example, IEEE 802.11a/b/g/n.

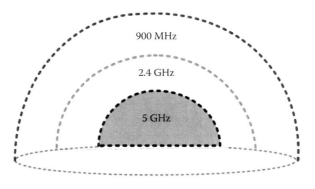

FIGURE 3.2
Extended range compared with older versions of Wi-Fi.

TABLE 3.2

IEEE 802.11ah Modulation and Coding Schemes for 2-MHz Bandwidth Channels for Different Guard Intervals Used with OFDM-Based Physical Layer and Spatial Stream

Index	Modulation Technique	Code Rate	Data Rate (Mbps)	
			Moderate GI	Short GI
0	BPSK	1/2	0.65	0.72
1	QPSK	1/2	1.3	1.44
2	QPSK	3/4	1.95	2.17
3	16-QAM	1/2	2.6	2.89
4	16-QAM	3/4	3.9	4.33
5	64-QAM	2/3	5.2	5.78
6	64-QAM	3/4	5.85	6.5
7	64-QAM	5/6	6.5	7.22
8	256-QAM	3/4	7.8	8.67
9	256-QAM	5/6	–	–

Note: Moderate GI = 8 μs, Short GI = 4 μs (Similar table are there for 4, 8, and 16 MHz channels).

For, IEEE 802.11ah, a new PHY and MAC design is suggested to maintain extended range of Wi-Fi and the IoT. PHY design improves link budget compared with 2.4-GHz technology. In order to handle the trade-off between range, throughput, and energy efficiency, IEEE 802.11ah found the solutions as follows:

- Different sets of modulation and coding schemes with mandatory coding schemes include binary conventional coding and optional coding scheme low-density parity check.
- Number of spatial streams.
- Different duration of the Guard Interval.

Table 3.2 shows the 2 MHz mode with PHY requirements.

To support M2M communication apart from the abovementioned physical layer aspects, the following requirements are specified by IEEE 802.11ah:

- 8191 devices at most can be associated with an AP in a tree structure (hierarchy).
- At least 100 kbps data rate.
- One-hop network topologies.
- Short and sporadic data transmissions
  - Packet data size ~100 bytes
  - Packet interarrival time >30 s
- Lowered energy consumption by adopting power-saving strategies. These may include the adoption of modified IEEE 802.11v power-saving features as well as enhanced ultra-low-power consumption strategies, such as Radio On Demand (ROD).

MAC layer management is done by AP by sending power-save poll frames and enhanced distributed channel access (EDCA) mechanisms (Aust et al., 2012; Tian et al., 2016) (IoT protocols you need to know about) (IEEE 802.11ah-sub GHz WiFi).

3.4.2 Bluetooth Low Energy

The Bluetooth protocol is enhanced with low power as the primary design consideration. Its major role is in wearable IoT designs, which are system on chip (SOC) type. To address the limitations of earlier versions and for interoperability, the Bluetooth Special Interest Group (SIG) introduced Bluetooth Low Energy (BLE) for short-range communication, also known as *Bluetooth Smart*. BLE continues to operate in 2.4-GHz ISM band with a bandwidth of 1 Mbps. Few main features of BLE are as follows:

- Robust architecture supporting adaptive frequency hopping with a 32 bit CRC.
- A low data rate makes it suitable for applications in which only state information has to be exchanged.
- Due to burst transmission of small blocks of data, ignoring duplicate packets, due to client/server mode along with master/slave, and so on, the host processor can operate in a low power mode frequently.
- Reduced time for connection setup to data exchange to within a few ms.
- It supports broadcaster mode—just transmits the advertising packets—without undergoing a connection procedure.
- The link layer is optimized for quick reconnections, thereby reducing power.

A BLE device is not compatible with conventional Bluetooth device as it is a different technology. However, Bluetooth dual-mode devices support both BLE and classic/conventional Bluetooth. Figure 3.3 represents the major differences in the Bluetooth protocol layered architecture for conventional, dual-mode and low-energy Bluetooth devices. Here, the major change in MAC layer is observed along with changes in PHY.

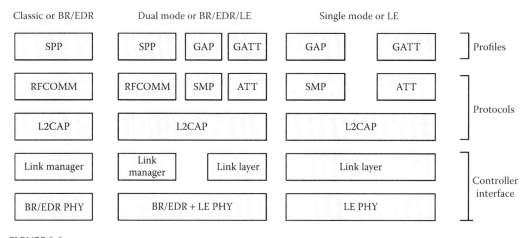

FIGURE 3.3
Protocol architecture difference in various versions of Bluetooth technology. GAP—Generic Access Profile, SPP—Serial Port Profile, GATT—Generic Attribute Profile, RFCOMM—Serial cable emulation protocol based on ETSI TS 07.10, SMP—Security Manager Protocol, ATT—Attribute Protocol, BR/EDR—Basic Rate/Enhanced Rate, LE—Low Energy.

Standard Bluetooth 4.2 core specifications are as follows:

- Range: 50–150 m (Smart/BLE)
- Data Rates: 1 Mbps (Smart/BLE)
- Low latency connection (3 ms)
- Low power (15 mA peak transmit, 1-μA sleep)
- Designed to send small packets of data (opposed to streaming) via Connect→transmit→disconnect→sleep states
- Security (128 bit AES CCM)
- Modulation (GFSK @ 2.4 GHz)
- Adaptive frequency hopping
- 24 bit CRC
- Output power: ~10 mW (10 dBm)
- 40 Channels—2 MHz spacing
- Frequency hopping in connections: pseudo-random, set in the connection request
- Transmit power (−20 to +10 dBm)
- Receive sensitivity (−70 dBm)

The various states and hence the processes of link layer are shown in Figure 3.4 in form of link layer state machine for BLE (Gomez et al., 2012; Decuir, 2010).

Bluetooth low-energy beacons can be studied separately.

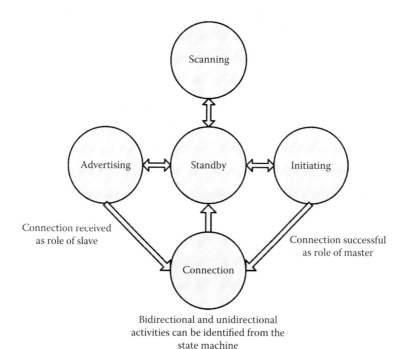

FIGURE 3.4
State diagram of link layer state machine.

3.4.3 IEEE 802.15.4e

The technologies that allow the location and identification of physical objects will be important in the context of IoT. IEEE 802.15.4 (2003) is a standard specifying physical layer and media access control for Low-Rate Wireless Personal Area Networks (LR-WPANs). It forms the basis for the ZigBee, WirelessHART, and many other specifications. It can also be used with 6LoWPAN and standard IP protocols to build a wireless-embedded Internet, for example, wireless sensor network (WSN) providing an autonomous and intelligent connection between the physical and virtual worlds.

IEEE 802.15.4 supported two physical layers originally, one with 868 and 915 MHz frequency bands and the other with 2.4 GHz band. IEEE 802.15.4 standard has well-defined frame format, headers and deals with the techniques how nodes can communicate with each other. The physical layer is based on Direct Sequence Spread Spectrum (DSSS) techniques. Twenty-seven channels were supported across three bands. One channel in 868 MHz band 10 channels in 915 MHz band and 16 channels in 2.4 GHz band are supported. The three bands provide a transmission rate of 20, 40, and 250 kb/s respectively. In the later versions, many additional bands were added.

The IEEE 802.15.4 supports two classes of devices for the MAC layer: full function devices (FFDs) and the reduced function devices (RFDs). All Personal Area Networks (PAN) consist of at least one FFD that acts as the PAN coordinator to maintain the PAN. RFDs are responsible for obtaining data directly from the environment and sending them to a PAN coordinator. As the traditional frame formats are not suitable for low-power multihop networking in IoT due to their overhead. In 2008, IEEE802.15.4e was created to extend IEEE802.15.4 and support low-power communication to meet IoT requirements. This is especially the *MAC amendment* for industrial application. By time synchronization and channel hopping, it enables high reliability and low cost. In contrast, IEEE 802.15.4g Smart Utility Networks (SUN) Task Group is given the task of *PHY amendment* in 802.15.4 to provide a standard that facilitates very large-scale process control applications.

Specific MAC features of IEEE 802.15.4e can be summarized as follows:

- *Frame structure and power saving*: IEEE 802.15.4e frame structure is designed for scheduling and telling each node what to do. A node can sleep, transmit, or receive information.
    - *Sleep* mode: The node turns off its radio to save power and to store all messages that it needs to send at the next transmission opportunity.
    - *Transmit* mode, it sends its data and waits for an acknowledgment.
    - *Receive* mode: The node turns on its radio before the scheduled receiving time, receives the data, sends an acknowledgment, turn off its radio, delivers the data to the upper layers, and goes back to sleep.
- *Synchronization*: Necessary to maintain nodes connectivity to their neighbors and the gateways. Two approaches can be used for synchronization: acknowledgment based or frame based.
    - *Acknowledgment-based* mode can be used for maintaining the connectivity/communication as well as guaranteed reliability simultaneously.
    - *Frame-based* mode: The nodes are not communicating, and hence, they send an empty frame at prespecified intervals of about 30 s typically.

- *Scheduling*: There is no standardization for how the scheduling is done. It can be centralized by a manager node that is responsible for scheduling, informing others about the schedule. Other nodes have to just follow the schedule.
- *Channel hopping*: IEEE802.15.4e introduces channel hopping for time slotted access to the wireless medium and hence frequency diversity. It requires changing over the frequency channel using a predetermined random sequence. This reduces the effect of interference and multipath fading.
- *Network formation*: Network formation phase includes advertisements and joining components. A new device will listen to advertisement commands and upon receiving at least one such command, it will send a join request to the advertising device.
    - In a centralized system, this join request is routed to the manager node and processed.
    - In distributed systems, join requests are processed locally.

Once a device joins the network, and it is fully functional, the formation phase is disabled and will be activated again if it receives another join request.

The MAC layer of IEEE 802.15.4 has the following features: association and disassociation, channel access mechanism, acknowledged frame delivery, frame validation, guaranteed time slot management, and beacon management. The MAC frame format of IEEE 802.15.4e is shown in Figure 3.5 that is almost the same as the general MAC of IEEE 802.15.4. There are four types of MAC frames: data frames, MAC command frames, acknowledgment frames, and beacon frames. The size of frame has to be less than 127 bytes considering the size constraint of the physical layer payload. The sequence number field is used to match the received acknowledgment frames. The frame check sequence is a 16 bit Cyclic Redundancy Checks (CRC). In the multipurpose format size of source, PAN identifier is varied. The multipurpose frame structure provides flexibility for the various purposes. Format supports short and long form of frame control field and allows for all MHR fields to be present or elided as specified by the generating service (IEEE 802.15.4, 2011; Montenegro et al., 2007; Patil and Lahudkar, 2016).

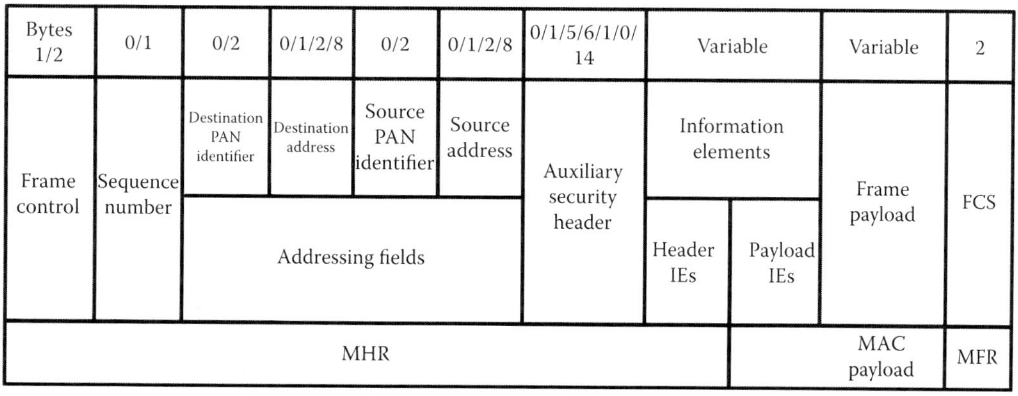

Bytes 1/2	0/1	0/2	0/1/2/8	0/2	0/1/2/8	0/1/5/6/1/0/14	Variable		Variable	2
Frame control	Sequence number	Destination PAN identifier	Destination address	Source PAN identifier	Source address	Auxiliary security header	Information elements		Frame payload	FCS
		Addressing fields					Header IEs	Payload IEs		
MHR									MAC payload	MFR

FIGURE 3.5
IEEE 802.15.4e MAC layer frame—general and multipurpose formats.

The IEEE standard supports two types of channel access mechanisms: (1) in the case of non-beacon networks, the devices use unslotted Carrier Sensing Multiple Access with Collision Avoidance (CSMA-CA); (2) in the case of beacon-enabled networks, a slotted version of CSMA-CA is used.

3.4.4 ZigBee and ZigBee Low-Energy Smart above IEEE 802.15.4

ZigBee, like Bluetooth, is a well-established standard protocol. ZigBee PRO and ZigBee Remote Control are among other available ZigBee profiles, are based on the IEEE 802.15.4 protocol operating in the unlicensed frequency range.

ZigBee has some significant advantages such as low-power operation, high security, robustness, and high scalability with more number of nodes. It can take advantage of both wireless control and sensor networks in M2M and IoT applications.

The latest version of ZigBee is version 3.0, which is essentially the unification of the various ZigBee wireless standards into one forming the foundation for IoT. This standard provides seamless interoperability. ZigBee supports mesh routing based on AODV protocol (Patil and Lahudkar, 2016).

ZigBee 3.0 certification is now available. Standard: ZigBee 3.0 based on IEEE802.15.4

- Frequency: 2.4 GHz
- Range: 10–100 m
- Data Rates: 250 kbps
- Maximum number of nodes in the network: 1024
- Encryption: 128 bit AES

Few plus points for ZigBee are reliability and robustness, low power, scalability, and security.

3.4.4.1 *ZigBee Network and Management*

As shown in Figure 3.6, ZigBee network comprises coordinator, router, and end devices.

3.4.4.1.1 *Coordinator*

In the beginning, the coordinator needs to be installed for establishing ZigBee network service. It starts a new PAN; once started, other ZigBee components, router, and end devices can join PAN. The coordinator is responsible for selecting the channel and PAN ID. It can help in routing the data through the mesh network and allows join request from router and end device. It is mains powered (AC) and support child devices. It will not go to sleep mode.

3.4.4.1.2 *Router*

A first router needs to join the network; then it can allow other routers and end devices to join the PAN. It is mains powered and support child devices. It will not go to sleep mode.

3.4.4.1.3 *End-Devices*

They neither allow other devices to join the PAN nor can assist in routing the data through the network. They are battery powered and do not support child devices. These may sleep;

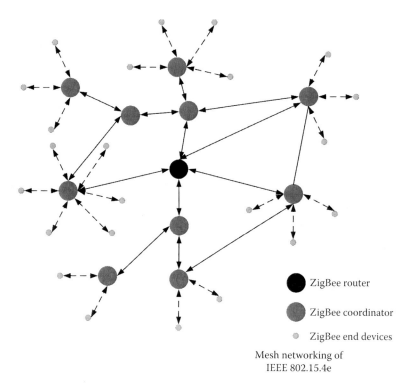

ZigBee router

ZigBee coordinator

ZigBee end devices

Mesh networking of
IEEE 802.15.4e

FIGURE 3.6
ZigBee network components.

hence, battery consumption can be reduced to a great extent. Star and mesh topologies are supported in ZigBee mesh routing. PAN ID (a 16 bit number) is used to communicate between ZigBee devices. The Coordinator will have the PAN ID set to zero always, and all other devices will receive a 16 bit ID address. Thus, there are two main steps in completing ZigBee Network Installation—forming the network by the Coordinator and joining the network by Routers and End devices.

3.4.4.2 ZigBee Network Formation Procedure

Formation procedure indirectly exhibits the capability of the MAC. Also, indirectly, it exhibits scalability features as well. Coordinator searches for suitable free RF channel, not interfering with Wireless LAN frequencies in use. This is done on all the 16 channels. It is also referred to as energy scan. Coordinator then starts the network formation by assigning a PAN ID to the network. Assignment is done in two ways—manual (pre-configured) and dynamic (obtained by checking other PAN IDs of networks already in the operation nearby to avoid PAN ID conflict). Here Coordinator also assigns a network address to itself, that is, 0×0000. Now coordinator completes self-configuration and is ready to accept request queries from routers and end devices who wish to join the PAN. In addition to the above, the coordinator sends broadcast beacon request frame on remaining quiet channels. This is beacon scan or PAN scan. By this, coordinator receives PAN ID of routers and end devices present nearby. It also comes to know whether routers or end devices allow join or not. Now routers or end devices can join by sending association request to the coordinator. The coordinator will respond to association response.

There are two ways of ZigBee network joining

1. MAC association—implemented by the device underlying MAC layer
2. Network rejoin—implemented by the network layer

Few more protocols related to data link layer are mentioned in brief.

3.4.5 WirelessHART

HART is an acronym for highway addressable remote transducer. The HART Protocol makes use of the Bell 202 Frequency Shift Keying (FSK) standard to superimpose digital communication signals at a low level on top of the 4–20 mA. One can say that it is digital over analog. Due to phase continuous digital FSK, no interference with the 4–20 mA signal. HART technology is a master/slave protocol. The HART Protocol communicates at 1200 bps without interrupting the 4–20 mA signal and allows a host application (master) to get two or more digital updates per second from a smart field slave device. Approximately 30 million HART devices are in use worldwide, and HART technology is the most widely used field communication protocol for intelligent process instrumentation. Due to capability of wireless communication, it becomes competent for IoT application.

WirelessHART is a data link protocol that operates above IEEE 802.15.4 PHY and adopts time-division multiple access (TDMA) in its MAC. It is secure and reliable and uses advanced encryption techniques for messages calculating the integrity to offer reliability. The architecture, as shown in Figure 3.7, consists of a network manager, a security manager, a gateway to connect the wireless network to the wired networks, wireless devices as field devices, APs, routers and adapters. The standard offers end-to-end, per-hop, or peer-to-peer security mechanisms.

The specifications and characteristics of WirelessHART are as follows:

- Radios comply with IEEE 802.15.4-2006 and use 2.4 GHz license free frequency band. They employ DSSS technology and channel hopping for security and reliability, as well as TDMA-synchronized, latency-controlled communications between devices on the network.

- It can coexist with other wireless networks. Provision of blacklisting avoids frequently used channels; thus, network optimizes bandwidth and radio time. Time synchronization is incorporated for on-time messaging.

- Self-healing type of property of these networks can be exhibited. This is because the network
  - Adjusts communication paths for optimal performance.
  - Monitors paths for degradation and repairs itself.
  - Finds alternate paths around obstructions.

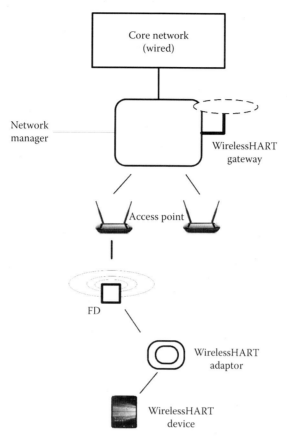

FIGURE 3.7
WirelessHART architecture. FD—Field Device, AP—Access point.

Mesh network topology and multiple APs are supported.

Each WirelessHART network includes three main elements as shown in Figure 3.7:

- *HART-enabled wireless field devices* (with a WirelessHART adapter attached to it) are connected to process or plant equipment.
- *Gateways* enable communication between wireless field devices (through AP) and host applications connected to a core network or other plant communications network.
- *Network manager* is responsible for network configuration, scheduling communications between devices, managing routes, and monitoring network functionality. The network manager can be integrated into the gateway, host application, or process automation controller.

TCP/IP   WirelessHART

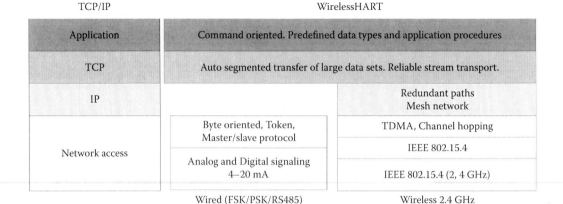

FIGURE 3.8
WirelessHART-layered architecture in comparison with TCP/IP stack.

This technology has been proven in field trials and real plant installations across various process control industries.

- Each device in the mesh network can serve as a router for messages from other devices. This extends the range of the network and provides redundant communication routes.
- The network manager determines the redundant routes based on latency, efficiency, and reliability. Consequently, data are automatically rerouted to follow a known good, redundant path with no loss of data in the case of nonsuitable path.
- The mesh design also makes adding or removing devices easy.
- For flexibility to meet different application requirements, the WirelessHART standard supports multiple messaging modes: (1) one-way publishing of process and control values, (2) spontaneous notification by exception, (3) ad-hoc request/response, and (4) autosegmented block transfers of large data. Figure 3.8 represents the protocol stack (Kim et al., 2008).

3.4.6 LoRaWAN

Long-range wide area network (LoRaWAN) is a new wireless technology designed for long range, low power wide area networks (LPWANs) with low cost, mobility, security, and bidirectional communication features for future needs of IoT applications. It is based on Semtech's long range (LORA) radio modulation technique and a specific MAC protocol. It is optimized for low power consumption and designed for scalability feature with millions of devices. It is location-free and energy-harvesting technology and supports redundant operation. In the OSI Reference Model, LoRa represents PHY, and LoRaWAN roughly maps primarily a data link (specifically MAC) layer, with only some components of a network layer.

LoRa transmissions work by chirping, breaking the chips in different places in terms of time and frequency to encode a symbol. In LoRaWAN, Spreading Factor (SF) refers to chirp rate. Different SFs can be decoded in the same frequency channel at the same time. The fact that LoRa transmissions jump from one place to another at a particular time means bit strings are changing.

3.4.6.1 LoRaWAN Architecture and Operation

LoRaWAN networks use a star of stars topology (as shown in Figure 3.9) with a hierarchy, in which *gateways* are a transparent bridge to relay messages between *end-devices* and a central *network server* in the backend. Gateways are connected to the network server via standard IP connections, whereas end devices use single-hop wireless communication to one or many gateways. Gateways forward messages between nodes and applications in the backend. Nodes use a single hop wireless connection to one or more gateways. Many end devices in the network could be trying to communicate with the gateway at the same time, but the gateway would not be able to hear or understand them all at a time. To support a wide range of applications, LoRaWAN has several different classes of end point devices, the capabilities of each are shown in Figure 3.10.

1. *Bidirectional end devices (Class A)*: They allow for bidirectional communications whereby the uplink transmission of each end device is accompanied by two short downlink receive windows. The end device can transmit on the basis of its own needs with a small variation based on a ALOHA. Downlink communications from the server at any other time will have to wait until the next scheduled uplink.

2. *Bidirectional end devices with scheduled receive slots (Class B)*: In addition to the Class A random receive windows, Class B devices open extra receive windows at scheduled times. Network may send downlink packet to nodes at any receive slot. To open its receive window at the scheduled time end device receives a time synchronized Beacon from the gateway. This allows the server to know when the end-device is listening.

3. *Bidirectional end devices with maximal receive slots (Class C)*: End devices of Class C have nearly continuously open received windows, only closed when transmitting.

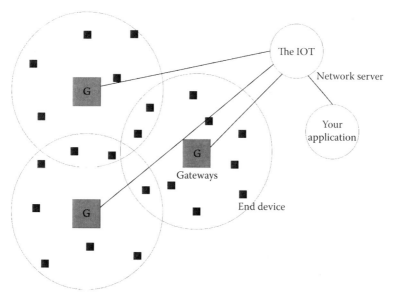

FIGURE 3.9
LoRaWAN topology.

Class A: Receiver Initiated Transmission (RIT) strategy

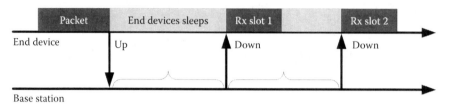

Class B: Coordinated Sampled Listening (CSL)

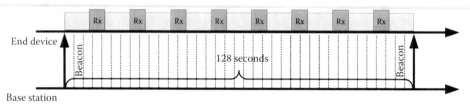

Class C: Continuous Listening (CL)

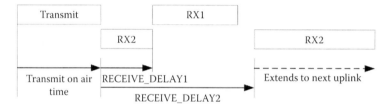

FIGURE 3.10
LoRaWAN devices with operative capabilities.

Similar to all wireless systems, the gateway cannot find out anything when it is carrying. So, when the gateway sends a message, it disconnects the receivers and is momentarily off the air. For this reason, transmissions and downlinks do not happen often in LoRaWAN. On account of this, LoRaWAN is best suited for uplink focused networks. Nearly all LPWAN systems, including LoRaWAN, have multiple receive channels. Most of the LoRaWAN systems can receive eight messages simultaneously, across any number of frequency channels.

The selection of the data rate is a trade-off between communication range and message duration. Due to the spread spectrum technology, communications at different data rates create a set of virtual channels (without interference between channels) increasing the capacity of the gateway. LoRaWAN data rates range from 0.3 to 50 kbps with an adaptive data rate (ADR) technique. It supports a coverage range of 2–5 km (urban environment), 15 km (suburban environment). Devices and applications have a 64 bit unique identifiers, DevEUI and AppEUI, respectively. When a device joins the network, it receives a dynamic (nonunique) 32 bit device address DevADDR (Tian et al., 2016) (LoRaWAN) (What is LoRaWAN?).

3.5 Other MAC Layer Protocols

3.5.1 Z-Wave

Z-wave is intended to provide a simple and reliable method to wirelessly control lighting, security systems, home cinema, automated window treatments, swimming pool, garage, home access controls, and suitable for IoT applications using a Z-Wave gateway or central control device. Z-Wave is a low power MAC protocol. It covers about 30–100 m range. The protocol-layered architecture is shown in Figure 3.11.

Z-Wave is designed to provide reliable, low latency transmission of small data packets at data rates up to 100 kbps. Message ability to hop is up to four times between nodes. It uses 15 unlicensed ISM band. It operates at 868.42 in Europe, at 908.42 MHz in the United States and Canada and uses other frequencies in other countries depending on their regulations. Modulation is by Manchester channel encoding. The MAC and PHY are described by ITU-T G.9959 and fully backward compatible. It uses CSMA/CA for multiple access and ACK messages for reliable transmission. It follows a master/slave architecture in which the master sends commands to slave and handles scheduling of the whole network.

Z-wave uses source-routed mesh network architecture. The devices can communicate by using intermediate nodes to route around overcoming various obstacles or radio dead spots that might occur in the multipath environment. A message from one to another node can be delivered successfully even if the two nodes are not within range, providing that a third node can communicate with both these nodes.

The simplest Z-wave network is a primary controller and one controllable device. Additional devices can be added at any time, may be as secondary controllers such as traditional hand-held controllers, key-fob controllers, wall-switch controllers, and PC applications specifically designed for management and control of a Z-wave network. A Z-wave network can include up to 232 devices with the option of bridging networks if more devices are required.

Each Z-wave network is identified by a NetID (also called HomeID), and each device is identified by a NodeID/an address. The NetID is the common identification of all nodes in

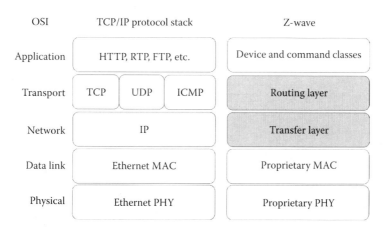

FIGURE 3.11
Z-wave protocol stack.

a logical Z-wave network. The NetID is 4 bytes (32 bits) long and is assigned to each device by the primary controller, when the device is included into the network. Nodes with different NetIDs cannot communicate with each other. The NodeID is 1 byte (8 bits) long and unique in its network.

Z-wave units can operate in power save mode reducing power consumption. Z-wave units to be able to route unsolicited messages cannot be in sleep mode. So, battery-operated devices are not designed as repeater units. As Z-wave assumes that all devices in the network remain in their original detected position, mobile devices, such as remote controls, are excluded (Bojkovic et al., 2016; The IoT powered by Z-wave).

3.5.2 G.9959

G.9959 is a short-range narrowband communication technology with PHY and MAC layer specifications and protocol from ITU, designed for low bandwidth and low cost, half-duplex reliable wireless communication. The reference model is shown in Figure 3.12. It is designed for critical real-time applications to deal with reliability and low power consumption.

The PHY and MAC layers of G.9959 are as follows, and also MAC indicates how to use these for IPv6 transport. G.9959 PHY recommendations specify sub-1-GHz RF transceivers. The PHY layer provides bit rate adaptation, that is, data flow control between the MAC and PHY. It encapsulates transmit MAC PDUs into the PHY frame and adds PHY-related control and management overhead. This is all via PMI (Figure 3.12). The PHY layer provides encoding of the PHY frame content including header and payload and modulates these encoded frames for transmission over the channel.

PHY is responsible for (1) assignment of an RF profile to a physical channel, (2) activation and deactivation of the radio transceiver, (3) data frames transmission and reception, (4) clear channel assessment, (5) frequency selection, and (6) link quality for received frames.

The MAC layer of G.9959 includes (1) unique network identifiers that allow 232 nodes to join one network, (2) unique channels access mechanism, (3) collision avoidance mechanism and backoff time in the case of collision, (4) acknowledgements and automatic

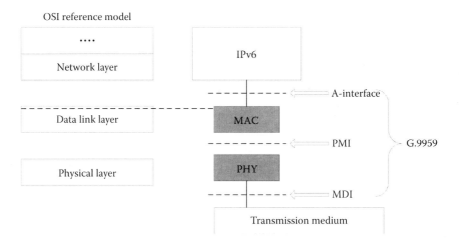

FIGURE 3.12
Protocol reference model of an ITU-T G.9959 transceiver. PMI—Physical medium-independent interface, MDI—Medium-dependent interface.

TABLE 3.3

M Flag Support and Interpretation for RA

M Flag Support	M Flag Value	Required Node Behavior
No	ignore	Node must use linklayer-derived addressing
Yes	0	Node must use linklayer-derived addressing
	1	Node must use DHCPv6 based addressing Node must comply fully with RFC6775

retransmissions of MAC PDUs to guarantee reliability, (5) dedicated wakeup pattern that allows nodes to sleep when they are out of communication and hence saves their power, and (6) frame validation. G.9959 defines how a unique 32 bit HomeID network identifier is assigned by a network controller and how an 8 bit NodeID host device identifier is allocated to each node. This forms the basis for IPv6 support. Specific M-Flag support may or may not be there for node addressing. By skipping support for the M flag, a cost-optimized node implementation may save memory. The M flag must be interpreted as defined in Table 3.3.

The G.9959 Segmentation and Reassembly (SAR) layer does the same for larger datagrams as usual and routing layer also provides a similar feature for routed communication. An IPv6 routing stack communicating over G.9959 may utilize link-layer status indications such as data delivery confirmation and ACK timeout from the MAC layer (Brandt and Buron, 2015).

3.5.3 Weightless

Weightless is another wireless low-power WAN (LPWAN) technology for IoT applications designed by a nonprofit global organization Weightless Special Interest Group (SIG). Weightless is both the name of a group and the technology. It is to note that the technology is an open global standard rather than a proprietary technology in order to guarantee low cost and low risk, and to maximize user choice and ongoing innovation. It is capable of long range, long battery life, QoS, ultra-low cost, and high capacity IoT products to be deployed rapidly and economically.

Three different weightless open standards are available: Weightless-N, Weightless-P, and Weightless-W. Weightless-N was first developed to support low cost and low-power M2M communication. P and N versions use ultra-narrow bands in the sub-1-GHz ISM frequency band, whereas Weightless-W provides the same features but uses white spaces of TV frequencies. The modulation scheme is differential binary phase shift keying (DBPSK) to transmit within narrow frequency bands using a frequency hopping algorithm for interference mitigation. It provides encryption and implicit authentication using a shared secret key to encode transmitted information via a 128-bit AES algorithm. Each Standard is designed to be deployed in different use cases depending upon a number of main priorities and summarized in Table 3.4.

Similar to other wireless systems, weightless system consists of two main parts, namely network subsystems and user terminals. It consists of base station, Internet backbone systems on the network side and terminal having weightless radio module along with a protocol stack. Weightless system frame structure is similar to WiMAX system frame structure having TDD topology. The weightless system operates with single carrier modulation, broadband downlink, narrowband uplink, direct sequence spreading, long frame size, and frequency hopping at the frame rate.

TABLE 3.4

Comparison of Weightless Protocol Versions

	Weightless-N (for Cost Effective Solution and One-Way Communication)	Weightless-P (for High Performance and Two-Way Communication)	Weightless-W (Where TV White Space is Available, Two-Way Communication)
Feature set	Simple	Full	Extensive
Range	>5 km	>2 km	>5 km
Battery life	9–10 years	3–8 years	3–5 years
Terminal cost	Very low	Low	Moderate
Network cost	Very low	Medium	Medium

Weightless frame consists of a preamble followed by downlink part and uplink part. Downlink refers to transmissions from base station to terminals, whereas uplink refers to transmissions from terminals to the base station. Usually weightless frame is of duration of about 2 seconds. The frame is divided into slots that are the smallest possible length of resource allocation to the terminals. Mobility is supported by the network by automatic routing terminal messages to the correct the destination.

Other PHY/MAC technologies suitable for IoT applications are DECT/ULE, HomePLUG, IEEE 802.3 Ethernet, IEEE 802.16 WiMAX, ultra-wide band (UWB), infrared data association (IrDA), near-field communication (NFC), and so on (What is weightless?).

3.6 Network Layer Protocols with Key Features and Elimination of Challenges

The main functions of the network layer are to enable the correct use of the MAC sub-layer and provide a suitable interface for the upper layer, mainly transport and application layer. Routing and encapsulation of IP are the two basic aspects of the network layer. It also needs to control fragmentation as per the capabilities of the network and exhibits Internetworking.

On the one hand, it is necessary to manage network layer data units coming from the payload of the application layer and their routing according to the existing topology. On the other hand, layer control is required to handle configuration of new devices and establish new networks. It is necessary to determine a neighboring device belongs to the particular network and also discovering new neighbors and routers. The control also detects the presence of a receiver to allow direct communication and MAC synchronization.

Generally, the routing protocol used by the network layer of IoT is based on Ad Hoc On-Demand Distance Vector (AODV) protocol, and algorithm indirectly highlights functions associated at network layer. To find the path to destination, source device broadcasts a route-request to all of its neighbors in the dynamic topology. The neighbors in turn broadcast the route-request to their neighbors until the destination is reached. Once the destination is reached, destination device sends route-reply via unicast transmission following the lowest cost path back to the source. Once the source receives the route-reply, it will update its routing table for the destination address of the next hop in the path and total path cost. There are few routing protocols described soon (Patil and Lahudkar, 2016; Salman, 2008).

3.6.1 Challenges in IoT Networking

IoT system may be deployed in harsh environments such as forests, underwater, battle fields, and consequently has to rely on wireless mesh technologies for communication, hence the following challenges:

- IoT nodes must not be always *ON* to conserve the battery power. This brings more challenges to the TCP/IP protocol architecture. Mesh networks typically adopt the multilink subnet model that is not supported by the original IP addressing scheme.

- Broadcast and multicast are expensive on a battery-powered network because single multicast will involve a series of multihop forwarding, resulting in waking up of many nodes.

- A scalable routing mechanism is necessary for IP communications to happen over the mesh networks. Also, communication between non-IP and IP networks is required.

- The reliable and in-order byte stream delivery as in TCP is often not suited for applications that require customized control and prioritization of their data (Salman, 2008).

3.6.2 Network Layer Routing Protocols

3.6.2.1 RPL

Routing Protocol for Low-Power and Lossy Networks (RPL) provides a mechanism to control different traffic inside the Low-Power and Lossy Networks (LLNs) supporting specifications on routing metrics, objective functions, and security. In LLN, both the routers and their interconnections are constrained. LLN routers typically operate with constraints on processing power, memory, and battery power, whereas interconnects are constrained by high loss rates, low data rates, and instability. LLNs comprise tens to thousands of routers.

Supported communications are as follows:

- Point-to-point (between devices inside the LLN)
- Point-to-multipoint (from a central control point to a subset of devices inside the LLN)
- Multipoint-to-point (from devices inside the LLN toward a central control point)

RPL is developed by routing over low power and lossy networks (ROLL) working group of IETF, who did a detailed analysis of the routing requirements focusing on varieties of IoT applications. The objective of the group was to design a routing protocol for LLNs, supporting a variety of link layers without specific assessment on the link layer, sharing the common link characteristics of being low bandwidth, lossy and low power. Link layer could either be wireless such as IEEE 802.15.4, IEEE 802.15.4g, IEEE 802.11ah, or Power Line Communication (PLC) using IEEE 802.15.4 such as IEEE P1901.2.

Routing issues are very challenging especially when the low-power and lossy radiolinks, the battery-powered nodes, the multihop mesh, and the dynamic topologies due to mobility. The solutions should take into account the specific application requirements, along with IPv6 behavior and its adaption mechanisms. LLNs must be able to build up network routes quickly, to distribute routing knowledge among nodes, and to adapt the topology efficiently.

In RPL, typically, the nodes of the network are connected through multihop paths to a small set of root devices, which are usually responsible for data collection and coordination duties. RPL is a distance vector routing protocol for LLNs that makes use of IPv6, and network devices running the protocol are connected like a spanning tree. For devices, an optimized Destination Oriented Directed Acyclic Graph (DODAG) is created taking into account link costs, node attributes/status information, and an Objective Function (OF) that defines the way of routing metric computation. The graph is routed at a single destination. OF specifies how routing constraints and other functions are taken into account during topology construction (Al-Fuqaha et al., 2015).

In some cases, a network has to be optimized for different application scenarios, situations, and deployments, and in order a DODAG may be constructed as per the expected number of transmissions or the current status of battery power of a node. RPL allows building a logical routing topology over an existing physical infrastructure. RPL *Instance* defines an OF for a set of one or more DODAGs. The protocol tries to avoid routing loops and compute a node's position relative to other nodes with respect to the DODAG root. This position is called a *Rank* that increases if nodes move away from the root and vice versa. The Rank may be in terms of a simple hop-count, or it may be calculated with respect to other constraints.

The RPL protocol defines four types of control messages with respect to DODAG for topology maintenance and data exchange.

1. *DODAG information object (DIO)*: It is the main source of routing control information. It may store information such as the current *Rank* of a node, the current RPL *Instance*, the IPv6 address of the root, and so on.

2. *Destination advertisement object (DAO)*: It enables the support of down traffic and is used to propagate destination information upward along the DODAG.

3. *DODAG information solicitation (DIS)*: It makes possible for a node toward requiring DIO messages from a reachable neighbor.

4. *DAO-ACK*: It is sent by a DAO recipient in response to a DAO message.

The RPL specification defines all four types of control messages as ICMPv6 information messages with a requested type of 155.

Another important fact about the protocol's design is the maintenance of the topology. As most of the devices in LLNs are typically battery powered, it is crucial to restrict the number of sent control messages over the network. Many routing protocol broadcast control packets at a fixed time interval that causes waste of energy when the network is stable. RPL adapts the sending rate of DIO messages by extending the Trickle algorithm. Number of control messages will be less over stable links whereas RPL is made to send control information more often in dynamic topologies.

3.6.2.1.1 RPL—Mode of Operation

RPL specifies how to build a DODAG sometimes referred to as a graph in the rest of this section using an OF. The objective function operates on a combination of metrics and constraints to compute the *best* path. The objective function does not necessarily specify the metric/constraints but dictates some rules to form the DODAG (e.g., the number of parents, backup parents, use of load-balancing). There could be several OFs in operation on the same node in the mesh network because deployments vary greatly with different objectives and a single mesh network may need to carry traffic over different path quality. For example, several DODAGs may be used with the objectives

1. Find paths with best Expected Transmissions (ETX) values (metric) and avoid non-encrypted links
2. Find the best path in terms of latency (metric) while avoiding battery-operated nodes

The graph built by RPL is a logical routing topology. The network administrator may decide to have multiple routing topologies (and hence graphs) active at the same time to carry different traffic with different requirements. A node in the network can join one or more graphs (in this case we call them RPL *Instances*) and mark the traffic according to the graph characteristic to support QoS aware and constraint-based routing. The marked traffic flows up and down along the edges of the specific graph.

Each RPL message has a level of security (32 bit and 64 bit MAC and ENC-MAC modes are supported). The algorithms (CCM and AES-128) in use are indicated in the protocol messages (Patil and Lahudkar, 2016).

3.6.2.2 CORPL

An extension of RPL is cognitive and opportunistic RPL (CORPL) that is designed for cognitive networks and uses DODAG topology generation technique but with two modifications to RPL. In cognitive networks the primary and secondary users are supported. Secondary users are called opportunistic users. An opportunistic forwarding approach is used in CORPL to meet the utility requirements of secondary network to forward the packet by choosing set of multiple forwarders and choosing best next hop out of them in coordination among the nodes. DODAG is built in the same way as RPL. Each node maintains a *forwarding set* instead of its parent only and updates its neighbor with its changes using DIO messages. Based on latest information, each node dynamically updates its neighbor priorities in order to construct the forwarder set (Patil and Lahudkar, 2016).

3.6.3 Network Layer Encapsulation Protocols

3.6.3.1 6LoWPAN

6LoWPAN is the first and most commonly used encapsulation protocol. 6LoWPAN is an open standard defined in RFC 6282 by IETF and represents an IPv6 adaptation layer allowing short packet networks to exchange IP packets. Core protocols for 6LoWPAN architecture have already been specified. It efficiently encapsulates IPv6 long headers in small packets. The specification/protocol supports different address lengths, low bandwidth, different topologies, low cost, scalable networks, mobility, unreliability, and long sleep time. The standard provides header compression to reduce transmission overhead, fragmentation to meet the required frame length (Patil and Lahudkar, 2016). 6LoWPAN uses four types of headers in various frames:

- No 6LoWPAN header (00): Any frame that does not follow 6LoWPAN specifications is discarded.
- Dispatch header (01): It is used for multicasting and IPv6 header compressions.
- Mesh header (10): These headers are used for broadcasting.
- Fragmentation header (11): These headers are used to break long IPv6 header to fit into fragments of maximum say 128 byte length.

3.6.3.2 6LoWPAN Use Case

To send IPv6 packets over, for example, 802.15.4 efficiently, one needs to concentrate on the issues arising from the underlying MAC and PHY protocols as discussed earlier. In the case of Ethernet links, a packet with the size of the IPv6 maximum transmission unit (1280 bytes) can be sent easily as single frame over the link. In the case of 802.15.4, because it supports a maximum packet size of 127 bytes, 6LoWPAN acts as an adaptation layer between the IPv6 networking layer and the 802.15.4 link layer. It solves the issue by fragmenting the IPv6 packet at the sender and reassembling it at the receiver. 6LoWPAN also provides a compression mechanism to reduce the IPv6 headers sizes to reduce transmission overhead, which, in turn, reduces the energy consumption as the fewer bits are sent over the wireless channel. It provides a very efficient low overhead mechanism for forwarding multihop packets in a mesh network.

A typical example of IoT network based on 6LoWPAN is shown in Figure 3.13. The 6LoWPAN network is connected to the IPv6 network using an edge router. The edge router handles three actions:

1. The data exchange between 6LoWPAN devices and the Internet/IPv6
2. Local data exchange between devices within the 6LoWPAN
3. The generation and maintenance of the 6LoWPAN network

One 6LoWPAN network may be connected to other IP networks through one or more edge routers that forward IP datagrams between different media (Al-Fuqaha et al., 2015).

Two other device types are included inside the typical 6LoWPAN network shown in Figure 3.13: routers and hosts. Routers route the data destined to another node in the 6LoWPAN network. Routers form the distributed network scenario. Hosts are the end devices (or nodes in case of sensor networks) and are not able to route data to other devices in the network. Host can also be a sleepy device, waking up periodically to check its parent (a router) for data. This enables very low power consumption. 6LoWPAN network protocol architecture is represented in Figure 3.14 in comparison with TCP/IP protocol stack.

6LoWPAN connects more things to the cloud. A powerful feature of 6LoWPAN is that although originally designed to support IEEE 802.15.4 in the 2.4 GHz band, it is now being adopted over a variety of other networking media including sub-1 GHz low-power RF, Bluetooth Smart, Power Line Control (PLC), and low power Wi-Fi (Patil and Lahudkar, 2016; Lora Alliance: Wide Area Networks for IoT) (Thread usage of 6LoWPAN).

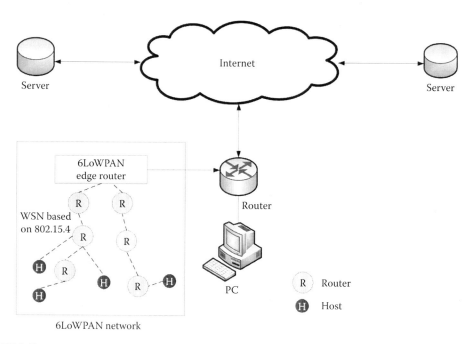

FIGURE 3.13
An example of an IPv6 network with a 6LoWPAN mesh network.

TCP/IP Protocol stack				
HTTP		RTP		Application
TCP	UDP	ICMP		Transport
IP				Network
Ethernet MAC				Data link
Ethernet PHY				Physical

6LoWPAN protocol stack	
Application	
UDP	ICMP
IPv6 with LoWPAN	
IEEE 802.15.4 MAC	
IEEE 802.15.4 PHY	

FIGURE 3.14
Protocol stack comparison of TCP/IP and 6LowPAN.

3.7 Conclusion

After studying the various challenges of MAC and Network layer, one has the scope to optimize or enhance the various protocols described in this chapter for IoT applications. The new protocols for IoT are developed on the basis of the situation and requirements. Major developments required are up to IP level; thereafter, everything is set in the Internet. This shows the importance of developing PHY-MAC and relevant network layer along with adaptation layer. These protocols may be non-IP. Internet of local low power networks is a future of this whole world that will bring the revolution in human lives. Of course upper layers are equally important along with the security aspects over these basic aspects. They add quality of services in the application.

References

Al-Fuqaha, A., Guizani, M., Mohammadi, M., Aledhari, M., and Ayyash, M. (2015), Internet of things: A survey on enabling technologies, protocols and applications. *IEEE Communications Surveys Tutorials*, 17: 2347–2376. doi:10.1109/comst.20152444095

Aust, S., Prasad, R. V., and Niemegeers, I. G. (2012), IEEE 802.11 ah: Advantages in standards and further challenges for sub 1 GHz Wi-Fi. In *2012 IEEE International Conference on Communications (ICC)*, pp. 6885–6889.

Bojkovic, Z., Bakmaz, B., and Miodrag, B. (2016), *Security Aspects in Emerging Wireless Networks*, 1: 158–165.

Brandt, A. and Buron, J. (2015), Transmission of IPv6 Packets over ITU-T G.9959 Networks. *IETF RFC 7428*, Retrieved from http://www.ietf.org/rfc/rfc7428.txt

Decuir, J. (2010), Bluetooth 4.0: Low Energy, Presentation slides, Retrieved from: http://chapters.comsoc.org/vancouver/BTLER3.pdf

Gomez, C., Oller, J., and Paradells, J. (2012), Overview and evaluation of Bluetooth low energy: An emerging low-power wireless technology. *Sensors*, 12(9), 11734–11753.

Home of RF and Wireless Vendors and Resources, Retrieved from http://www.rfwireless-world.com

IEEE 802.11ah-subGHz Wi-Fi, Retrieved from http://www.radio-electronics.com/info/wireless/wi-fi/ieee-802-11ah-sub-ghz-wifi.php

IEEE 802.15.4, (September 5, 2011), IEEE Standard for Local and metropolitan area networks-Part 15.4: Low-Rate Wireless Personal Area Networks (LR-WPANs), http://standards.ieee.org/getieee802/download/802.15.4–2011.pdf

IOT protocols you need-to know about Retrieved from http://www.rs-online.com/designspark/electronics/knowledge-item/eleven-internet-of-things

Kim, A., Hekland, F., Doyle, P., and Petersen, S. (2008), When HART goes wireless: Understanding and implementing the Wireless HART standard in *IEEE International Conference on Emerging Technologies and Factory Automation (ETFA 2008)*, pp. 899–907.

Lora Alliance: Wide Area Networks for IOT, Retrieved from: https://www.lora-alliance.org/What-Is-LoRa/Technology

LoRaWAN, Retrieved from: https://www.thethingsnetwork.org/wiki/LoRaWAN/Overview

Montenegro, G., Hui, J., Kushalnagar, N., and Culler, D. (2007), Transmission of IPv6 Packets over IEEE 802.15.4 Networks. *IETF RFC 4944*, 1–30, https://tools.ietf.org/html/rfc4944

Patil, B., Nieminen, J., Isomaki, M., Savolainen, T., Shelby, Z., and Gomez, C. (2015), IPv6 over Bluetooth low energy. *IETF RFC 7668*, http://www.ietf.org/rfc/rfc7668.txt

Patil, K. and Lahudkar, S. L. (2016), A survey of MAC layer issues and application layer protocols for machine to machine communications, 3(2), 742–747

Rajandekar, A. and Sikdar, B. (2015), A survey of MAC layer issues and protocols for machine-to-machine communications. *IEEE Internet Things J.*, 99: 175–186. http://dx.doi.org/10.110g/JIOT.2394438

Reiter, G. (2012), White paper "Wireless connectivity for IoT", Texas Instruments. http://www.ti.com/lit/wp/swry010/swry010.pdf

Salman, T. (2008), Networking protocols and standards for internet of things. http://www.cse.wustl.edu/~jain/cse570–15/ftp/iot_prot

The Internet of Things powered by Z-wave, Retrieved from http://z-wavealliance.org/

Thread Usage of 6LoWPAN, Retrieved from https://www.silabs.com/SiteDocs/white-papers/Thread-Usage-of-6LoWPAN.pdf

Tian, L., Famaey, J., and Latre, S. (2016), Evaluation of the IEEE 802.11ah restricted access window mechanism for dense IoT networks. *IEEE Conference world of wireless mobile and multimedia networks*, Coimbra, Portugal, pp. 21–24. doi:10.1109/WoWMoM.2016.7523502

Transmission of IPV6 Packets over ITU-T g.9959 Networks, Retrieved from http://www.ietf.org/rfc/rfc7428.txt

What is LoRaWAN? Retrieved from http://www.link-labs.com/what-is-lorawan/

What is Weightless? Retrieved from http://www.weightless.org/about/what-is-weightless

4

Privacy and Security of Data in IoT: Design Principles and Techniques

K.V. Arya and Vishnu Prabhakaran

CONTENTS

4.1 Introduction

The Internet of Things (IoT) is moving from a unified structure to a perplexing system of decentralized shrewd gadgets. This movement guarantees altogether new administrations and business opportunities. An inexorably associated world will see the developing systems administration and cloud enablement of a wide range of physical gadgets from machines through cars to home apparatuses.

The IoT includes the expanding predominance of articles and elements—referred to, in this setting as things—furnished with one-of-a-kind identifiers and the capacity to consequently exchange information over a system. A significant part of the expansion in IoT correspondence originates from figuring gadgets and inserted sensor frameworks utilized as a part of mechanical machine-to-machine (M2M) correspondence, vehicle-to-vehicle communication and wearable processing gadgets, and so on.

The principle issue is that on the grounds that systems administration machines and different articles are moderately new, security has not generally been considered in item plan. IoT items are frequently sold with old and non-patched implanted working frameworks and programming. Besides, buyers frequently neglect to change the default passwords on shrewd gadgets—or in the event that they do transform them, neglect to choose adequately solid passwords. To enhance security, an IoT gadget that should be specifically open over the Internet ought to be fragmented into its own particular system and has system access limited. The system section ought to then be checked to distinguish potential bizarre activity, and move ought to be made if there is an issue.

4.2 Issues in IoT

A quarter century, on the off chance that you let me know my telephone could be utilized to take the secret word to my e-mail account or to take a duplicate of my unique mark information, I would have snickered at you and said you observe a lot of James Bond. Be that as it may, today, in the event that you let me know that programmers with malevolent expectations can utilize my toaster to break into my Facebook account, I will freeze and rapidly pull the fitting from the detestable machine.

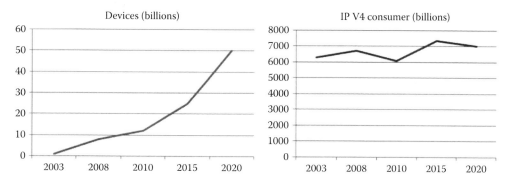

FIGURE 4.1
Evolution of IoT (in terms of number of devices and number of consumers) over the years.

Welcome to the period of the IoT, in which digitally associated gadgets are infringing on each part of our lives, including our homes, workplaces, cars, and even our bodies. With the appearance of IPv6 and the wide arrangement of Wi-Fi systems, IoT is developing at a perilously quick pace, and scientists assess that by 2020, the quantity of dynamic remote-associated gadgets will surpass 40 billion as observed in Figure 4.1.

In any case, IoT brings up essential issues and presents new difficulties for the security of frameworks and forms and the protection of people. Some IoT applications are firmly connected to delicate foundations and vital administrations, for example, the dispersion of water and power and the reconnaissance of advantages. Different applications handle delicate data about individuals, for example, their area and developments, or their well-being and buying inclinations. Trust in and acknowledgment of IoT will rely on upon the insurance it gives to individuals' protection and the levels of security it sureties to frameworks and procedures.

The upside is that we can do things we at no other time envisioned. In any case, as with each good thing, there is a drawback to IoT: It is turning into an undeniably alluring focus for cybercriminals as indicated in Figure 4.2. More associated gadgets mean more assault vectors and more potential outcomes for programmers to target us; unless we move quickly to address this rising security concern, we will soon be confronting an inescapable fiasco. The IoT is based on a wide range of semiconductor advances, including power administration gadgets, sensors, and chip. Execution and security necessities change significantly

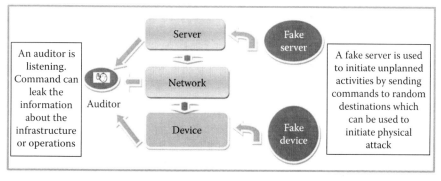

FIGURE 4.2
A typical information leakage scenario.

starting with one application then onto the next. One thing is consistent, nonetheless. The more prominent the volume of delicate information we exchange over the IoT, the more prominent the danger of information and fraud, gadget control, information adulteration, IP robbery, and considerably server/system control. The issues in IoT can be classified in different ways in the following sections.

4.2.1 Security Issues

In all the business processes, the required degree of reliability is very high. The following security and privacy requirements are supposed to be the part of secure system (Fabian and Gunther, 2007):

Power to handle attacks: The system is designed in such a way that it is capable of avoiding any point of failure and should reconfigure itself to avoid node failures.

Data authentication: In principle, the system has the mechanism to authenticate the retrieval of address and information.

Access control: System must be capable of implementing access control on the given data.

Privacy of Client: By very design, the measures to be taken so that the information provider can only interpret the information based on the need of particular customer.

Security issues in IoT are classified depending on the different layers of IoT architecture. In network layer, a number of security issues can arise. For example, Denial of Service, man-in-the-middle attacks, and so on. Due to an enormous heterogeneous volume of data exchange, a lot of network security issues come into picture. This apparently leads to network congestion or sometimes even data eavesdropping too. In the application layer, because of the integration between different technologies, the application may face extra issues like replay attacks or data tampering. The key exchange mechanism should be secure enough in order to prevent it from any intruder. Apart from these, the attacker might gain the encryption key based on the time taken to encrypt the data, which is known as timing attack, must be addressed. The attacker can take control of all the nodes in the system which is called Node Capture attack.

4.2.2 Communication Issues

IoT communication can be stated in two different ways, namely, inter- and intra-domain. In both the cases, a lot of issues can be identified. To be precise, there is no standard protocol or framework for IoT communication. This essentially means incapability to merge different IoT technologies together. The lack of standard universal agreement makes the system difficult to realize in all practical aspects. The existing protocols do not provide the actual meaning to the IoT, as the IoT is much different, in which different technologies of different functionalities are combined, which makes the system heterogeneous. As IoT systems experience heterogeneity, intracommunication process becomes an unrealizable task. Moreover, effective mechanisms should be handled to take care of the traffic explosion caused by exchange of different IoT devices. There is a possibility to maintain interdomain communication for a long period, thus energy efficient protocols have to be considered.

4.2.3 Security Implementation in IoT

A light-weight protocol, Message Queue Telemetry Transport, which is especially designed for IoT can be implemented. The key feature here is to implement *a broker*. There are different sorts of brokers available like HiveMQ, IBM Message Sight, and so on. The purpose of these brokers is to help in obtaining the information from the *things*. All these brokers vary in their features and some of them implement additional features on top of standard Message Queue Telemetry Transport functionality. A notable implementation has been provided by using Wired Drone Docking system that can perform an assortment of capacities careful cooperation with gadgets in existing IoT environment, and it can be connected to industrial security.

4.3 Internet of Insecure Things

The Rapid7 exploration is the recent one to find the alarming security issues. This is connected to IoT similar to regular gadgets—including clothes washers, indoor regulators, and cars—that have processing and system capacities implanted into them. The Rapid7 scientists concentrated on electronic devices as they are generally utilized and given the seriously individual uses IoT gadgets could serve. The scientists went ahead to caution that the bugs they found could do substantially more than permit eavesdroppers to attack the proprietors' close to home protection. A portion of the additionally terrifying vulnerabilities found on IoT gadgets has brought IoT security further up the pile of issues that should be tended to rapidly. The *Internet of Things* is a term not without buildup and metaphor, but rather at its heart is an idea that is as of now picking up energy. Innovation and information transfers firms are connecting *things* as various as cell phones, cars, mechanical sensors, and family apparatuses to the web, empowering intercommunication and independent M2M information exchange. The quantity of associated buyer items as of now counts in the billions, and investigators gauge it will hit 6.4 billion overall this year. There is a lot of business potential in these extending associations, yet there is a major drawback: the quantity of conceivable IoT security vulnerabilities is becoming exponentially increasing too. The IoT makes basically every part of data administration more confounded: there are more gadgets, more vulnerability, and more data that is streaming. That makes IoT information security more muddled, as well.

> It just opens up new pathways

Experts say that there are IoT security vulnerabilities along the entire chain: Someone could spoof a sensor connecting to transmit data, an organization with poor authentication could inadvertently allow a hacker onto its network, and an insecure connection could allow a cybercriminal in.

4.3.1 New IoT Threats and Attacks

As the technology develops, there are much more threats that come into limelight. Thanks to numerous researches that have provided a way to distinguish the new IoT threats. The following gives an insight of how IoT threats can be classified based on different factors.

- Vehicles or human body can be manipulated
- Possibility of wrongly diagnosing and treating patients

- Possibility to gain physical access to any building
- Loss of vehicle control
- Critical information such as warning of floods can be blocked
- Critical infrastructure damage can occur
- Sensitive information such as credit card can be stolen
- Unauthorized tracking of people's locations
- Manipulation of financial transactions

Let us see how a car could be controlled by a hacker. We assume the car has the ability to make use of GPS and judge the route based on the traffic information. This can also mean that hackers can gain control over your car. To be precise, a hacker can control the door opening, flashing headlights, and so on. The scenario has been explained in Figure 4.3.

Apart from all these threats, there are many threats that are being discovered every day or minute rather.

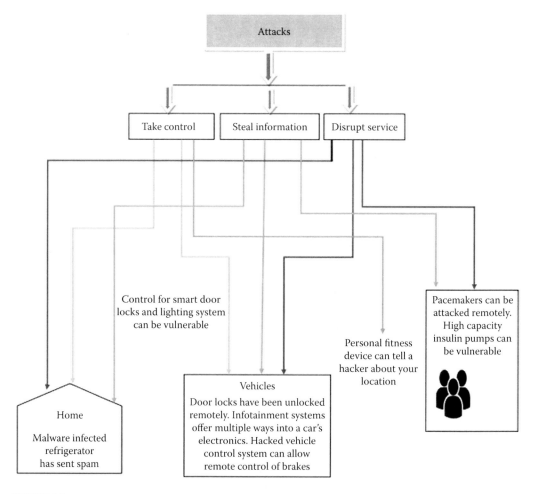

FIGURE 4.3
IoT dangers—car attacks.

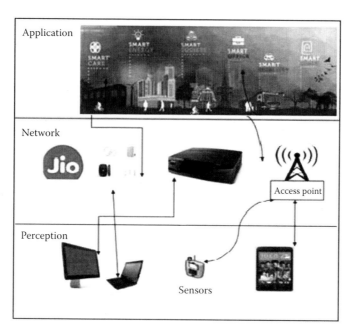

FIGURE 4.4
Ensuring security in trilayer of IoT architecture.

4.3.2 IoT Security Countermeasures

Security measures have to be implemented in all three layers starting from physical, network, to application layers as shown in Figure 4.4. Proper authentication measures have to be provided such as the one discussed in Zhao et al. (2011). A feature extraction method has to be combined with hash function to avoid collisions between IoT terminal nodes and platforms. This method emphasizes the authentication process when the data are exchanged from IoT platform to terminal nodes.

To provide an accurate access control is as vital as authenticating for security. This is given in the form of Identity Authentication and Capability based Access Control (Mahalle et al., 2013). This method uses timestamp in the authentication messages between various gadgets. Thus, it prevents one of the most popular attacks *man in middle*.

4.4 Challenges in Securing IoT Technologies

As indicated by an extraordinary number of researchers, gadgets in the IoT world get an assorted cluster of potential security dangers. Some of these dangers are customary dangers for standard PCs. For example, new smart TVs that empower clients to surf the Internet buy other *things* and offer photographs by means of social media could put information away. More sensitive data such as card data, passwords, and individual information can encourage data tampering. The idea of *Multiple Points of Vulnerability* in the security of IoT framework, actually, incorporates the IoT item, installed programming, information inside the gadget, correspondence channels, information collection stage, to examine the sensor

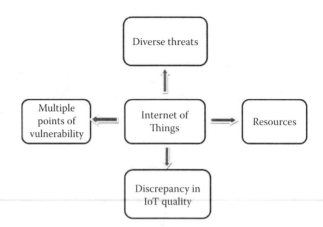

FIGURE 4.5
Common security characteristics of IoT products.

information. Giving security to these surfaces is a genuine test for organizations here. The common characteristics of IoT security products are given in Figure 4.5.

In all actuality, there is an incredible inconsistency in the nature of IoT gadgets overall, as some of them are modern and costly and others are shabby and expendable. Security vulnerabilities in one gadget might be utilized to encourage assaults on different gadgets sometime later in a given customer's system. There can be many challenges that arise in deploying security in IoT architecture. For example, some of the IoT systems are badly designed. These are done with the help of protocols that give rise to complex configurations. There is a lack of mature IoT technologies and business processes. There is no proper guidance for lifecycle maintenance and management of IoT devices. Furthermore, the IoT presents one-of-a-kind physical security concerns. IoT security concerns are mind boggling and not generally promptly apparent. Some security concerns are not promptly identifiable and some worries are not feasible by essentially implementing secrecy assurances, character, or area to exchanges. Numerous IoT designers are not yet acquainted with secure improvement best practices. The race to make new IoT-based capacities will probably bring about restricted spotlight on the security of the new usefulness being made. There is an absence of models for confirmation and approval of IoT edge gadgets. Review and logging principles are not characterized for IoT parts. Confined interfaces are accessible to collaborate with IoT gadgets onto security gadgets and applications. There is no emphasis yet on recognizing techniques for accomplishing situational consciousness of the security stance of an association's IoT resources. A security standard for stage setups including virtualized IoT stages supporting multi-tenancy is juvenile.

4.4.1 Types of Threats in IoT

The threats in IoT can be classified into many types.

4.4.1.1 Spoofing Attacks

When a malicious user impersonates the authorized user, he falsifies the data. Many of the existing IoT protocols do not authenticate the messages from a source or destination, results of which make the system more vulnerable to spoofing attacks. The spoofing attacks can be of different types like IP spoofing, e-mail spoofing, or network spoofing. To understand the concept better, let us go to reality where we get some of the phishing mails asking to update the profiles like Paypal. Many of us are aware of the fact that it is from the spoofed site. But some of us may fall into the trap of these websites and click the link to navigate to the spoofing site. This spoofing site is well designed such that it looks exactly similar to the original site. Thus, the attackers have the higher probability of stealing the sensitive information like credit or debit card details. There is a very popular spoofing occurs in the name of IP spoofing. This necessarily implies that the attacker tries to steal the information from a trusted computer network. IP spoofing is always carried out by the hacker for malicious actions. The attacker first modifies the source address field in the transmitted packet. When the destination opens up the connection for the forged source address, it can lead to a large number of attacks. This IP spoofing can be classified as many.

4.4.1.2 Non-Blind Spoofing

This is a kind of attack that takes place when the hacker and the victim share the same subnet. This hacker can bypass any authentication measures with the help of a mechanism called session hijacking. Further, the stream of data can be corrupted. With the help of any advanced tool, the hacker can reestablish the correct stream of data.

4.4.1.3 Blind Spoofing

The attackers can target the host with a bunch of data packets to obtain the correct sequence numbers. These sequence numbers are really useful to assemble the data packets for the attacker. However, modern systems use random numbering for sequences hence it is a bit difficult for the attacker to guess the correct one. With the help of obtaining the sequence numbers, the attacker can inject the data into the packets as it requires nonauthentication connection establishment.

4.4.1.4 Man in the Middle Attacks

The purpose of this attack is to interrupt the legal communication between two legal parties. Without having the exact knowledge about the machine, the hacker can send spoofed packets into the stream of data. The attacker can fool the host and can modify or delete the packets by forging the original identity.

4.4.1.5 Denial of Service Attacks

Within zero time, the attacker can flood the host machine with a large number of requests. The attacker can spoof the IP address of the victim which in turn makes it difficult to be traced. First, the attackers can create a list of more vulnerable hosts which is obtained with the help of scanning the Internet and convince them to install any type of attack programs which can then be exploited to obtain the full access.

4.4.1.6 Data Tampering

This is the deliberate modification of data. The interesting fact is that the tampered data will always not be detected in the first place. These data can be tampered in any ways like it can be tampered when it is in the channel or while it is being stored in the system.

The main motive of the attackers is to target the data packets that are not encrypted. A good measure is to encrypt the data whenever and wherever possible.

4.4.1.7 Repudiablility

It is the action of denying something that has already occurred. It is like denying the reception of some receipt or service and asking for another one. This can be prevented with the help of digital signatures and hashes.

4.4.1.8 Information Disclosure

The severity of this depends on the type of sensitive data are being disclosed. For example, medical or credit card data will attract high severity. This could be a serious threat in case of IoT applications. This is somewhere related to data tampering as disclosure is the very first step in data tampering. So keeping the data tempering out of the context will help us to get rid of information disclosure attacks.

4.4.1.9 Network Intrusions

IoT gadgets bring the guarantee of business streamlining, remote patient checking, help with discovering parking spots, and a large group of different advantages, some not yet even imagined. Most of the devices are built using low-cost hardware which makes them vulnerable to security risks. One of such risks is network intrusions. Due to the inexpensive nature of the IoT devices, it becomes really difficult for them to report or log any information about whether the hacker has penetrated into the network or when the network security has been compromised. The first phase to overcome this problem is to use detection software for the network intrusion. For instance, let us consider a hacker tries to gain access control of any IoT device. After discovering the IoT device on the network, this hacker can use brute force attack to find out the correct log in information. When the hacker floods with multiple login attempts, most embedded systems should just process and drop the request each time when the login information is matched with the available data. These embedded systems just consider the next request as a normal request. This essentially means that the embedded system is not aware of the fact that it is under network intrusion attack. Hence, it cannot report to the security management system. When the users have a strong password which cannot be easily guessed by attacker, he may not succeed in any of his attempts. Unfortunately, users have the habit of setting the easily guessed passwords. For example, many of the passwords set by the users have *user* or *admin* in it. This opens up many security risks. Intrusion-detection software would usually alert the system after the flooding of the failed login requests and generates report for such requests.

4.4.1.10 Social Engineering Attacks

Social engineering attacks have become very common against enterprises. With programmers formulating perpetually astute techniques for tricking workers and people into giving over significant organization information, ventures must use due industriousness with an end goal to stay two stages ahead of cyber criminals. These types of attacks commonly include some type of mental control, tricking generally clueless clients or workers into giving over classified or delicate information. Normally, social engineering includes e-mail or other correspondence like asking to click a malicious link or to open an infected file.

It is important to analyze the severity levels of each of these threats and document it before IoT deployment.

4.4.2 Security Requirements

The security requirements can be classified into three broad categories.

4.4.2.1 Data Security Requirements

A data security system is an arrangement to mitigate dangers while submitting to with legitimate, statutory, inside, and legally binding created requests. Common strides to building a system incorporate the meaning of control destinations; the evaluation and recognizable proof of ways to deal with meet the targets, the determination of controls, measurements, the foundation of benchmarks, and the arrangement of usage and testing plans. The decision of controls is ordinarily relies on upon cost examination of various key ways to deal with minimize the danger. The cost correlation normally differentiates the expenses of various methodologies with the potential pick up money-related establish-ment could understand as far as expanded availability, confidentiality, or integrity of the information is concerned. Whatever be the approach, it should consider each and every aspect of the following entries.

- There should be a standard policy or procedures
- Technology should be designed in a way that is compatible to all IoT products
- A typical data security system should support every gadget

4.4.2.2 System Security Requirements

Providing security nowadays has turned into an intense errand for all the business and the distinctive associations. Security must be given to the clients and the vital informa-tion to shield them from the vindictive and automatic leaks. Information is imperative for each undertaking; it might be the custom records or licensed innovation. To keep the trust between the IoT products and the consumers, it is vital to provide the security in the system.

4.4.2.3 Network Security Requirements

As the technology builds up, the utilization of Internet has turned out to be progressively increasing. Furthermore, the use of various advancements turns out to be more. In the meantime, security risk additionally turns out to be increasing and offers opportunity to do more faults. Network attacks have become more common. When the network is more targeted by the attackers, it is sometimes possible to lose data too. The Internet should be the safest thing to do business. When the trust is broken, IoT does not have any meaning. To avoid the network being targeted by the hackers, security policies should be taken that safeguard the network from any such vulnerability.

4.4.3 Privacy Principles in IoT Technologies

There should be a mandate to let the IoT users know that kind of information that are being collected from them. The users should rather have the flexibility of sharing limited information. Before deploying the system in operational state, it is required to decide the privacy principles. A proper analysis should be done on what kind of data should be shared over the network. Proactive measures rather than reactive measures should be taken. The security should be deployed as a part of IoT system development process. The customers should be notified about their unexpected use of data. IoT organization that employs this should ensure that the required privacy controls are installed perfectly. A few innovative protections are usually talked about or executed: information minimization, encryption, access control, and advances giving the people upgraded control over their own information, the way it is traded and secured. As in a legitimate examination of the dangers and effects, one should not just concentrate on the technical aspects. Corporate frameworks will have huge amount of data derived from the IoT sensors. Take the case of organizations naturally gathering readings from client smart meters. They could parody messages being sent from the meter to the service organization and send false information. Yet, in the IoT that security capacity does not exist in a hefty portion of the gadgets that will all of a sudden get to be associated. Security must be incorporated with the outline of these gadgets and frameworks to make trust in both the equipment and trustworthiness of the information. Further, the organization should collect the data about threat analysis when the information is shared over the Internet. Application-specific privacy controls should be provided based on the type of application. Companies should review their privacy policies regularly in order to cope up with the technology changes. A better trade-off should be taken between security and privacy to achieve the objectives. Customers must be provided with the glimpse of what kind of their data is shared with the third parties by the IoT organizations as the data collected in IoT have a longer lifetime. This data protection should be available to secondary data also. For instance, if the data about when and where your car moves and some other information like shopping habits; it becomes easy for anyone to track you and do harm. Granular level data exemption should be provided to the customers. Maintaining all the sensitive data becomes a very serious threat, as there is a strong chance that the data could be viewed by anyone. IoT organization should ensure that only authorized persons have the access in order to avoid mishandling data.

4.4.3.1 *Privacy from the IoT Organization's Perspective*

The following principles help in achieving the privacy with respect to the organizations:

- Recognize the essential data streams in the associations and, particularly, the arrangements of information that nourish center computational frameworks, regardless of the possibility that information is not gathered or transmitted electronically today. Wherever information gathering or information utilization makes business esteem, you ought to hope to one day see associated (IoT) gadgets.

- Arrange the sorts of IoT gadgets expected and their benchmark administration necessities, for instance, gadget discovery, remote configuration, inventory, and programming update. Organize these data by timeframe.

- Characterize and organize the new threats of information leakage, particularly new vectors that develop because of the fragmentation of working frameworks, systems, and interfaces.

- Evaluate the danger of unapproved access to these gadgets. For instance, when the medical devices are compromised, the negative impacts of this system should be analyzed.

- Characterize the related security activities to be triggered. For example, the circumstances under which a gadget that is compromised would be removed the associated system.

- Characterize your big data technique for IoT. By what method will you secure the huge measure of business basic information that is delivered by the sensors in these gadgets? Imagine a scenario in which monstrous measure of sensor information exuding from a business basic gadget is traded off or spilled. Information arranged security with element information relationship, investigation, and insight is a center prerequisite for IoT.

- Create protection arrangements for sensor information. The expansion of sensors will bring about more individual information being produced. The entrance to and the security of this information will have numerous protection suggestions with constrained direction from existing case law.

- Ensure these new associated gadgets against system interruptions and foreswearing of-administration assaults. Endeavors have instruments to do this today; however, now they will need to do it over a much more extensive arrangement of gadgets.

4.4.4 Privacy at the Granular Level

Securing an IoT foundation requires a thorough security inside and out system. Beginning from securing information in the cloud, to ensuring information safe while in travel over the general population web, and giving the capacity to safely procurement gadgets, every layer assembles more noteworthy security confirmation in the general framework. Secure these new associated gadgets against system interruptions and dissent of-administration assaults. Ventures have systems to do this today, yet now they will need to do it over a much more extensive arrangement of gadgets.

This security strategy can be produced and executed with dynamic cooperation of different players required with the assembling, advancement, and installations of IoT gadgets and framework. For example, an *IoT device manufacturer* should make sure that following activities are to be carried out:

- The equipment configuration ought to incorporate least elements required for operation of the equipment and nothing more. A case is to incorporate USB ports just required for the operation of the gadget. These extra components open the gadget for undesirable assaults, which ought to be kept away from.

- There should be mechanism to identify physical altering of equipment, for example, opening of gadget spread, expelling a part of the gadget, and so on. These alter signs might be a piece of the information stream transferred to the cloud empowering cautioning of these occasions to the administrators.

- Overhauling firmware amid lifetime of the gadget is unavoidable. Building gadgets with secure ways for redesigns and cryptographic certification of firmware form will permit the gadget to be secured amid and after updates.

An *IoT solution developer* should ensure the following activities:

- Creating secure programming requires a ground-up thinking on security from the initiation of the undertaking the way to its usage and testing. The decision of platform and apparatuses is affected with this procedure.

- Open-source programming gives a chance to rapidly create solutions. At the point when picking open-source programming, consider the movement level of the group for every open-source segment. A dynamic group guarantees that programming will be bolstered, issues will be found and tended to. On the other hand, a dark and idle open-source programming will not be bolstered and issues will most presumably not be found.
- A hefty portion of the product security imperfections exist at the limit of libraries and APIs. Ensuring that all interfaces of segments being incorporated are secure guarantees general security.
- As far as the IoT solution deployment is concerned, one has to follow the best practices in the industry to make sure that IoT information is secured over the Internet.
- IoT organizations may oblige equipment to be conveyed in unsecure areas, for example, out in the open spaces or unsupervised regions. In such circumstances, make sure that equipment sending is carefully designed to the greatest degree. On the off chance that USB or different ports are accessible on the equipment, guarantee that these are secured safely.
- *Keep confirmation keys safe*: amid arrangement, every gadget requires gadget IDs and related verification keys created by the cloud administration. Keep these keys physically safe even after its deployment. Any traded off key can be utilized by a noxious gadget to take on the appearance of a current gadget.

The *IoT solution operator* should make sure that following best practices are to be made:

Stay up with the latest: Make sure that gadget OS and all gadget drivers are redesigned to the most recent renditions. Windows 10, with programed updates turned on, has stayed up with the latest by Microsoft, giving a protected working framework to IoT gadgets. For other working frameworks, for example, Linux, staying up with the latest guarantees they are additionally secured against malignant assaults.

Secure against malignant movement: If the OS licenses, put the most recent hostile to infection and against malware capacities on every gadget working framework. This can alleviate most outside dangers. Most cutting edge working frameworks, for example, Windows 10 IoT and Linux, can be ensured against this danger by making fitting strides.

Review every now and again: Evaluating IoT base for security-related issues is the key when reacting to security episodes. Most working frameworks, for example, Windows 10, in occasion logging that ought to be looked into every now and again to ensure that no security break has happened. Review data can be sent as a different telemetry stream to the cloud benefit and then inspected.

Physically ensure the IoT base: the most exceedingly terrible security assaults against IoT base are propelled utilizing physical access to gadgets. Ensuring against vindictive utilization of USB ports and other physical access is a vital well-being and security rehearses. Logging of physical access, for example, USB port utilization, is critical to revealing any rupture that may have happened. Once again, Windows 10 empowers logging of these occasions.

Ensure cloud certifications: cloud validation accreditations utilized for arranging and working an IoT sending are potentially the most effortless approach to obtain entrance and trade off an IoT framework. Ensure the qualifications by changing the secret key every now and again, and not utilizing these accreditations on open machines.

4.5 Methods of Providing Security

Securing IoT is similar to a protocol and has to be implemented accordingly. Figure 4.6 provides an illustration of different layers of security that can be provided to an IoT system. Apart from this there are five different steps of achieving perfect secrecy of an IoT system. They are to be dealt in following sections.

4.5.1 Authenticity

When the gadget is connected to the system, it ought to confirm itself before accepting or transmitting information. Profoundly implanted gadgets frequently do not have clients sitting behind consoles, holding up to enter the certifications required to get to the system. How, then, would we be able to guarantee that those gadgets are recognized accurately before approval?

Pretty much as client verification permits a client to get to a corporate system in light of client name and secret word, machine verification permits a gadget to get to a system in light of a comparable set of accreditations put away in a protected stockpiling range.

4.5.2 Redesigns and Fixes

Once the gadget is in operation, it will begin accepting hot patches and programming upgrades. Administrators need to take off patches, and gadgets need to verify them, in a way that does not expend data transfer capacity or weaken the useful well-being of the gadget. It is one thing when Microsoft sends upgrades to Windows clients and ties up their tablets for 15 minutes. It is very important when thousands of gadgets in the field are performing basic capacities, on the other hand benefits are subject to security patches

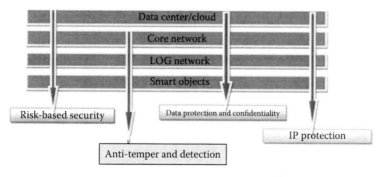

FIGURE 4.6
Different layers of security.

to ensure against the inescapable helplessness that departures into nature. Programming upgrades and security patches must be conveyed in a way that preserves the restricted data-transfer capacity and irregular network of an implanted gadget and totally wipes out the likelihood of trading off practical security.

4.5.3 Firewalling and IPS

The gadget additionally needs a firewall or profound bundle investigation ability to control movement that is foreordained to end at the gadget. Why is a host-based firewall or IPS required if system based machines are set up? Profoundly installed gadgets have remarkable conventions, unmistakable from big-business IT conventions. For example, smart energy grid has its own arrangement of conventions representing how gadgets converse with each other. That is, the reason business particular convention sifting and profound parcel review capacities are expected to distinguish pernicious payloads covering up in non-IT conventions. The gadget need not concern itself with sifting more elevated amount, basic Internet movement—the system devices ought to deal with that—however, it needs to channel the particular information bound to end on that gadget in a way that makes ideal utilization of the restricted computational assets accessible. A firewall, in any case, is not sufficient to ensure the IoT against programmers. That is on the grounds that control is not the main issue—there is additional spying. Information encryption is required too.

4.5.4 Secure Booting

When the device is turned on, it is required to verify the software in the device with the help of digital signatures. This has to be done with the help of cryptography. In addition to this, the device has to be protected from different run-time and malicious threats.

4.5.5 Access Control

Next, various types of asset and access control are connected. Required or part based access controls incorporated with the working framework restrain the benefits of gadget segments and applications so they get to just the assets they have to carry out their employments. In the event that any part is bargained, access control guarantees that the interloper has a negligible access to different parts of the framework as could be expected under the circumstances. Gadget-based access control systems are undifferentiated from system-based access control frameworks, for example, Microsoft Active Directory: even on the off chance that somebody figured out how to take corporate accreditations to pick up access to a system; traded off data would be constrained to just those regions of the system approved by those specific certifications. The guideline of slightest benefit directs that exclusive the negligible access required to perform a capacity should be approved keeping in mind the end goal to minimize the viability of any rupture of security.

4.6 Building Safer IoT

Building a safer IoT is not just a single day's task. A lot of parameters have to be considered in order to provide the IoT system with a perfect security.

4.6.1 Cloud Security and the IoT

Securing the information in-transit from the edge of the system will likewise be the key. Reportedly, medical device manufacturers and home automation are among the most endangered industry segments, as far as cyber security is concerned. Encryption ought to guarantee security and information honesty, yet associations will likewise require the capacity to identify when these remote gadgets could be utilized as method for assaulting their center system. Once more, it is impossible that this will be executed as something besides a cloud administration.

As it is so often in modern IT, the cloud is likely to be the savior here. There are many organizations that set up individual identities and credentials for connected devices and revoke them again to maintain the integrity of the system. The IoT will require the ability to segment the corporate network and to assume that many of the devices attached to it are at-best vulnerable and at-worst actually a security risk (Gou et al., 2013). Yet, these administrations will take years to develop—it could be close to the end of the prior decade they are working at scale and with the required productivity level. In the meantime, expect IoT security—or more likely the lack of it—to be making headlines for the next few years at least.

4.6.2 Security versus Efficiency

Security versus efficiency is a major problem in today's IoT. It is misty right now about what is the right answer for actualizing efforts to establish safety without affecting network resources adversely. However, devices used in today's technology cannot be secured because of technical or business reasons. They are highly error-prone as it uses wireless technology that is highly insecure in nature. Public key infrastructure (PKI) is a great method to exchange secret key but in the case of IoT, this seems to be obsolete.

4.6.3 Defend against IP Spoofing

The following measures can be helpful to fight against IP spoofing:

- Applying authentication measures when exchanging keys between two parties.
- Prepare a list of authorized users to control the access. This will hinder making Private IP addresses.
- Make different filtering techniques to control incoming or outgoing traffic. This will make sure that a source is authenticated when incoming packets are approaching, or it controls the outbound traffic.
- Routers should be configured not to accept the packets from the outside network.

Remember, these are the simple measures that you can follow to make sure that you are free from IP spoofing. However, this spoofing needs an intelligent solution as it emphasizes the inbuilt TCP/IP design.

4.7 Security Issues in IoT Products

The specific security issues in IoT products can be addressed in multiple ways. Refer to Figure 4.7 for a better understanding. Elements of the security solutions can be given in the further sections.

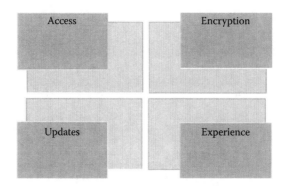

FIGURE 4.7
Solution to IoT security issues.

4.7.1 Unauthorized Access

A standout amongst the most difficult security issues with brilliant gadgets is the strength against presentation to physical attacks. IoT items are regularly left unattended and hence are simple prey for a pernicious performing artist who can catch them and afterward remove cryptography insider facts, interfere with or adjust their programming, or replace these devices with preexploited ones under his control. Gadgets physically worked to catch client login data at the source level provide reason for concern. Unauthorized access could result in creating perceptible risks to physical safety, but IoT can be accessed from a far as well.

4.7.2 Encryption

Another investigation by experts reveals that the attackers concentrate on the data that are transmitted without any kind of encryption. This can lead to leaking sensitive information like credentials, credit card information. So in order to protect the heterogeneous network, it is necessary to encrypt the data whenever possible.

4.7.3 Updates and Patches

Associated gadgets should be upgraded all the time to stay safe at any rate to a great degree unsophisticated threats. The danger of cyber attacks increments if patches are not regularly updated much of the time. Disregarding this, it is demonstrated that close to 49% of organizations offer remote updates for their smart things. There can be many reasons for this. The main concern to the firms involved in development of low end devices may lack economical support to provide updates. Many IoT products are built with the help of cheap rate chips. Then again, the reason could be that IoT sellers might not have the specialized mastery to grow such upgrades. In some examples, the fundamental issues is the absence of correspondence channels for the organization to remotely convey the patches, in light of the fact that it is more helpful for these organizations to give purchasers a chance to download and introduce them physically. Seen from the buyers' perspective, be that as it may, this may not function admirably for them as some of them experience challenges introducing the upgrades, or may not be at all mindful of their presence in any case.

4.7.4 Lack of Experience

In the first place, IoT gadgets are frequently designed by consumer goods manufacturers. On the contrary, some of the do not have the enough experience in providing the security to the IoT products. To design a secured product, one has to have excessive skills incorporated with the ability to look foresee the future problems that might arise. The fewer number of good security experts put a major obstacle in developing the secured products. New entries do not have the experience to adapt to security issues in IoT.

4.8 Conclusion

Security at both the gadget and system levels is basic to the operation of IoT. The same insight that empowers gadgets to perform their errands should likewise empower them to perceive and neutralize dangers. Luckily, this does not require a progressive approach, but instead an advancement of measures that have demonstrated effective in IT systems, adjusted to the difficulties of IoT and to the limitations of associated gadgets. Meanwhile, there are a couple of things you can do to secure your associated devices. In the event that you are on a PC or cell phone, ensure that any accessible firewall and anti-virus programming is actuated. Ensure you filter your framework routinely to recognize and evacuate any malware or conceivable interruptions. A PC tainted with malware or broken by a programer can be a starting point for assaults against the IoT hardware at your home, or it might store passwords for IoT items that the programer can utilize. At last, keep an eye out for phishing and social designing. Programers are exceptionally astute with regards to sending messages and messages that request client names and passwords. Be exceptionally suspicious. In the event that you get a telephone call requesting this secret data, do not give it. Ensure you hang up and afterward call the quantity of the business or association that the guest had guaranteed to speak to. Try not to utilize any telephone number the guest may give. What is more, customers are gradually growing practical insight. By consolidating rational strategies with state-of-the-art security technologies, we can keep programers from turning our gadgets against us.

To shield buyers from potential information security breaks, organizations need to create protection approaches that plainly detail how the information gathered from IoT items will be utilized, and these strategies ought to be effectively available to shoppers. Everyone is turning out to be more cognizant about where their information is being held, and an association attempting to show purchasers what their information is being utilized for will separate itself from the opposition. The IoT can possibly be huge for both purchasers and organizations; however, security must be at the heart of each phase of the procedure. We would not purchase a house that had no front entryway on it, so why might a shopper purchase an associated item with no security highlights set up? To rouse trust in the IoT as it assembles pace, organizations need to guarantee they are supplying the lock and additionally giving over the keys. Developing IoT solutions for security and protection are promising. These incorporate making clients' cell phones their security and protection *key* that can affirm gadget matching, utilizing cryptography rather than a console and passwords, and protection saving individual information stockpiling frameworks so clients control their private information shared crosswise over IoT frameworks. IoT can be made secure, and client protection can be safeguarded if sellers, government, and undertakings incorporate security with the IoT from the earliest starting point. The vision for the IoT is to make our regular lives simpler and help achieving the effectiveness and profitability of organizations and representatives. The information gathered will help us settle on more intelligent choices. Be that as it may, this will likewise affect security desires. On the off chance that information gathered by associated gadgets is compromised, it will undermine trust in the IoT. We are as of now seeing shoppers place higher desires on organizations and governments to defend their own data. Organizations ought to recognize the danger level for their present introduction to the IoT and where it is going later on furthermore consider the protection and security suggestions connected with the volume and sort of information the IoT will create. It is genuinely a bold new world that guarantees numerous energizing open doors. Trust is the establishment of the IoT and that should be supported by security and protection. Furthermore, it is a discussion we as a whole need to begin having now in the event that we are to profit from the associated world.

Bibliography

Baby monitors hacking issue description. Available: http://arstechnica.com/security/2015/09/9-baby-monitors-wide-open-to-hacks-that-expose-users-most-private-moments/ (accessed September 28, 2016).

Basic IoT eavesdropping illustration. Available: http://www.infineon.com/cms/en/applications/chip-card-security/internet-of-things-security/ (accessed September 28, 2016).

Car attacks description. Available: http://spectrum.ieee.org/telecom/security/how-to-build-a-safer-internet-of-things (accessed September 28, 2016).

Cloud Security Characteristics description. Available: http://www.zdnet.com/article/internet-of-things-finding-a-way-out-of-the-security-nightmare/ (accessed September 28, 2016).

Common Security Characteristics description. Available: http://resources.infosecinstitute.com/security-challenges-in-the-internet-of-things-iot/ (accessed September 28, 2016).

Fabian, B. and O. Gunther. 2007. Distributed ONS and its impact on privacy. *IEEE International Conference on Communications*, Glasgow, Scotland, pp. 1223–1228.

Gou, Q., L. Yan, Y. Liu, and Y. Li. 2013. Construction and Strategies in IoT Security System. *Green Computing and Communications (GreenCom), 2013 IEEE and Internet of Things (iThings/CPSCom), IEEE International Conference on and IEEE Cyber, Physical and Social Computing*, Beijing, China, pp. 1129–1132.

Growth and Challenges of IoT systems description. Available: http://www.cisco.com/c/en/us/about/security-center/secure-iot-proposed-framework.html (accessed September 28, 2016).

Mahalle, P. N., B. Anggorojati, N. R. Prasad, and R. Prasad. 2013. Identity authentication and capability based access control (IACAC) for the internet of things. *Journal of Cyber Security and Mobility*, 1: 309–348.

Privacy and Security issue description. Available: http://iotforum.org/wp-content/uploads/2014/09/D1.5-20130715-VERYFINAL.pdf (accessed September 28, 2016).

Zhao, G., X. Si, J. Wang, X. Long, and T. Hu. 2011. A novel mutual authentication scheme for internet of things. *International Conference on Modelling, Identification and Control (ICMIC)*, pp. 563–566.

5

Integration of Cloud Computing and IoT: Opportunities and Challenges

Sara Ghanavati, Jemal Abawajy, and Davood Izadi

CONTENTS

5.1 Introduction

Wireless body area network (WBAN) provides the majority of hardware as an infrastructure support, through providing access to sensors and actuators. From their origin, WBANs have been developed for specific monitoring purposes, for example, heart function of a patient. In reality however, collecting only vital sign cannot be considered as a comprehensive health care monitoring system. Regularly, a comprehensive monitoring system for telemedicine and ambient assisted living (AAL) needs to connect multiple medical devices such as blood pressure meter, weighting meter, blood-glucose meter, and pulse oximeter (Fei et al., 2010). Information about nurses, doctors, and also location of patients could also be significantly important. In some situations, input–output devices and sensors such as camera, microphone, speaker, can be extremely helpful in enhancing functionality of a health monitoring system as patients can be remotely visited by their doctors. However, WBANs are not capable to collect the information as biological sensors are very sophisticated, sensitive with significant limitations.

To overcome the existing challenges, IoT technology can be applied to work in parallel with WBANs in health care monitoring systems (Islam et al., 2015). IoT-based services are ubiquitous and personalized and will speed up the transformation of health care from career- to patient-centric (Klasnja & Pratt, 2012; Liu et al., 2011). In the application, various medical devices include biomedical sensors, and diagnostic devices can be considered as smart devices or objects representing core element of IoT technology. The main purpose behind the innovation is to reduce costs, while enhancing quality of life. IoT provides also an efficient scheduling of limited resources by ensuring their best use and service of more patients (Islam et al., 2015). IoT is characterized by a very high heterogeneity of devices, technologies as well as procedures. Therefore, ensuring about acceptance level of scalability, interoperability, reliability, efficiency, availability, and security are considerably important in establishing IoT-based systems (Islam et al., 2015).

Cloud Computing has the capability to resolve the majority of the aforementioned problems (Ghanavati, 2015). Cloud Computing is in health care informatics that provides unlimited recourse of data that can be accessed anytime and anywhere in the world (Rosenthal et al., 2010; Schweitzer, 2012). Health care systems require continuous and systematic innovation in order to remain cost and time effective and efficient while providing high-quality services (Rosenthal et al., 2010). Cloud Computing can ease data processing collected from wireless sensors that need complicated computational tasks to reveal patients' health status (Fortino et al., 2014). In this case, real-time processing of huge amounts of received data streams from WBAN and other related devices is a memory- and energy-intensive task (Chen et al., 2011). Cloud infrastructure can help one to run numerous processing tasks simultaneously in a real-time manner (Sultan, 2014). Cloud Computing also provides additional features such as ease-of-access, ease-of-use, and efficiency of deployment costs (Dash et al., 2010). However, integrating Cloud, WBAN and IoT imposes several challenges such as security and privacy for health care applications that are currently receiving attentions by the research communities.

The remainder of this chapter is organized as follows: Section 5.2 briefly shows the motivation and objective of the chapter. Section 5.3 introduces WBAN with advantages and possible challenges. Section 5.4 displays IoT technology and the related IoT health care services. Section 5.5 presents Cloud Computing and its related advantages. In Section 5.6, integration of Cloud-WBAN-IoT is presented. Lastly, Section 5.7 concludes the chapter.

5.2 Motivation and Objectives

The motivation for this chapter comes from the fact that integrating the three Cloud Computing, WBAN and IoT technologies can provide a high quality and scalable health monitoring systems. Although there are many approaches that use sensor Cloud-based monitoring systems, there is still a gap in the development of a comprehensive and scalable common platform for patient monitoring systems (Löhr et al., 2010; Misra & Chatterjee, 2014).

The main objectives of this chapter are to investigate the three technologies individually and explore advantages as well as challenges in detail. Furthermore, the integration of the three technologies is also studied in details.

5.3 Wireless Body Area Network

WBAN technology is a subfield of existing research in the field of WSNs and can be considered to be a specialization of biomedical engineering. WBANs consist of only less than few dozen sensor nodes and are generally linked with the human body in which fewer nodes are deployed than traditional WSNs. In WBANs, sensor nodes send their gathered information to a base station (BS) known as coordinator. Due to the low transmission rate, the network area of WBAN communication is normally within a few meters. Figure 5.1 shows a WBAN that is consisted of several body sensors such as Electrocardiogram (ECG), blood pressure, Glucose, and motion sensor. For example, ECG (Ghanavati et al., 2016b) is

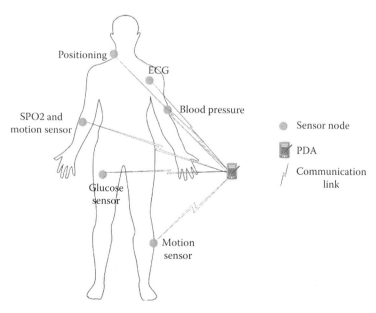

FIGURE 5.1
Network size in a WBAN.

one of the most popular biomedical sensing systems these days. As has been investigated in Nemati et al. (2012), Sullivan et al. (2007), ECG can monitor human heart activities (Ghanavati et al., 2015). Another type of body sensor is blood pressure sensor that can monitor flow and pressure in a blood vessel (Ghanavati, 2015). A glucose sensor, as it has been extensively used in many projects (Li et al., 2015; Liao et al., 2012), can monitor Glucose level in real time.

WBAN provides the majority of hardware as an infrastructure support, through providing access to sensors and actuators. From their origin, WBANs have been developed for specific monitoring purposes for example heart function of a patient. That is due to the reality, collecting only vital sign cannot be considered as a comprehensive health care monitoring system. Regularly a comprehensive monitoring system for telemedicine needs to connect multiple medical devices such as blood pressure meter, weighting meter, blood–glucose meter, and pulse oximeter (Fei et al., 2010). Information about nurses, doctors, and also location of patients could also be significantly important. Hence, ambulances can be called by the monitoring system to pick up patients and somehow find their specific doctors. In some situations, input–output devices and sensors such as camera, microphone can be extremely helpful in enhancing functionality of a health monitoring system as patients can remotely be visited by their doctors. However, WBANs are not capable of collecting the information as biological sensors are very sophisticated, sensitive with significant limitations.

5.3.1 Challenges in WBANs

In the following sections, some of the challenges in developing a comprehensive health care monitoring system are discussed.

5.3.1.1 Node Characteristics

Small size of sensor nodes is usually preferred in most of the WBAN applications. As patient comfort is an important factor in the applications, size of implantable or wearable sensor nodes must be as small as possible. Apart from the size, accuracy of data provided by sensor nodes is an extreme challenge in WBANs (Latré et al., 2011). That is because the number of nodes is extremely limited in WBAN. Hence, nodes need to be very accurate and sophisticated to reimburse the accuracy without data redundancy (Li et al., 2006). For example, in ECG application that measures the activity of a human heart, sensors are attached close to the heart to sense information with high accuracy (Latré, 2008). Therefore, we cannot expect the network to collect different information such as location of patients, due to their unique features of WBAN technology.

5.3.1.2 Data Transmission Capability

Due to the diversity of WBAN scenarios, data transmission parameters are different from each application to another. The required values for data transmission parameters in different applications are already investigated in Ghanavati et al. (2015), Latré (2008), Latré et al. (2011), Patel and Wang (2010).

As can be seen from the given information, data rates vary from simple with low rate to more complicated data such as video streams. If any of WBANs applications use several of these sensors together, data rate will reach over Mbps. In the case, as most of the current existing low power radios are not able to pass such transmission rate, WBANs technology is not a reliable paradigm to transfer different data such as sound and video of patients to health centers. In fact, due to different data rate, congestion could be a possible significant issue in such networks (Ghanavati, 2015).

5.3.1.3 Limitation of Resources

In WBANs, integrated batteries in sensor nodes provide the required energy for network tasks. Due to the small size of sensor nodes in WBANs, batteries are also small and are able to save limited amount of power. Mainly, there are three categories for energy consumption in a WBAN: sensing information, wireless communications, and data processing (Latré, 2008). As have been investigated in Hanson et al. (2009), wireless communications such as Wi-Fi, wireless USB, ZigBee, and Bluetooth consume higher amount of saved energy compared with sensing part and processing circuits in a typical WBAN.

Battery usage is a critical issue in such networks. This is because for those sensors implanted in the human body, changing or replacing the sensors is not feasible (Latré, 2008). Hence, expecting the sensor nodes with limited energy capacity to transfer large amount of different information, such as GPS and voice data as well as their specific collected data with high accuracy, is not realistic.

Apart from battery capacity, available memory, bandwidth, and processing capabilities are among resource limitations of a WBAN that are because of very small size of the nodes. Therefore, not all the collected information, such as voice and GPS data, can be processed. These resource limitations cause a number of difficulties in implementing WBANs in a comprehensive monitoring health care system (Ghanavati, 2015).

5.3.1.4 WBANs and IoT Technology

To overcome the described challenges, IoT technology can be applied to work in parallel to WBANs in health care monitoring systems (Islam et al., 2015). WBANs can exist without IoT however; IoT cannot exist without WBANs. That is because WBANs provide the majority of hardware as an infrastructure support, through providing access to sensors and actuators. From their origin, WBANs have been developed for specific monitoring purposes, for example, heart function of a patient. In contrast, IoT is not focused on specific applications as it would not be expected to collect specific types of sensor data.

5.4 The Internet of Things

In the late 1960s, Computer Network was invented, and communication between two computers was made possible (Natalia & Victor, 2006). Then, TCP/IP stack was introduced in the early 1980s. In the late 1980s, the World Wide Web (WWW) became available and made the Internet very popular to be used commercially by 1991. That led people to become connected to the Internet through different devices, including mobile devices, and form the mobile-Internet (Associati, 2011). Finally, with the emergence of social networking, users are now able to become connected through the Internet. The next evolution in the IoT is to let the entire devices be able to communicate with each other via the Internet (INFSO, 2008). Figure 5.2 presents five typical phases in the evolution of the Internet.

FIGURE 5.2
Five phases in the evolution of the Internet.

Recently, IoT has become a significant technology in academia as well as industry. This technology provides the opportunity of connecting the entire smart objects around us to the Internet to communicate with each other and enables many applications in various domains. Some objects have very limited capabilities of processing data, whereas there is limitation in storing data. On the contrary, there are objects with larger memory and higher processing capabilities and able to connect to the Internet intelligently. The devices could be Internet-enabled computers and mobile phones, which already connected 2.5 billion of them to the Internet. This rate is also increasing, and it is estimated that it will reach 50 to 100 billion devices connected to the Internet by 2020 (Sundmaeker et al., 2010). Thus, IoT technology is necessarily required to facilitate the communication among the online devices to meet the required QoS in handling of massive number of interactions.

The goal behind that is to reduce humans' interactions to the devices and act to meet people's needs by bringing the capability of interacting autonomously to the monitoring systems (Dohr et al., 2010b). This can be achieved by developing a proper architecture based on applications (Alam et al., 2010; Patel et al., 2009). Application domain in IoT generally can be mainly divided into industry environment such as Supply chain management (Chaves & Decker, 2010) and society categories (Sundmaeker et al., 2010) such as Telecommunication, medical technology (Wang et al., 2011) that this chapter mainly focuses on.

5.4.1 Health-IoT

In the near future, the current form of health care systems is expected to change from the hospital-centric model to hospital–home-balanced in 2020 to the home-centric form of health care in 2030 (Koop et al., 2008). To achieve that, IoT technology has been invented to establish the in-home health care service. In fact, health care represents one of the most attractive application areas for the IoT (Pang, 2013).

Health-IoT-based services are universal and personalized that enhance health care data communication from career centric to patient centric (Klasnja & Pratt, 2012; Liu et al., 2011). IoT has the potential to be applied in many medical applications such as remote health monitoring that could include treatment and medication usage monitoring at home by health care providers. In these applications, various medical devices such as biomedical sensors and diagnostic devices are considered smart devices or objects that play core roles in the IoT.

The main purpose behind the innovation is to reduce costs, while enhancing quality of life. IoT paradigm can also efficiently schedule limited available resources to ensure their best services in a real time (Islam et al., 2015).

5.4.2 IoT Health Care Services

IoT-based health care systems can be applied to variety of applications such as health monitoring, supervising, and managing elderly patients as well as private health and fitness of others. The services and applied procedures, however, require modifications and justifications to ensure properly function in the scenarios. These include different services such as notification, resource sharing, and Internet services. For health care scenarios with major connectivity with heterogeneous devices, cross-connectivity protocols and link protocols are also required to be fully taken into account. Security, low-power and fast discovery of devices and services can also be added to this list (Islam et al., 2015). In this section, popular IoT services are explained in detail.

5.4.2.1 Ambient-Assisted Living

An IoT platform equipped with artificial intelligence can address health care of aging and incapacitated individuals, that is, called AAL. The main purpose of this service is to allow elderly individuals living in their places in a convenient and safe manner. This service ensures them a greater autonomy and providing them with a human-servant-like assistance in the case of any problem.

Applying the central AAL paradigm over the IoT has been investigated in (Dohr et al., 2010b). The authors explored combination of smart objects and health care services to facilitate AAL. Such infrastructure enables communication between service users such as elderly individuals, health care givers, and maybe family members. For instance, a modular architecture for automation, security, control, and communication is proposed for IoT-based AAL in Shahamabadi et al. (2013). This approach proposes a framework for providing health care services to elderly and incapacitate individuals. The authors used 6LoWPAN for active communications, and radio-frequency identification (RFID) and near-field communications (NFCs) for passive communications. They applied medical-based knowledge algorithms to detect issues faced by elderly people and patients.

5.4.2.2 Adverse Drug Reaction

An adverse drug reaction (ADR) is an injury that might occur because of consuming a single dose of a drug or a combination of two or more medicines. As this injury is not a particular disease with a specific solution, a separate monitoring system is necessarily required to be developed. An IoT-based ADR is proposed in Jara et al. (2010). In this approach, authors developed a monitoring system by cooperating a pharmaceutical intelligent information algorithm and sensor network. In this approach, the patient's terminal identifies the drug by applying barcode/NFC-enabled devices.

The collected information from the devices and patients' information is processed to ensure that the drug is compatible with their allergy profile and electronic health record. The iMedPack has also been developed as part of the iMedBox to address the ADR (Yang et al., 2014) by applying of RFID and controlled delamination material (CDM) technologies.

5.4.2.3 Community Health Care

Community health care (CH) monitoring essentially comes with the idea of establishing a network covering areas such as hospitals and rural communities. IoT paradigm can play a backbone role in connecting several required networks to establish a cooperative network structure.

A cooperative IoT platform for rural health care monitoring has been proposed in Rohokale et al. (2011) to enhance energy efficiency. A community medical network has also been proposed in You et al. (2011). This network applied WBANs to materialize CH. The proposed platform shares the collected data from the networks to other medical facilities. Thus, medical staff can have access to the information for making decisions. In fact, the structure of a community medical network can be viewed as a virtual hospital (Wang, Li et al., 2011).

5.4.3 IoT Health Care Challenges

To provide the best health care services, there are numbers of challenges to overcome. In this section, we are addressing the challenges in detail.

5.4.3.1 Storage Resources

IoT involves a large amount of information as Things produce a considerable amount of data. The data in different scenarios could be semistructured or even not-structured in variety of sizes collected in different frequencies (Aguzzi et al., 2014). This collected information then needs archiving and sharing, which cannot be accomplished by IoT paradigm due to its capabilities (Rao et al., 2012). This integration apprehends a new convergence scenario with newly arisen opportunities such as data aggregation (Fox et al., 2012) and sharing with third parties (Zaslavsky et al., 2013).

Hence, a technology is required to offer the most convenient solution to deal with data produced by IoT. It should capable of virtually offering unlimited, low cost, and on-demand storage capacity to collect data (Rao et al., 2012). It should be capable of treating data in a homogeneous manner through specified standards (Fox et al., 2012). Furthermore, it should apply top-level security (Dash et al., 2010) and directly access to and visualize from different locations (Rao et al., 2012).

5.4.3.2 Computational Resources

Due to the very limited processing resources of IoT paradigm, resources are not permitted to process data on-site. Data processing is usually accomplished in powerful nodes in which aggregation and realizing is possible (Rao et al., 2012). To address the issue, a third party technology can be applied to provide unlimited real-time processing capabilities and on-demand model to allow IoT processing be properly satisfied and enable analyses of unprecedented complexity (Dash et al., 2010).

Another paradigm is also required to offer data-driven decision-making and prediction algorithms to reduce risks and costs while enhancing performances with higher energy efficiency (Zaslavsky et al., 2013). The performance needs to be real-time processing (Rao et al., 2012), to implement scalable, collaborative, sensor-centric applications in complex events (Rao et al., 2012).

5.4.3.3 Communication Resources

The main concept of IoT is enabling IP-enabled devices communicating through different dedicated hardware. However, providing the capability for different communication purposes can be considerably expensive. Therefore, a technology is required to offer an effective and cheap solution to connect, track, and manage requests from anywhere at any time using customized portals and built-in applications (Rao et al., 2012).

5.4.3.4 IoT and Cloud Computing

IoT is characterized by a very high heterogeneity of devices that are applying different protocols to collect and communicate with each other. Thus, QoS parameters such as scalability, interoperability, and reliability as well as security necessarily need to be fully considered. However, the integration with the Cloud overcomes most of these concerns as proved in Suciu et al. (2013). Cloud Computing also provides additional features such as ease-of-access, ease-of-use, and efficiency of deployment costs (Dash et al., 2010).

5.5 Cloud Computing

The general concept of Cloud Computing refers to a new technique capable of delivering enterprise IT. Cloud-based services provide flexible deployment of applications to adjust their resources to changes in client requirements. Cloud Computing provides virtual machines (VMs) instances for servers, and other Internet-faced services. Thus, Cloud Computing based models can be applied to any type of data centers regardless of its operational goals (Schweitzer, 2012).

One interesting application of Cloud Computing is in health care informatics, which provides unlimited recourse of data, can be accessed anytime from anywhere in the world (Ghanavati et al., 2016a). Health care systems require continuous and systematic innovation in order to remain cost and time effective and efficient while providing high-quality services (Rosenthal et al., 2010). Cloud Computing can ease data processing collected from wireless sensors that need complicated computational tasks to reveal patients' health status (Fortino et al., 2014). In this case, real-time processing of huge amounts of received data streams from WBAN and other related devices is a memory and energy-intensive task (Chen et al., 2011). However, Cloud infrastructure can help one to run numerous processing tasks simultaneously and in a real-time manner (Sultan, 2014).

Another challenging issue is managing collected data from the devices that include WBANs as the data might be distributed in time and or space (Borradaile, 2013). Time distribution means occurrence of several activities in different times, whereas they have been managed to reach the same goal. Space distribution refers to activities that take place in different locations, whereas they are managed by a same data network (Fortino et al., 2014). In this case, a Cloud Computing solution could simplify the management of distributed data by using methods such as data fusion in a scalable and low cost manner (Chatman, 2010).

5.5.1 Mobile Cloud Computing Technology

Mobile Cloud Computing (MCC) technology was invented not long after Cloud Computing was introduced. The paradigm has been attracted attentions as a profitable business option that reduces development and running cost of mobile applications (Ali, 2009).

Alternatively, MCC can be defined as a combination of mobile web and CC (Christensen, 2009), which is the most paradigm for stakeholders to access applications and services on the Internet. MCC provides an infrastructure in which both the data storage and data processing happen outside of the mobile device. The infrastructure moves the computing power and data storage into powerful and centralized platform located in Cloud. The centralized platform and applications are then, accessed over wireless connection via not just smartphones but mobile subscribers (Suciu et al., 2013). Thus, with this technology, the mobile devices do not need a powerful configuration. The devices are not required high CPUs speed and memory capacity either. That is because the entire complicated computing modules are processed in the Clouds.

5.5.1.1 Advantages of Mobile Cloud Computing

In this section, we explore the advantages of MCC in detail.

5.5.1.1.1 Extending Battery Lifetime

Efficiency of energy consumption is a significant concern in developing mobile devices. Literature shows that there are a various number of developed approaches to enhance CPU performances (Ghanavati, 2015) and to screen in an intelligent manner (Steinfeld et al., 2015) to reduce power consumption. However, these solutions did not fully address the issue as they require to investigate in the structure of mobile devices or new hardware, which is not feasible for all mobile devices using in different systems. Then, the idea of using offloading techniques is proposed (Ghanavati et al., 2016a). This avoids taking a long application execution time on mobile devices and consequently enhance large amount of power consumption.

For instance, (Barbera et al., 2013) evaluate the effectiveness of offloading techniques such as applying MCC through several experiments. They proved remote application executions that can save energy significantly up to 25%. Task migration and remote processing is also another advantage of applying offload techniques. Mobile Cloud for Assisting Healthcare (MoCAsH) (Hoang & Chen, 2010) is MCC-based information management system that collects and utilizes vital sings. Users interact with the system via mobile agents to the Cloud to communicate and manage the collected data.

5.5.1.1.2 Improving Data Storage Capacity and Processing Power

MCC is developed mainly with the purpose of ensuring mobile users are capable of storing and accessing a large data on the Cloud through wireless networks. Amazon Simple Storage Service (S3) (Vartiainen & Väänänen-Vainio-Mattila, 2010) utilize a large storage space in Clouds for authorized users. Apart from storage, running cost of compute-intensive applications is also considerable. That is because of required long time and large amount of energy especially when performed on the limited-resource devices. MCC enhances the expected cost of running the applications. Amazon offers Elastic Compute Cloud (EC2) that is developed to perform data computation on the stored data in S3. That provides a number of advantages for users, including lower maintenance and operation cost with higher utilization through virtualization. This type of computing model reduces start-up time for new services as well as easier disaster recovery (Kumar & Lu, 2010).

5.5.1.1.3 Improving Reliability

Storing data and running applications on Clouds are an effective way to enhance data-processing reliability. That is because of storing and backing up data and applications on a number of computers or servers. This reduces the possibility and risk of data and application lost on the mobile devices. MCC can also be developed with a fully comprehensive data security model for both service providers and users. For instance, the Cloud is expected to remotely provide mobile users with security services such as malicious code detection and authentication (Oberheide et al., 2008). Also, these Cloud-based security services are capable of enhancing usage efficiency of the collected data and records from different stakeholders to improve the effectiveness of the services (Kumar & Lu, 2010).

5.6 Cloud-WBAN-IoT Integration

To develop a comprehensive monitoring health care system, Cloud Computing can be considered as the missing technology in the WBAN-IoT integration scenario. In fact, some people believe that Cloud fills some gaps such as limited storage of WBAN-IoT integration (Jeffery, 2014).

Health care applications usually generate a vast amount of data that require to be managed efficiently. The collected information is also required to be analyzed to abstract technical details while eliminating the need for expertise over the technology. In general, the entire collected patients' vital data via a WBAN and information from things (devices) connected to the Internet, such as recovery, medication, logistics, finance, and even daily activities, are managed, utilized, and processed through a Cloud Computing system (Domingo, 2012).

The technology simplifies health care process and enhances QoS by improving the cooperation among different entities. Moreover, the paradigm helps one effectively to overcome existing challenges such as privacy and reliability of the applications (Nkosi & Mekuria, 2010).

Figure 5.3 depicts an example of architecture for Cloud-WBAN-IoT for acquiring and managing information including biomedical sensor data on Cloud. This monitors general health situation of patients by collecting and forwarding requested information to Cloud for further analysis, computational process, and storage. Cloud technology provides this opportunity to adjust the system's tasks and performance based on the users' needs. Specifically, the framework composes three subsystems that provide real-time monitoring of patients. Figure 5.3 shows a general overview of the integrated framework:

In the following, we explain the key components of our proposed framework:

1. WBAN-IoT integration module: This module is the lowest layer of a health care monitoring system composed of two main procedures described in the following:

   a. Data collection phase: In this phase, patient's vital signs such as blood pressure, blood glucose, temperature, and others are monitored using body senor nodes. There are adequate sensor nodes such as ECG (ElectroCardioGraphy) and temperature sensor, and so on. These senor devices are able to sense, process, and transmit health data (Abawajy et al., 2013; Zhang & Xiao, 2009). There are also Things that can connect to the Internet to monitor patients' health situations. Thus, information such as diet and drug consumptions of patients also can be collected. In emergency situations, the patients can use specific cameras

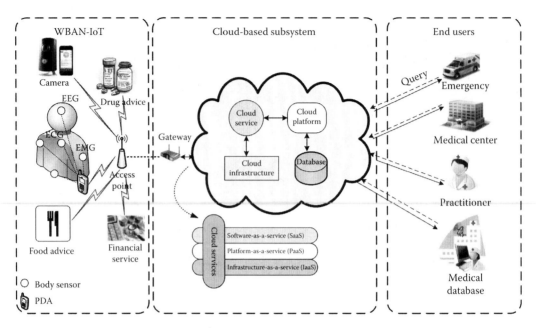

FIGURE 5.3
Cloud-WBAN-IoT integration for remote health care monitoring.

to be virtually visited by authorized doctors and discuss about their health conditions.

b. Transmission phase: In this phase, the monitored data will be transferred via wireless communication. The data are forwarded to the Cloud data centers for monitoring and storage. There are several transportation protocols available for WBAN such as Bluetooth over IEEE 802.15.1, ZigBee over IEEE 802.15.4, UWB over IEEE 802.15.6 and radio-frequency identification (RFID) (Bhatia et al., 2007).

2. Cloud-based module: The Cloud subsystem allows patients to submit their personal and medical information through a gateway. The medical records are accessible by doctors and other health care givers as well as patients from a client-server web service at any time. Authentication is needed for each time accessing Cloud. This module provides the three types of Cloud services, namely SaaS, PaaS, and IaaS. SaaS service offers various applications that are necessary for working with health data. The PaaS includes tools such as operating system (OS), Database Management System (DBMS), and virtualization as well as servers, storage, and networking. Lastly, IaaS provides the physical infrastructure needed for fundamental computing recourses in a heterogeneous health care system. Cloud service also provides a platform for development, testing, and execution of software needed by both patients and medical staff. It suits to say that the Cloud provides the standard interfaces for application integration.

3. End-user module (monitoring phase): This module offers patients with intelligent monitoring services in a real-time manner. It consists of a variety of applications that are used for analyzing-recorded health information. This module can also have different terminals in different buildings such as emergency centers or family

members of patients, based on the service purposes and expectations. End-users are also able to subscribe to the service to become updated automatically about the recent transferred medical data and health situations.

5.6.1 Advantages of Cloud-WBAN-IoT Integration

In this section, we are investigating advantages of integrating Cloud Computing, WBAN, and IoT.

5.6.1.1 Communication

The availability of high-speed networks of WBAN-IoT integration enables effective monitoring and control of Things, their coordination, their communications, and real-time access to the produced data (Rao et al., 2012). In fact, WBAN-IoT is characterized by a very high heterogeneity of devices, technologies, and protocols. Therefore, scalability, reliability, and efficiency as well as availability and security need to take into account. The integration with the Cloud resolves most of these challenges (Suciu et al., 2013). Furthermore, Cloud-WBAN-IoT paradigm brings the advantage of enhancing data and application sharing. The integration personalizes ubiquitous applications, whereas it reduces expenses of both data collection and distribution (Rao et al., 2012). In this integration, Cloud offers an effective and cheap solution not just to connect any IP enabled devices from any location at any time it provides opportunities to track and manage them by using customized portals and built-in applications.

5.6.1.2 Storage

Integration of WBAN-IoT is basically identified by a large amount of information sources from both biomedical sensors and Things that are required to work together in monitoring systems. Thus, a large amount of data including nonstructured or semistructured are expected to be produced and transferred (Aguzzi et al., 2014). In some cases, vast variety types of data in different sizes and frequencies are collected and are required to be stored. Therefore, a large-scale and long-lived storage is required.

To address the concern, Cloud Computing represents an import Cloud-WBAN-IoT driver as it provides virtual low cost and on-demand storage capacities. In fact, Cloud Computing generates new opportunities for data aggregation, integration, and sharing with third parties (Rao et al., 2012). The data then can be directly accessed and visualized from any places (Zaslavsky et al., 2013). Cloud also offers well-protected technology to protect data by applying top-level security modules (Rao et al., 2012).

5.6.1.3 Computation

Data processing in the integration of WBAN-IoT needs to be properly satisfied for performing real-time data analysis, implementing scalable and real time collaborative to manage complex events, and supporting task offloading for energy saving (Yao et al., 2013). However, the integration of WBAN-IoT has limited processing and energy resources thus; complex and on-site data processing cannot be occurred. Cloud Computing, however, offers virtually unlimited processing capabilities and an on-demand usage model. This represents another important Cloud-WBAN-IoT driver (Rao et al., 2012).

5.6.2 Challenges of Cloud-WBAN-IoT

Integrating Cloud, WBAN, and IoT imposes several challenges for health care applications that are currently receiving attentions by research communities. In this section, we deeply investigate the challenges.

5.6.2.1 Communications

Integration of Cloud-WBASN-IoT involves various communications among heterogeneous devices with different protocols (Zaslavsky et al., 2013). For instance, smartphones that usually capable of providing a wide range of mobile applications in different domains of the integrations. This requires several heterogeneous network technologies continually transferring data from Things such as wearable and implantable biosensors. Therefore, dealing with this heterogeneity to manage source nodes including the Things and biomedical sensor nodes to enhance expected performance of the monitoring system is a considerable challenge (Botta et al., 2008).

There are wireless network communication protocols, for example, Bluetooth (IEEE 802.15.1) or WiFi (IEEE 802.11a/b/p) technologies that are using for data transmissions. However, there is still no agreement on the network architecture on managing the Things in a uniform fashion of a heterogeneous scenario, whereas providing required performance still represents an open issue (Botta et al., 2008).

Moreover, continuity in transmission of data increases the overall bandwidth consumption dramatically whereas the current bandwidth limitation cannot support the increasing trend (Xu et al., 2011). Thus, efficiency of bandwidth access and also its optimization is still a significant challenge. In addition, in health care applications, providing and acceptable rate of fault tolerant and reliable continues data transfer from Things to the Cloud is a challenge (Biswas et al., 2010).

5.6.2.2 Scalability and Flexibility

Cloud-WBAN-IoT requires efficient mechanisms to match collected data and tasks to appropriate applications and services. As captured information about events and relevant access services for different applications are distributed timely and geographically, collection and dissemination becomes critical challenges. Therefore, providing flexible subscription schemas and events management, while guaranteeing scalability with respect to WBAN-IoT integration and users, is still considered an open issue (Le et al., 2010).

5.6.2.3 Security and Privacy

Critical monitoring health care applications requires a very high level of trust in services. That is to provide a very high level of privacy by authorizing individuals to access specific information.

In such applications, there is a high possibility of exposing several attacks such as session riding, SQL injection, cross site scripting, and side-channel (Le et al., 2010). Multitenancy can also compromise security and lead to sensitive information leakage (Kapadia et al., 2011). Moreover, data integrity is an important factor that necessarily needs to be provided from security and privacy-related aspects especially on outsourced data (Liu et al., 2015).

Therefore, security and privacy of health care services is considered a challenge in Cloud-WBAN-IoT to cope with different threats from hackers (Li et al., 2013).

5.6.2.4 Heterogeneity

Cloud services usually require resource integration to be efficiently customized for particular purposes and applications. The requirements become significant in enhancing applications with multi-Cloud services (Grozev & Buyya, 2014). To achieve the Cloud customizations and facilitate applications deployment particularly in health care applications, there are significant challenges such as sophisticated information selection and fusion mechanisms of heterogeneity to be taken into considerations (Botta et al., 2016). In fact, a model in Cloud-WBAN-IoT in real-time data collection process from heterogeneous devices needs to be developed to enhance decision-making capabilities. Although research efforts have been investigated in this field, maximizing the intelligence in this context is also still an open challenge (Botta et al., 2016; Pedersen et al., 2013).

5.6.2.5 Performance

Cloud-WBAN-IoT based health care applications usually require specific performance and QoS requirements for different aspects such as communication and computation. However, meeting the expected level of QoS in different scenarios may not be simply achievable. Generally, obtaining acceptable network performance of the Cloud-WBAN-IoT integration is a main challenge (Rao et al., 2012) for the applications. That is because health care scenarios are real-time applications, and they necessarily need to provide services with extremely high QoS performances. For example, in scenarios using mobility of data resources, provisioning of data and services need to be performed with high reactivity (Perera et al., 2014). However, timeliness may be heavily impacted by unpredictability issues; real-time applications are mainly susceptible to performance challenges (Suciu et al., 2013).

5.7 Conclusion

This study explored WBANs, IoT, and Cloud Computing technologies individually and explore advantages as well as challenges in details. This study proved that merging the three technologies can overcome many of the existing challenges in each individual paradigm by explaining an example architecture for eHealth monitoring systems. In fact, we justified that a WBAN-IoT-Cloud paradigm provides a comprehensive health care monitoring system with a number of enhancements such as location independency, reliability, and acceptable security. In fact, it improves cooperation among different requests due to the capability of processing different applications while sharing resources. However, there are challenges such as required flexibility and privacy in processing heterogeneous data need to be considered accurately.

References

Abawajy, J. H., Kelarev, A., & Chowdhury, M. (2013). Multistage approach for clustering and classification of ECG data. *Computer Methods and Programs in Biomedicine, 112*(3), 720–730.

Aguzzi, S., Bradshaw, D., Canning, M., Cansfield, M., Carter, P., Cattaneo, G., & Stevens, R. (2014). Definition of a research and innovation policy leveraging cloud computing and IoT combination. *European Commission, Directorate-General of Communications Networks, Content & Technology,* Brussels, Belgium.

Alam, S., Chowdhury, M. M., & Noll, J. (2010). *Senaas: An event-driven sensor virtualization approach for internet of things cloud.* Paper presented at the Networked Embedded Systems for Enterprise Applications (NESEA), 2010 IEEE International Conference on.

Ali, M. (2009). *Green cloud on the horizon.* Paper presented at the IEEE International Conference on Cloud Computing.

Associati, C. (2011). The evolution of internet of things. *Focus. Milão, fev.* Retrieved from http://www.casaleggio.it/pubblicazioni/Focus internet of things v1.81%20-%20eng.pdf (accessed on August 6, 2015).

Atzori, L., Iera, A., & Morabito, G. (2010). The internet of things: A survey. *Computer Networks, 54*(15), 2787–2805.

Barbera, M. V., Kosta, S., Mei, A., & Stefa, J. (2013). *To offload or not to offload? The bandwidth and energy costs of mobile cloud computing.* Paper presented at the INFOCOM, 2013 Proceedings IEEE.

Bhatia, D., Estevez, L., & Rao, S. (2007). *Energy efficient contextual sensing for elderly care.* Paper presented at the Engineering in Medicine and Biology Society, 2007. 29th Annual International Conference of the IEEE.

Biswas, J., Maniyeri, J., Gopalakrishnan, K., Shue, L., Phua, J. E., Palit, H. N., … Li, X. (2010). *Processing of wearable sensor data on the cloud-a step towards scaling of continuous monitoring of health and wellbeing.* Paper presented at the 2010 Annual International Conference of the IEEE Engineering in Medicine and Biology.

Borradaile, G. J. (2013). *Statistics of earth science data: Their distribution in time, space and orientation.* Berlin, Germany: Springer Science & Business Media.

Botta, A., de Donato, W., Persico, V., & Pescapé, A. (2016). Integration of cloud computing and internet of things: A survey. *Future Generation Computer Systems, 56,* 684–700.

Botta, A., Pescapé, A., & Ventre, G. (2008). Quality of service statistics over heterogeneous networks: Analysis and applications. *European Journal of Operational Research, 191*(3), 1075–1088.

Chatman, C. (2010). How cloud computing is changing the face of health care information technology. *Journal of Health Care Compliance, 12*(3), 37–70.

Chaves, L. W. F., & Decker, C. (2010). *A survey on organic smart labels for the internet-of-things.* Paper presented at the Networked Sensing Systems (INSS), 2010 Seventh International Conference on, Kassel, Germany.

Chen, M., Gonzalez, S., Vasilakos, A., Cao, H., & Leung, V. C. (2011). Body area networks: A survey. *Mobile Networks and Applications, 16*(2), 171–193.

Christensen, J. H. (2009). *Using RESTful web-services and cloud computing to create next generation mobile applications.* Paper presented at the Proceedings of the 24th ACM SIGPLAN conference companion on Object oriented programming systems languages and applications, Penn Plaza, NY.

Dash, S. K., Mohapatra, S., & Pattnaik, P. K. (2010). A survey on applications of wireless sensor network using cloud computing. *International Journal of Computer science & Engineering Technologies, 1*(4), 50–55.

Dohr, A., Modre-Opsrian, R., Drobics, M., Hayn, D., & Schreier, G. (2010). *The internet of things for ambient assisted living.* Paper presented at the 2010 Seventh International Conference on Information Technology, IEEE.

Dohr, A., Modre-Osprian, R., Drobics, M., Hayn, D., & Schreier, G. (2010). The internet of things for ambient assisted living. *Information Technology: New Generations, 10*, 804–809.

Domingo, M. C. (2012). An overview of the Internet of Things for people with disabilities. *Journal of Network and Computer Applications, 35*(2), 584–596.

Fei, D.-Y., Zhao, X., Boanca, C., Hughes, E., Bai, O., Merrell, R., & Rafiq, A. (2010). A biomedical sensor system for real-time monitoring of astronauts' physiological parameters during extra-vehicular activities. *Computers in Biology and Medicine, 40*(7), 635–642.

Fortino, G., Parisi, D., Pirrone, V., & Di Fatta, G. (2014). BodyCloud: A SaaS approach for community body sensor networks. *Future Generation Computer Systems, 35*, 62–79.

Fox, G. C., Kamburugamuve, S., & Hartman, R. D. (2012). *Architecture and measured characteristics of a cloud based internet of things.* Paper presented at International Conference on the Collaboration Technologies and Systems (CTS), Denver, CO.

Ghanavati, S. (2015). *Congestion control mechanism for sensor-cloud Infrastructure.* Retrieved from http://dro.deakin.edu.au/view/DU:30083703.

Ghanavati, S., Abawaji, J., & Izadi, D. (2015). *A congestion control scheme based on fuzzy logic in wireless body area networks.* Paper presented at IEEE 14th International Symposium on the Network Computing and Applications (NCA).

Ghanavati, S., Abawajy, J., & Izadi, D. (2016a). *An alternative sensor cloud architecture for vital signs monitoring.* Paper presented at the Neural Networks (IJCNN), 2016 International Joint Conference on, IEEE.

Ghanavati, S., Abawajy, J., & Izadi, D. (2016b). *ECG rate control scheme in pervasive health care monitoring system.* Paper presented at the Fuzzy Systems (FUZZ-IEEE), 2016 IEEE International Conference on.

Grozev, N., & Buyya, R. (2014). Inter-Cloud architectures and application brokering: Taxonomy and survey. *Software: Practice and Experience, 44*(3), 369–390.

Hanson, M. A., Powell Jr, H. C., Barth, A. T., Ringgenberg, K., Calhoun, B. H., Aylor, J. H., & Lach, J. (2009). Body area sensor networks: Challenges and opportunities. *Computer, 42*(1), 58.

Hoang, D. B., & Chen, L. (2010). *Mobile cloud for assistive healthcare (MoCAsH).* Paper presented at the Services Computing Conference (APSCC), IEEE Asia-Pacific.

INFSO, E. (2008). Internet of Things in 2020: Roadmap for the Future. *INFSO D, 4.*

Islam, S. R., Kwak, D., Kabir, M. H., Hossain, M., & Kwak, K.-S. (2015). The internet of things for health care: A comprehensive survey. *IEEE Access, 3*, 678–708.

Jara, A. J., Belchi, F. J., Alcolea, A. F., Santa, J., Zamora-Izquierdo, M. A., & Gómez-Skarmeta, A. F. (2010). *A Pharmaceutical Intelligent Information System to detect allergies and Adverse Drugs Reactions based on internet of things.* Paper presented at the Pervasive Computing and Communications Workshops (PERCOM Workshops), 8th IEEE International Conference on.

Jeffery, K. (2014). *Keynote: CLOUDs: A large virtualisation of small things.* Paper presented at the 2nd International Conference on Future Internet of Things and Cloud (FiCloud-2014).

Kapadia, A., Myers, S., Wang, X., & Fox, G. (2011). *Toward securing sensor clouds.* Paper presented at the Collaboration Technologies and Systems (CTS), International Conference on, IEEE.

Klasnja, P., & Pratt, W. (2012). Healthcare in the pocket: Mapping the space of mobile-phone health interventions. *Journal of Biomedical Informatics, 45*(1), 184–198.

Koop, E., Mosher, R., Kun, L., Geiling, J., Grigg, E., Long, S., & Rosen, J. M. (2008). Future delivery of health care: Cybercare. *Engineering in Medicine and Biology Magazine, IEEE, 27*(6), 29–38.

Kumar, K., & Lu, Y.-H. (2010). Cloud computing for mobile users: Can offloading computation save energy? *Computer, 43*(4), 51–56.

Latré, B. (2008). *Reliable and energy efficient network protocols for wireless body area networks.* Belgium: Ghent University.

Latré, B., Braem, B., Moerman, I., Blondia, C., & Demeester, P. (2011). A survey on wireless body area networks. *Wireless Networks, 17*(1), 1–18.

Le, X. H., Lee, S., True, P. T. H., Khattak, A. M., Han, M., Hung, D. V., Lee, & Y.-K. (2010). *Secured WSN-integrated cloud computing for u-life care*. Paper presented at the Proceedings of the 7th IEEE conference on Consumer communications and networking conference.

Li, B., Wang, Q., Yang, Y., & Wang, J. (2006). *Optimal distribution of redundant sensor nodes for wireless sensor networks*. Paper presented at the Industrial Informatics, 2006 IEEE International Conference on.

Li, H., Guo, C.-Y., & Xu, C.-L. (2015). A highly sensitive non-enzymatic glucose sensor based on bimetallic Cu-Ag superstructures. *Biosensors and Bioelectronics, 63*, 339–346.

Li, W., Zhong, Y., Wang, X., & Cao, Y. (2013). Resource virtualization and service selection in cloud logistics. *Journal of Network and Computer Applications, 36*(6), 1696–1704.

Liao, Y.-T., Yao, H., Lingley, A., Parviz, B., & Otis, B. P. (2012). A 3-CMOS glucose sensor for wireless contact-lens tear glucose monitoring. *IEEE Journal of Solid-State Circuits, 47*(1), 335–344.

Liu, C., Yang, C., Zhang, X., & Chen, J. (2015). External integrity verification for outsourced big data in cloud and IoT: A big picture. *Future Generation Computer Systems, 49*, 58–67.

Liu, C., Zhu, Q., Holroyd, K. A., & Seng, E. K. (2011). Status and trends of mobile-health applications for iOS devices: A developer's perspective. *Journal of Systems and Software, 84*(11), 2022–2033.

Löhr, H., Sadeghi, A.-R., & Winandy, M. (2010). *Securing the e-health cloud*. Paper presented at the Proceedings of the 1st ACM International Health Informatics Symposium, ACM: New York.

Misra, S., & Chatterjee, S. (2014). Social choice considerations in cloud-assisted WBAN architecture for post-disaster healthcare: Data aggregation and channelization. *Information Sciences, 284*, 95–117.

Natalia, O., & Victor, O. (2006). Computer networks, principles, technologies and protocols for network design. Hoboken, NJ: John Wiley & Sons.

Nemati, E., Deen, M. J., & Mondal, T. (2012). A wireless wearable ECG sensor for long-term applications. *Communications Magazine, IEEE, 50*(1), 36–43.

Nkosi, M., & Mekuria, F. (2010). *Cloud computing for enhanced mobile health applications*. Paper presented at the Cloud Computing Technology and Science (CloudCom), IEEE Second International Conference on.

Oberheide, J., Veeraraghavan, K., Cooke, E., Flinn, J., & Jahanian, F. (2008). *Virtualized in-cloud security services for mobile devices*. Paper presented at the Proceedings of the First Workshop on Virtualization in Mobile Computing, ACM: New York.

Pang, Z. (2013). *Technologies and architectures of the Internet-of-Things (IoT) for health and well-being*.

Patel, M., & Wang, J. (2010). Applications, challenges, and prospective in emerging body area networking technologies. *Wireless Communications, IEEE, 17*(1), 80–88.

Patel, P., Jardosh, S., Chaudhary, S., & Ranjan, P. (2009). *Context aware middleware architecture for wireless sensor network*. Paper presented at the Services Computing, IEEE International Conference on.

Pedersen, T. B., Pedersen, D., & Riis, K. (2013). On-demand multidimensional data integration: Toward a semantic foundation for cloud intelligence. *The Journal of Supercomputing, 65*(1), 217–257.

Perera, C., Zaslavsky, A., Christen, P., & Georgakopoulos, D. (2014). Sensing as a service model for smart cities supported by internet of things. *Transactions on Emerging Telecommunications Technologies, 25*(1), 81–93.

Rao, B., Saluia, P., Sharma, N., Mittal, A., & Sharma, S. (2012). *Cloud computing for Internet of Things & sensing based applications*. Paper presented at the Sensing Technology (ICST), Sixth International Conference on, IEEE.

Rohokale, V. M., Prasad, N. R., & Prasad, R. (2011). *A cooperative Internet of Things (IoT) for rural healthcare monitoring and control*. Paper presented at the Wireless Communication, Vehicular Technology, Information Theory and Aerospace & Electronic Systems Technology (Wireless VITAE), 2nd International Conference on, IEEE.

Rosenthal, A., Mork, P., Li, M. H., Stanford, J., Koester, D., & Reynolds, P. (2010). Cloud computing: A new business paradigm for biomedical information sharing. *Journal of Biomedical Informatics, 43*(2), 342–353.

Schweitzer, E. J. (2012). Reconciliation of the cloud computing model with US federal electronic health record regulations. *Journal of the American Medical Informatics Association, 19*(2), 161–165.

Shahamabadi, M. S., Ali, B. B. M., Varahram, P., & Jara, A. J. (2013). *A network mobility solution based on 6LoWPAN hospital wireless sensor network (NEMO-HWSN)*. Paper presented at the Innovative Mobile and Internet Services in Ubiquitous Computing (IMIS), Seventh International Conference on.

Steinfeld, L., Ritt, M., Silveira, F., & Carro, L. (2015). Optimum design of a banked memory with power management for wireless sensor networks. *Wireless Networks, 21*(1), 81–94.

Suciu, G., Vulpe, A., Halunga, S., Fratu, O., Todoran, G., & Suciu, V. (2013). *Smart cities built on resilient cloud computing and secure Internet of Things*. Paper presented at the 19th International Conference on Control Systems and Computer Science.

Sullivan, T. J., Deiss, S. R., & Cauwenberghs, G. (2007). *A low-noise, non-contact EEG/ECG sensor*. Paper presented at the Biomedical Circuits and Systems Conference, IEEE.

Sultan, N. (2014). Making use of cloud computing for healthcare provision: Opportunities and challenges. *International Journal of Information Management, 34*(2), 177–184.

Sundmaeker, H., Guillemin, P., Friess, P., & Woelfflé, S. (2010). *Vision and challenges for realising the Internet of Things* (Vol. 20). EUR-OP, Brussels, Germany.

Vartiainen, E., & Väänänen-Vainio-Mattila, K. (2010). *User experience of mobile photo sharing in the cloud*. Paper presented at the Proceedings of the 9th International Conference on Mobile and Ubiquitous Multimedia, New York.

Wang, W., Li, J., Wang, L., & Zhao, W. (2011). *The internet of things for resident health information service platform research*. Paper presented at the Communication Technology and Application, IET International Conference on, IEEE.

Wang, Y.-W., Yu, H.-L., & Li, Y. (2011). *Notice of retraction internet of things technology applied in medical information*. Paper presented at the Consumer Electronics, Communications and Networks (CECNet), International Conference on, IEEE.

Xu, Y., Helal, S., & Scmalz, M. (2011). *Optimizing push/pull envelopes for energy-efficient cloud-sensor systems*. Paper presented at the Proceedings of the 14th ACM international conference on Modeling, analysis and simulation of wireless and mobile systems, Gainesville, FL.

Yang, G., Xie, L., Mäntysalo, M., Zhou, X., Pang, Z., Da Xu, L., Zheng, L.-R. (2014). A health-IoT platform based on the integration of intelligent packaging, unobtrusive bio-sensor, and intelligent medicine box. *IEEE Transactions on Industrial Informatics, 10*(4), 2180–2191.

Yao, D., Yu, C., Jin, H., & Zhou, J. (2013). Energy efficient task scheduling in mobile cloud computing. *Network and parallel computing* (pp. 344–355). Springer: New York.

You, L., Liu, C., & Tong, S. (2011). *Community medical network (cmn): Architecture and implementation*. Paper presented at the Mobile Congress (GMC), IEEE.

Zaslavsky, A., Perera, C., & Georgakopoulos, D. (2013). Sensing as a service and big data. *arXiv preprint arXiv:1301.0159*.

Zhang, Y., & Xiao, H. (2009). Bluetooth-based sensor networks for remotely monitoring the physiological signals of a patient. *Information Technology in Biomedicine, IEEE Transactions on, 13*(6), 1040–1048.

Section II

Frameworks for the Internet of Things
An Architectural Perspective

6

IoT Framework Based on WSN Infrastructure with Different Topological Distributions

Marcel Stefan Wagner, Diogo Ferreira Lima Filho,
Miguel Arjona Ramírez, and Wagner Luiz Zucchi

CONTENTS

6.1 Introduction

The internet of thinks (IoT) is based on the idea that devices, objects, and living beings have interaction within the environment to achieve a particular purpose. These elements may be distributed and connected across the network. The environment can comprehend different areas, such as terrestrial, air, space, underwater, or underground. The network can be designed to work in different topologies and the network nodes can be fixed or mobile.

It is possible to provide an environment with monitoring and control using sensors and actuators for the interactions amongst the different IoT network mechanisms. The use of sensors helps the network to collect data about environment features. The use of actuators serves to modify the environment or to carry out alterations in order to meet a main purpose, according to the readings taken from the sensors and in accordance with the network policy.

A wireless sensor network (WSN) consists basically of autonomous sensors distributed in an environment with specific addressing number for monitoring physical or environmental conditions such as temperature, pressure, lightness, proximity, and movement, among others. Another important activity of a node is the relaying or routing of information received from other nodes.

Usually, the data collected by the sensors are sent over the network to a central location, which is responsible for processing the information. The most modern networks have nodes with high processing power and these nodes can operate with bidirectional communication. In this scenario, they can also control sensor activity and manage devices optimally.

As well as in many other branches of technology, initially the development of sensor networks was motivated by military applications regarding data collection. Currently, these networks are also used for monitoring industrial processes, hospital management, commercial issues, traceability, and other applications in various fields of activity. The nodes may have different dimensions and because of the progress in nanotechnology, there are chips or integrated circuits (ICs) of reduced dimensions, maintaining the same or providing increased capability processing.

6.2 General Sensor Description

Each IoT node normally is comprised by a power supply, which is a battery or an energy harvesting module, an electronic circuit for interfacing sensors, microcontroller or microprocessor and radio transmitter with an internal or external antenna. In Figure 6.1, the blocks refer to the sensor internal modules for medium reading and actuation. From the left to the right: analog-to-digital converter is responsible for the analog to digital conversion (it can be represented by pulse width modulation) to a low-power microcontroller or microprocessor. In Figure 6.1, around the microprocessor, there is a memory group composed by random access memory/read-only memory for data processing and storage, radio frequency (RF) transceiver for the chosen operating band as can be seen in Section 6.4. In the reception (RX) path, the sensor node receives data from other sources to perform some tasks in the network, such as setting an actuator to turn on a motor, valve, or gate. It happens by a command from the microcontroller to provide the actuators with digital or analog data.

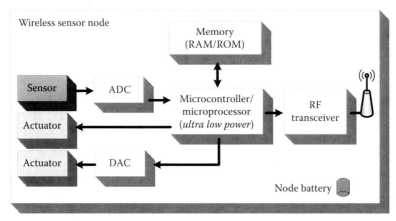

FIGURE 6.1
Basic structure in blocks of a sensor node.

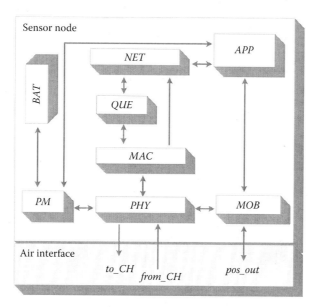

FIGURE 6.2
Basic internal structure of a wireless sensor node.

As Sohraby et al. (2007) pointed, the cost of the sensors varies according to the application and size, that is, it depends on the environment in which they will be used, including climatic conditions. For instance, it may be important to know whether the system is vulnerable to rain, if what the water level is, and whether the temperature has reached a critical value. It should be considered for sensor nodes, resource constraints, such as energy, memory, computational speed, and communication bandwidth (BW).

According to Dargie and Poellabauer (2010), the topology of sensor networks can vary from a simple network with few static nodes till an advanced WSN with mesh distribution and multihop feature, including mobility. The propagation technique used among the receivers, which influences the number of hops to the destination depending on the application purpose, can be done by forwarding packets or broadcasting information to network nodes.

In Figure 6.2, node mobility is assigned to the MOB module (related to the mobility) and the backbone communication with the environment is assigned to PHY, which interfaces with the channel (to_CH/from_CH) and outputs a position vector (pos_OUT). There is an internal flow, in which APP corresponds to the application, NET to the network layer, QUE to the data queue (if necessary it may perform part of the node internal processing), ACC to network access layer, and PHY to the physical layer. Although in the internal structure of the sensor, the BAT is related to energy storage unit and PM to the power management.

It is through the PHY layer that the node interacts with the network, generating the transmission (TX) and RX of data. This layer is directly related to the RF transceiver block of Figure 6.1.

The sensor functions are associated with the application and are suited to the type of environment to be monitored, analyzed, and managed. Sensors use physical entities that provide the electronic circuits with data collected from the environment. To provide it, the sensors can use changes in resistance, capacitance, inductance, frequency, amplitude, phase, and signal intensity.

The connectivity with sensors can be implemented using microcontrollers with TX connection interfaces. These connections are essentially based on the communication technology chosen, such as Ethernet, Fiber Optic, and Twisted Pairs for wired connections, and Infrared (IR), Bluetooth (BT), Wireless-Fidelity (Wi-Fi), ZigBee, radio-frequency identification (RFID), near-field communication (NFC), LoRa, 3G, 4G, and 5G for wireless communications.

6.2.1 Types of Sensors and Applications

The sensors can be classified into different groups for measurement or environment reading according to the application, such as: electric current, potential, magnetic, radio, humidity, fluid velocity, liquid flow, pressure, thermal, heat, temperature, gas, proximity, optical, position, chemical, and others.

A general vision about possible types of sensors that can be used in the smart city scenarios is provided as follows:

- *Speed*: Speed is usually used for detecting speed of an object (or vehicle) and can be implemented with different types of sensors, such as speedometers, wheel speed sensors, ground speed, Doppler radar, and air speed indicators.

- *Temperature*: It gives the temperature measurement as an electrical signal in the form of electrical voltage, which, in turn, is proportional to the temperature value. There are different types of sensors to do it, classified as contact type and noncontact type, with subclasses as mechanical (thermometer and bimetal) and electrical (thermistor, thermostat, thermocouple, resistance thermometer, and silicon band gap).

- *Presence*: Passive IR (PIR) sensors are used for motion detection and are said to be passive because they do not emit any energy or radiation for sensing motion. These sensors are classified based on the angle over which they can detect motion of the objects in the monitored environment, such as 120°, 180°, and 360°.

- *Distance*: The basic principle is similar to sonar, in which the reading of echoes produced by a primary sound wave or RF is used to estimate the attributes of a target, such as the distance of the object from the source. A transducer is used for converting energy into ultrasound waves within bands above human hearing, typically around 40 kHz.

- *Humidity*: It actuates in the measurement of humidity in the environment monitored. They can convert humidity measures into electric voltage. Its measurement is determined by the amount of water vapor present in a specific gas that can be a mixture, such as air or a pure gas (nitrogen). Most commonly used units are: Relative Humidity that is a relative measurement and a function of temperature, Dew/Frost point as a function of the gas pressure but independent of the temperature (is also defined as an absolute humidity measurement), and parts per million that corresponds to an absolute measurement unit.

- *Gas and smoke*: The detector, flammable and smoke/smother gas sensor is able to detect combustible gases and smoke concentrations in the air. This sensor can detect gases of different types, including smoke, liquefied petroleum gas, butane, propane, methane, hydrogen, alcohol, and natural gas.

- *Luminosity*: Light Dependent Resistor (LDR) is a component whose resistance varies with the intensity of the light. The more light falls on the component, the lower the resistance is. These sensors are recommended for external use, which through photocells recognize the decrease in light and trigger the system.

- *Level*: It operates as a power switch which can trigger keys, pumps, and lamps or send a signal to a microcontroller. For liquids, they can be considered as level switches in the type of magnetic float.

- *Accelerometers*: They can operate with different physical effects; therefore, they have a wide range of acceleration values they are capable of measuring. These devices are mainly used in positioning systems, tilt, and vibration sensors. A well-known application of accelerometers is in mobile phone screens that adjust according to the angle they make by means of g-forces or proper accelerations, which are accelerations due to reactions to the pull of gravity or other equivalent forces.

- *Gyroscopes*: These are devices consisting of a rotor suspended by a support formed by two articulated circles. Its operation is based on the principle of inertia and the rotating shaft has a memory effect, eliminating the need for a geographical coordinate.

The use of the sensors described above can be made in conjunction and applied as parts of the IoT nodes. They work as arms for data acquisition from the monitored environment, to be allowed in microcontrollers or microprocessors for the decision-making process (enabling or not the actuators).

6.3 Node Interface Technologies

To understand the nature and location of the interfaces between nodes, a short description about the technologies involved in IoT environments is presented ahead, based on Wagner (2009) work. To visualize the differences among the various wireless TX technologies, Table 6.1 shows the application, coverage area, and operating frequency of the main technologies in the telecommunications market.

BT is a wireless communication technology for short distances, designated to replace the cables that connect many types of local devices, such as mobile phones, headsets, monitors, audio, and medical equipment, among others. The energy–efficiency ratio makes it ideal for use in devices that need to work from a small battery charge for long periods.

IR is a form of invisible electromagnetic radiation. The major advantages of IR are related to its immunity to interference and the possibility of being transmitted and received with low-cost and easy-to-operate devices.

The Wi-Fi term is a registered trademark of the Wi-Fi Alliance. The expression has become a synonym for IEEE 802.11 technology, which allows the connection between different wireless devices. The range of a Wi-Fi network is variable, depending on the router and the antenna used. This distance can vary from 100 m (indoor) up to 300 m (outdoor).

LoRa is a wireless technology developed to operate in the Low-Power Wide-Area Networks (LPWANs). It seems to be an important tool for machine-to-machine (M2M) and IoT applications, because the technology offers low power consumption, long range, and secure data TX, so it also can be connected with existing cellular/wireless networks. According to the LoRa Alliance, it is easy to plug in the LoRa to the existing cellular infrastructure and also allow them to offer a solution for more low-power IoT applications.

TABLE 6.1

Comparison among Various Wireless Transmission Technologies

Technology	Standard	Application	Coverage Area (m)[a]	Frequency (Hz)[a]
RFID	ISO/IEC 14443 and ISO/IEC 18092	WPAN	100 (active)/25 (passive)	13.56 MHz
IR	RECS-80, RC-5, CEA-931-A and CEA-931-B	WPAN	Up to 1.5 (targeted direct)	35–41 kHz
NFC	ISO/IEC 14443 A&B and JIS-X 6319-4	WPAN	0.2 (contact)	13.56 MHz
UWB	IEEE 802.15.3a	WPAN	About 10	3.1–10.6 GHz
Bluetooth	IEEE 802.15	WPAN	1 (class 3)/10 (class 2)/100 (class 1)	2.4 GHz
Wi-Fi	IEEE 802.11a	WLAN	25–100 (indoor)	5 GHz
Wi-Fi	IEEE 802.11b	WLAN	100–150 (indoor)	2.4 GHz
Wi-Fi	IEEE 802.11g	WLAN	100–150 (indoor)	2.4 GHz
Wi-Fi	IEEE 802.11n	WLAN	70 (indoor)/300 (outdoor)	2.4 or 5 GHz
ZigBee	IEEE 802.15.4	WPAN/WLAN	10–100	0.8, 0.9, or 2.4 GHz
WiMax	IEEE 802.16d	WMAN	6400–9600	11 GHz
WiMax	IEEE 802.16e	*Mobile*/WMAN	1600–4800	2–6 GHz
LoRa	Proprietary—Semtech (SX127X product family)	LPWAN	Greater than 15000	Sub-GHz band
WCDMA	3GPP 3G	WWAN	1600–8000	1.8, 1.9, or 2.1 GHz
CDMA2000	3GPP 3G	WWAN	1600–8000	0.4, 0.8, or 0.9 GHz
1xEV-DO	3GPP 3G	WWAN	1600–8000	1.7, 1.8, 1.9, or 2.1 GHz
LTE	3GPP 4G	WWAN	About 15000	0.7–2.6 GHz
MBWA	IEEE 802.20	WWAN	4000–12000	3.5 GHz

[a] Data collected according to Brazil information and/or specification.

RFID is an automatic identification method that uses radio signals, retrieving and storing data remotely using devices called RFID tags. This tag generally is a small object that can be placed on a person, animal, or things such as equipment, packaging, product, and others. It contains silicon chips and antennas that allow transponders to answer radio signals sent by a transmitter (passive tags) and there are labels equipped with battery which allows them to send the signal itself (active tags). It is an alternative to bar codes, enabling product identification at a short distance from the scanner or self-off position.

NFC is a technology that allows the exchange of information wirelessly and securely between compatible devices that are close to each other. Some examples are mobile phones, tablets, badges, electronic ticket cards, and any other device that has an NFC chip inserted. When the devices are sufficiently close, the communication is automatically established without the need for additional configuration.

ZigBee corresponds to a set of specifications that are intended as wireless communication between electronic devices. It is based on features of low power operation, reasonable data transfer rate, and low cost for implementation. This set of specifications can be correlated with the Open System Interconnection (OSI) layers, as established by the IEEE 802.15.4 standard. The network was designed to connect small units to collect local data, together with its control, using part of the radio spectrum. In Brazil, there is no need for licensing this RF use; however, the devices must be approved by the National Telecommunications Agency (Anatel) according with Brazil (2008).

The Inter-IC connection is used to communicate data with fewer wires, interfacing with a lot of devices, such as displays, cameras, and others. This protocol works with Serial Data and Serial Clock for bidirectional data transfer.

The 3G technology can be used to connect IoT nodes in a WAN manner, using high data TX. With this connection scheme, data collected from end sensors can be delivered to data centers or a respective node in a wide area, interfacing WSN with cellular networks. The Wide-Band Code-Division Multiple Access is the air interface technology applied. For 4G, two standards are attractive: WiMax and Long-Term Evolution. Both of them have high-data transfer performance in the end-to-end TX. The interconnection between WSN with other devices or WSN can be done using the approach WSN-cellular-end/WSN.

Global positioning system refers to a satellite service as the reference for location systems. It can operate with other services and devices to deliver a vast array of customized applications, such as in agricultural machinery, security applications, and traceability.

5G technology is a promise for very high data transfer (around 20 Gbps) with connectivity among things, network, and cloud. This theme is yet a subject of extensive discussion in the academic and market areas, providing scenarios for the future of communications and IoT applications.

According with the range of presented technologies as possibility to IoT environments, the technology chosen for describing in this work, at this section for a WSN scenario was the ZigBee communication standard.

6.3.1 Connectivity Approach for IoT

Silva et al. (2014) pointed to the use of MICA motes sensors (MICAz) as a ZigBee use solution, because of features as energy consumption, support to wireless communication, and increased storage capacity. These motes are standardized by IEEE (2011) on the IEEE 802.15.4 at a 2.4 GHz and BW of approximately 250 kbps, ensuring low cost, low power, and reliability.

Table 6.2 shows the description of the MICA motes sensor family features, linking the information necessary to select the sensor to be used in the WSN. Tiny OS is an operating system aimed at events that was developed for use in sensor nodes. It was designed to withstand intensive competition operations to access the channel that are common in sensor networks and use few hardware resources. Against this background, it is possible that a node not only acts on the environment phenomena detection, but also as a router, actuator, and handle cases relating to the application, serving in this way, as a node for IoT solutions.

Figure 6.3 shows the basic internal interaction of a generic node applied for the network environment. This node is provided with cognitive processes as a solution proposed by Wagner (2016) for IoT in the cognitive sensor networks scenario. Actually, nodes are provided with more processing power, higher speed, and storage, so its approach seems to be a great solution for the management of IoT networks.

In the air interface of WSN, it should be considered the fading effect applied over the signals transmitted for communication between nodes. In this scenario, for free space condition, only part of the energy transmitted through the electromagnetic waves is captured by the receiving antenna. This energy decreases with frequency and distance. This loss is called loss in free space.

$$L = 32.5 + 20\log(d) + 20\log(f) \qquad (6.1)$$

TABLE 6.2

MICA Motes Sensors Features

Features	MICA Mote Sensors		
	MICA Mote	**MICA2 Mote**	**MICAz Mote**
Model	MOT300–MPR300CA	MICA2–MPR400CB	MICAz–MPR2400CA
Radio transceiver	TR1000	CC1000	CC2420–IEEE 802.15.4
Frequency	916 MHz	868/916 MHz, 433 MHz (MPR410CB) or 315 MHz (MPR420CB)	2.4–2.48 GHz
Transmission rate	mode OOK (10 kbps) or ASK (115 kbps), with max 50 kbps	38.4 kbaud in Manchester	250 kbps in High Data Rate Radio
Processor type	Microcontroller Atmel Atmega 103. AT90LS8535 uses 0.75 mW, with 12 mA to transmission and 1.8 mA to receipt	Microcontroller Atmel Atmega128L, In transmission 27 mA, reception 10 mA and in Sleep <1 μA	Microcontroller Atmel Atmega128L. The radio MICAz (MPR2400), with hardware security (AES-128). Uses 8 mA in Active mode and less than 15 μA in Sleep
Communication	Radio transceiver RF in 916 MHz, for 10 kbps. For transmission uses 36 mJ/s and reception about 5.4 mJ/s–14.4 mJ/s. Range approx. 20 m	Radio transceiver RF in 916 MHz. Transmission power between 0.01 and 3.16 mW, in the range of 5–91 m	Radio transceiver RF in 2.4 GHz. The external range is 75–100 m and for internal purpose of 20–30 m
Operational System	TinyOS. 178 bytes of memory	TinyOS. 178 bytes of memory	TinyOS. 178 bytes of memory
Memory	128 kB of programmable memory, 512 kB SRAM for data and 32 kB for EEPROM	128 kB of programmable memory, 512 kB of *flash* and 4 kB of EEPROM for configuration	128 kB of programmable memory, 512 kB of *flash* and 4 kB of EEPROM for configuration.
Dimension	7.62 × 2.54 × 1.27 cm	5.71 × 3.17 × 0.63 cm. The mote weight is 18 g	5.8 × 3.2 × 0.7 cm. The mote weight is 18 g
Enabled type of sensors	Light, temperature, accelerometer, noise, and seismic	Light, temperature, accelerometer, noise, magnetometer, humidity, and barometric pressure	Light, temperature, accelerometer, noise, magnetometer, humidity, barometric pressure, PH, acoustic, magnetic, and other MEMSIC sensors
Battery	Utilizes 2 batteries AA 3V. Can operate about approximately 1 year in low consume mode	Utilizes 2 batteries AA working in 2.7–3.3V. Can operate about approximately 1 year in low consume mode	Utilizes 2 batteries AA working in 2.7–3.3 V. Can operate about approximately 1 year in low consume mode
Application	WSN with routing capacity in large scale networks (+1000 nodes), *Ad Hoc* topology covering and monitoring	WSN with routing capacity in large scale networks (+1000 nodes), *Ad Hoc* topology covering and monitoring	WSN with routing capacity in large scale networks (+1000 nodes), *Ad Hoc* topology covering, mesh networking and monitoring

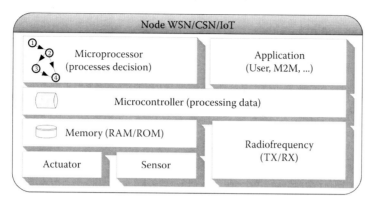

FIGURE 6.3
Basic internal block interaction of a generic node applied for the network environment.

where *d* is the distance in kilometers and *f* the frequency in MHz.

For selective fading, the Rayleigh, Rice, Log-normal, and other approaches can be applied. However, in general, environments for IoT implies in a multipath signal distribution; therefore, the Rayleigh fading is the approach chosen for explanation, because this effect can possibly impair the connection between IoT nodes depending on the distance of these end-to-end nodes or on the distance to the intermediate nodes in a multihop communication.

As mentioned, the antenna's range signal is directly linked to the fading effect in the TX medium, that is, the greater the distance between the nodes, the higher the signal decrease as the curve of the Rayleigh Probability Density Function, which is directly related to distance from the source node, representing:

$$R \sim \text{Rayleigh} (\sigma)$$ (6.2)

where σ is the scaling factor for the fading effect. The Rayleigh Probability Density Function of random variable *x* is

$$f\left(x \mid \sigma\right) = \frac{x}{\sigma^2} e^{\left\{\frac{-x^2}{2\sigma^2}\right\}}$$ (6.3)

where $f(x|\sigma)$ is the function that can be understood as an amplitude feature of the signal transmitted in the WSN environment with fading effect.

Figure 6.4 shows the sequence for 8σ simulations. It can be noted from the figure, that the higher is the σ value, the softer becomes the curve of the function. The general use approach in air interface for WSN is $\sigma = 1$; therefore, thinking in the IoT where the nodes will collect data from the environments events or in a periodic data acquisition for monitoring, it is reasonable to establish a threshold for the RX warranty, determining here the QoS for the service applied in delivering data from the source node to the destiny.

The generally applied topology for IoT scenarios include infrastructured distribution (it indicates that all group of nodes are under a direct control of a unique coordination function), *Ad Hoc* [also calling as independent mode, it indicates that devices can communicate directly among them without the necessity for an access point (AP)], and star (this topology generally is applied to LPWAN connectivity, where the devices are connected directly with an AP).

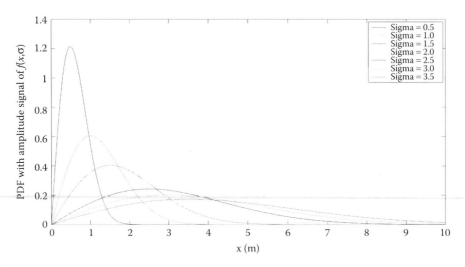

FIGURE 6.4
Probability density function (PDF) of Rayleigh functions.

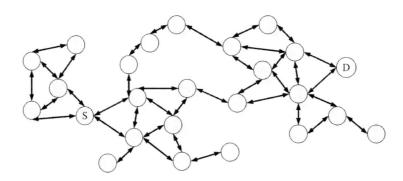

FIGURE 6.5
A simple scheme representation of connection among nodes in an *Ad Hoc* topology scenario.

Figure 6.5 shows a simple connection scheme in an *Ad Hoc* topology scenario, where the source (S) is in contact with the destination (D) node using a route determined by the range of each node's antenna, to the correct signal RX and re-TX between intermediate nodes. Therefore, the data is wirelessly delivered and the network connection is remade when the distance between nodes changes or if they lose their connection routes.

6.3.2 Data Sensing and Acquisition

The data acquisition is done by sensors present in the IoT nodes. According to Dias (2016), the sensors are necessary for IoT systems to convert the features or conditions of an environment, such as temperature, movement, and humidity in another element that can be readable for microcontroller processing, that is, converting the measurements in digital signals. These data will be carried to the application, in which it will be understood as information.

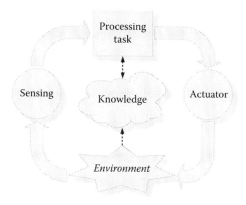

FIGURE 6.6
Node interaction when collecting data from the environment monitored.

A node is comprised by one or more sensors and transducers. The transducers are responsible by the conversion of a primary energy form in another one. These devices can transform any type of energy in electrical. The transducer can operate as an actuator in the monitoring environment.

In Figure 6.6, it can be seen that a node collect data from the environment monitored, and this cycle is executed in repetition according to the available sensing tools, such as the types of sensors used by node on the desired application.

As mentioned in previous sections, the data acquisition is done by sensors measuring the environment's phenomena and converting these values into electrical signals to carry it for the microcontroller's inputs, that in turn, transform these receipted signals in the information necessary to be, thus, forwarded to other network nodes or treated into the own node.

The physical medium corresponds to the environment where the sensors are inserted for the management and control tasks. There are many types of sensors, such as capacitive, inductive, and photoelectric, as described below.

The capacitive sensor (or proximity detector with capacitive effect) enables noncontact detection and linear measuring for small displacements (it happens on the order of almost zero to three centimeters with a resolution that can reach the nanometer). It uses capacitance, which is the ability of a material to store electric charge (i.e., thinking in the electronics field; this principle is generally associated to the storage device called capacitor).

Figure 6.7 shows a parallel plate capacitor made with two conductive materials positioned in parallel and between these plates there is an insulating material called as dielectric. The value of the mutual capacitance is proportional to the plate's area and the dielectric index material permittivity (property). It operates similarly to the capacitor; however, the capacitance of the sensor varies according to the distance between sensor-reading surface and material to be detected. So, there is a variation in capacitance according to the electrical signal emitted by device.

The inductive sensor is responsible for conversion of a physical measurement to an electric signal that can be understood by a programmable logic controller as information. The sensor has the ability to detect metallic objects at small distances. On account of lacking moving parts, the inductive sensor possesses a long service life in relation to the mechanical contacts sensors limit. In addition, they are very well-sealed, can work in dusty environments (nonmetallic), in contact with liquids (i.e., they can be inserted in a tank filled with water), and present great accuracy for small detection distance. Therefore, it can be

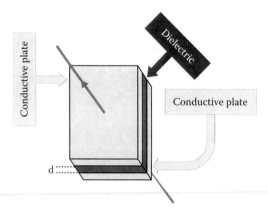

FIGURE 6.7
LDR sensor and the characteristic curve of the component.

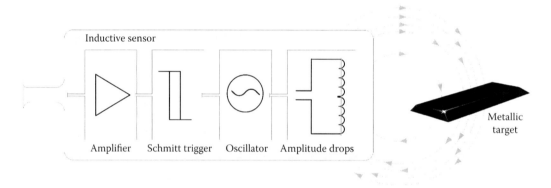

FIGURE 6.8
A simple operation of an inductive sensor.

used in proximity repeatability measurements and can operate at continuous (DC) or alternate (AC) output voltage level. In addition, the switching DC sensor can be found as normally open and normally closed and must be chosen according to the needs of the IoT project. In some DC models, transistors used in the amplifier circuit may be of PNP (positive potential is switched on) or NPN (connection should be made by negative potential) type. Figure 6.8 presents a simple operation of an inductive sensor.

The photoelectric sensors are used in a lot of applications, including: control, embedded electronics, security systems, industrial machinery, and medical equipment. The purpose of a photoelectric sensor is to convert a light signal (light or shadow) into an electrical signal that can be processed by an electronic circuit. LDR falls in the photo-sensors conductive type and is also known as photo-resistor. In them, the resistance to electric current flow depends on the intensity of light incident on a sensitive surface of cadmium sulphide material. The typical assembly of this sensor is intended to maximize the sensitive area of its electrodes, forming the structure shown in Figure 6.9.

As can be seen in the LDR sensor curve, the bigger the sensor, the greater is the ability to control more intense currents and in the dark, the typical resistance can reach more than 1 MΩ.

The data collected is generally sent to a microcontroller/microprocessor to run the embedded software (SW) and to attribute the referred task according to the data reading.

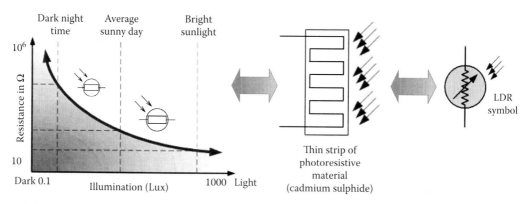

FIGURE 6.9
The characteristic curve of the LDR sensor.

There is a range of types of microcontrollers or microprocessors that can be applied in IoT environments, such as Raspberry PI, Arduino, PIC, BeagleBone, Galileo, and Edison. The most popular microcontroller family is Arduino and PIC. For the microprocessor, the most used are Raspberry PI and Field Programmable Gate Array. They have become popular because of their purchase facility and availability of source codes on the Internet. There is a vast content available for those who are interested in these technologies, but some knowledge is required about engineering, electronics, electricity, physics, and chemistry, besides a working knowledge of computer languages such as C, C++, and Java, in order to use most of them.

6.3.3 Data Transmission

The PHY layer is basically intended to be over wireless platform because of the emerging necessity for IoT nodes to be mobile and versatile. As it happens with other technologies for data sensing, the transceivers are the mechanisms that affect directly the TX. The basic structure of the general model block applied to the communication between two devices is shown in Figure 6.10.

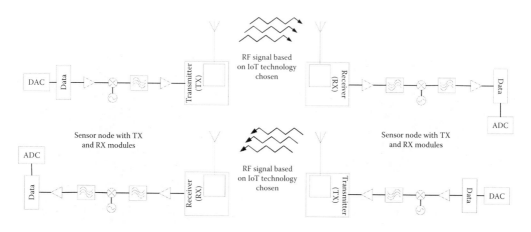

FIGURE 6.10
Basic general model block structure for the communication between two wireless devices.

The TX is done by the frequency related to the technology applied in the IoT environment, which are the RF TX, and RX between nodes. In Figure 6.10, the oscillator module (~) is responsible by the RF signal and the (►), (X), and (≈) modules are RF amplifiers and/or drivers with the intermediate frequency; signal mixer according to the oscillator; and filtering regarding to the low-pass, high-pass, band-pass, and/or band-stop filters, respectively.

The data is transmitted by the air interface after signal preparing stages mentioned as an analog signal to be received by the receiver node, which treats the signal received in the mentioned modules to finally turn the data able to be digitalized or used in analog manner by the nodes.

It was considered the use of DAC and analog-to-digital converter, because of the microcontroller/microprocessor modules utilization for the data management and possible intelligence improvement in the system for the IoT purpose.

6.4 Routing

As mentioned in previous sections of this chapter, the activity of an IoT node is not limited to the acquisition and actuation tasks, but it is also addressed with the advance in the applications and to the growing necessities. Thus the data routing among the devices is turning each day more important. This section will treat about some routing schemes and protocols that can be presented in the IoT scenarios, focusing in the newest Trellis Coded Network (TCNet) approach.

6.4.1 Current Challenges in Routing

Routing and forwarding packets remain a challenge in today's networks. In terms of the Internet backbone, the goal is to develop a shared understanding that the large operators are facing regarding the scalability of Internet routing system. It is recognized that the major contributing factors are the routing tables' growth, constraints in the routers technology and the limitations of Internet addressing architecture. Per year, the traffic volume is 50% (±5%) and the size of routing tables is around 15%–25%. Autonomous systems' growth rates are around 10%.

Between January 2006 and January 2009, Border Gateway Protocol (BGP) prefix update and withdrawal rates per day increased by a factor of 2.25 to 2.5, presenting an average value of 2–3 up to 1000 peaks per second. The research efforts can be classified as follows:

- Incremental improvements in BGP leading to a BGP++.
- Development of a new class of Path-Vector based routing protocols.
- Searching for new routing paradigms.

The routing protocols target space is characterized by distributed and adaptive requirements and the main objective is to determine if there is more than one possible routing scheme in that space besides BGP++ and Path-Vector.

On the other extreme of the Internet backbone lays the *Ad Hoc* and sensor networks. Such networks tend to proliferate according to the deployment of smart objects based services. They may be small or very large. In the latter case, the routing protocols also face the scalability challenges. Both kinds of networks in general consist of elements with limited resources in terms of processing power, storage space, and energy availability. The current employed routing protocols as, for example, the Ad Hoc On-Demand Distance Vector (AODV) suffers from a large signaling overload that entails large latencies. The situation is worsened by node failures and by the hidden nodes phenomenon. In addition, considering many of such networks may employ wireless TX; the spectral efficiency is also an issue.

To face the challenges presented by the described scenarios, geometric routing is being considered as an interesting solution. Geometric routing is based on topological network features that enable the implementation of greedy routing protocols.

For Papadimitriou and Ratajczak (2005), any planar three-connected graph can be embedded in the plane in such a way that for any nodes *s* and *t*, there is a path from *s* to *t* such that the Euclidean distance to *t* decreases monotonically along the path. As consequence of this, would be that in any *Ad Hoc* network containing such a graph as a spanning subgraph, two-dimensional virtual coordinates for the nodes can be found for which the method of purely greedy geographic routing is guaranteed to work. They present two alternative versions of greedy routing on virtual coordinates that probably work. Using Steinitz's theorem, it was shown that any three-connected planar graph can be embedded in three dimensions so that greedy routing works, albeit with a modified notion of distance. Experimental evidence that this scheme can be implemented effectively in practice has been presented by them. They also present a simple, but robust version of greedy routing that works for any graph with a three-connected planar spanning subgraph.

Krioukov et al. (2010) consider that at the same time, the nodes in the network are not aware of the global network structure, they analyzed if paths to specific destinations in the network can be found without such global topology knowledge and how optimal these paths can be. The salient feature of their model is that it allows one to study the efficiency of such path finding without global knowledge, because the analyzed networks have underlying geometry which enables greedy forwarding. As each node in the network has its address, that is, coordinates in the underlying hyperbolic space, a node can compute the distances between each of its neighbors in the network and the destination whose coordinates are written in the information packet or encoded in the signal. Greedy forwarding then accounts to forward the information to the node's neighbor closest to the destination in the hyperbolic space. As each node knows only its own address, the addresses of its neighbors and the destination address of the packet. In this scenario, no node has any global knowledge of the global network structure.

Comellas and Miralles (2011) propose a label-based routing for a family of scale-free, modular, planar, and nonclustered graphs. They provide an optimal labeling and routing algorithm for a family of scale-free, modular, and planar graphs with clustering zero. Relevant properties of this family match those of some networks associated with technological and biological systems with a low clustering, including some electronic circuits and protein networks. They suppose that the existence of an efficient routing protocol for this graph model should help when designing communication algorithms in real networks and also in the understanding of their dynamic processes.

All the presented approaches assume that the shortest path from the source to the destination node is the best path. In the realm of software defined networks, the best path may be defined considering a set of constraints, such as: cost, revenue, end-to-end delay, availability, packet loss rate, and so on. Based in the presented metrics, the use of the shortest path as the only metric is clearly not sufficient. In addition, it is important to assume that some metrics reflect technical performances whereas others are based on economic nature, in which some of them are linear. In the case of nonlinear, the geometric routing is not possible unless the metric space can be transformed into an equivalent underlying hyperbolic space in which the new source and destination nodes' coordinates can be used to evaluate the routing by means of a greedy algorithm. On the other hand, TCNet architecture was proposed by Lima Filho (2015) and Lima Filho and Amazonas (2013, 2014) presenting a different routing scheme to solve the pointed problems.

6.4.2 Fundamentals of the TCNet Architecture

The TCNet has been initially conceived for the scenario shown in Figure 6.11. This scenario considers a WSN in which a sink node queries a set of sensor nodes in a predetermined order. On the right side, the TCNet frame is presented.

In Figure 6.11, a node acts as an access point (AP) with internet protocol (IP) infrastructured networks and also as a sink to the information generated by other sensor nodes. A multicast scenario, where a sink node sends messages to a set of sensors and actuators, is a challenge for WSNs, and depending on the situation, the message has to be forwarded by several intermediate nodes before arriving at the desired destination. In such a case, each intermediate node has to consult a routing table to discover the next hop. If the next hop is not available in the table, a discovery procedure has to be initiated. The TCNet model changes this scenario completely. The sink node wishing to interrogate a set of nodes in the network sends a query q_i in a specific order defined by some optimization criteria. Each node receiving a frame undertakes the following actions:

- Recognizes that it is being queried.
- If the queried node concludes that it is an element of the route, it updates the payload in the frame with the value of its sensed variable and rebroadcasts the message.

The frame structure generated by the sink node, following the idea of the Multi-Protocol Label Switching shim header, which consists of the fields *WSN header|TCNet label|payload data* with the features described as follows:

- In the WSN header, a code word is loaded which enables metric calculation to be used by the routing algorithm.
- In the TCNet label, a specific sequence defined by a QoS criterion is loaded and transported along the route, allowing each node to recognize its position into the route and is used to update the code carried by the WSN header.
- Payload data is the field that stores the updated data sensed by the sensors during a query.

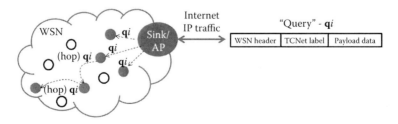

FIGURE 6.11
Illustration of a WSN in which a sink node queries a set of sensor nodes and in the right side the TCNet frame is shown.

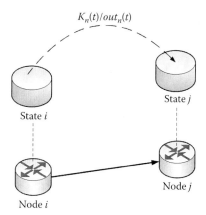

FIGURE 6.12
Network node modeled by a state of a Finite State Machine.

The process ends when all nodes of interest have been visited, the payload field carries up-to-date sensed data and the frame returns to the sink node.

The TCNet concept models the network as a digraph:

$$D = (V, E) \tag{6.4}$$

where V is the set of vertices v_i representing the network nodes and E is the set of edges v_i and v_j representing the physical or logical connections between the vertices v_i and v_j.

The model is based on finite automata or finite state machines (FSM) defined by a *cross* function (k_n/out_n) where a sequence of input symbols $\{k_n\}$ generates a sequence of output codes $\{out_n\}$ as shown in Figure 6.12. The $k_n(t)$ is the input symbol received at time t corresponding to the $out_n(t)$ code output. It is also assumed that at time t, a transition occurs at the FSM from state i to j. In TCNet, each state represents a network node and the transition state indicates that the frame information must be sent from node i to j.

A network node modeled as a state allows the application of FSM principles in order to exploit the analogy between networks and state diagrams to define paths between nodes, so as to enable each node to have full knowledge about the network by implementing a path generator, based on mealy machine of low complexity with the use of Exclusive-OR (XOR) gates and shift registers.

It is then possible to generate a specific route through a set of nodes, defined by a desired optimization criteria, such as latency, packet loss, BW and cost, by shifting an input sequence $\{k_n\}$ in the mealy machine and informing it in the frame as a TCNet label.

Using the described procedure, Figure 6.13 shows the route established by the protocol label using $\{k_n\} = \{1\ 1\ 0\ 0\}$. It can be observed that every node in the network is visited in the order [(10), (11), (01), (00)], that is, at the end, the frame returns to the sink node with the information collected from every other node.

The routing scheme helps in delivering data for network nodes. As described previously, a great effort has been done to improve the knowledge routing mechanisms. There is a wide field for network protocols optimization and should be considered that services and human necessities change with time.

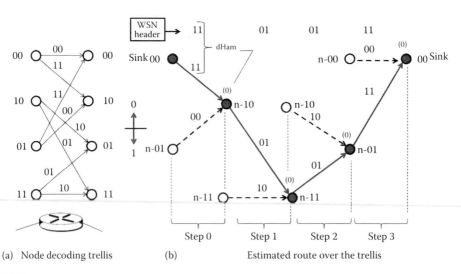

(a) Node decoding trellis   (b)   Estimated route over the trellis

FIGURE 6.13
TCNet. (a) Decoding trellis and (b) Route established by the input sequence $\{k_n\} = \{1\ 1\ 0\ 0\}$.

6.5 Data Management

One of the most important things to turn possible the IoT application in real environment or to enable the smart city development is the utilization of intelligence in the system, enabling the management optimization. Cognition and taking decision can be used to form a framework of IoT intelligence.

The use of cognition for the network management and the decision-making process for the actuators, not only in the way of reaction but aggregating learning to network nodes, is the most important tool for the evolution of WSN into the IoT.

As mentioned, the sensors collect data from the environment and in the actual network scenario, these data are sent to the microcontrollers or microprocessors for the decision based on the predefined rules established by the embedded SW or specific command lines, as a reaction form to the data collected, as, that is, if the lux measured in the environment falls to a predefined threshold, the microcontroller send to the respective pin (of the own IC) an information (binary 0_2 or 1_2) to turn on the light connected with this pin.

From the exposed, it can be seen that the actual mechanisms in the market yet do not have the necessary intelligence aggregated to the devices (unless the most powerful smartphone, which have the condition to use a more elaborated embedded SW and the availability of a great number of sensors) to create intelligence in network.

Figure 6.14 presents the actual relation of the issues applied to the standardization structure of IoT on the left side, and the increased blocks for the management and machine learning on the right side.

As pointed by Wagner et al. (2016), the cognitive processing allows the system to read just itself and at the same time perform the network monitoring tasks for the data acquisition and network management. This occur with the use of a control vector (V_C) that is interpreted by the adaptive cognitive system (ACS) to realize the Fuzzy logic (FL) approach in turn to deliver according to the defuzzification process output, the function to be adopted

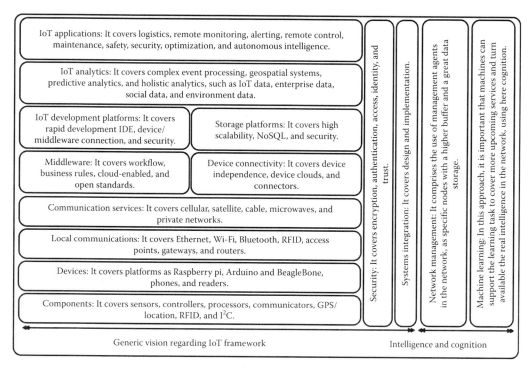

FIGURE 6.14
Relation of the issues applied to the standardization structure of IoT and the increased blocks for the management and machine learning of sensor nodes.

for some aspects readjust on the network nodes, which will direct the measures to be adopted in the network and accomplish the learning process for the network robustness.

Possible metrics involved with the measurement and V_C adoption are related with end-to-end delay, throughput, latency, and energy consumption. The ACS architecture was developed in order to seize control of WSN in multiple layers. For the ACS modeling, it was defined the block diagram present in Figure 6.15 in which, the ACS module has interaction with the WSN, Cognitive Process or Module (CPMod) and Fuzzy blocks.

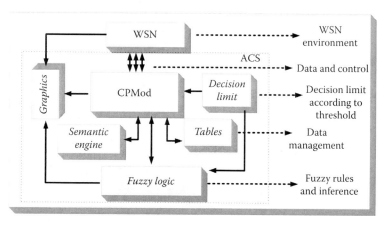

FIGURE 6.15
Adaptive cognitive system internal modules.

The Cognitive Network (CN) architecture, proposed by Thomas (2006) is considered in ACS, in the logical interaction between the blocks. In Figure 6.15, at CN, the Software Adaptable Network layer connects to the network medium, that is, the ACS performs the WSN interface with CPMod. The cognitive process layer of the CN, which includes cognitive elements in order to achieve the objectives outlined, is represented by CPMod block. Cognition selfish layer, which has the end-to-end objective and concentrate the network knowledge, is represented by ACS with the FL block.

In addition to the ACS representation as autonomous system along with the connections shown, thinking on the specific architecture of WSN, Figure 6.16 shows how the ACS is related to the different network layers, including FL and CPMod with availability to making adjustments in different layers.

For the Link and Physical layers, Mitola III (2006) presented the cognitive radio, a technology that is being applied for different reasons in the reading of channels available, because this technology enables the system to verify and optimize the use of radio spectrum. This radio system was idealized for the integration of substantial computational intelligence, especially machine learning, computer vision, and natural language processing in a Software-Defined Radio. Its application can be found in LoRa WAN radio technologies.

As the chosen sensor node for the study was the MICAz (family MICA motes), then the model of Figure 6.16a is more coherent because it specifically meets the IEEE 802.15.4 standard for attending the ZigBee technology. The model of 5 layers of Figure 6.16b is also valid for the ACS, for the IEEE 802.11 standard. The CPMod is the intelligent part for application in sensors, involving cognitive aspects to execute the decision-making process, acting as required on specific aspects of sensors, which are related to modeling of memory, type routing protocol, change processing queue, and it can apply energy saving.

Figure 6.17 shows the structure corresponding to a common sensing node with routing capability, along with the insertion of a cognitive module responsible for performing cognitive processes. In the scheme, the white blocks represent the traditional internal interface node, the yellow blocks are the representation of the physical wireless environment and green block represents the cognitive module. In CPMod, the solid red interfaces were initially implemented for processing, memory, and routing, and the dashed corresponds to

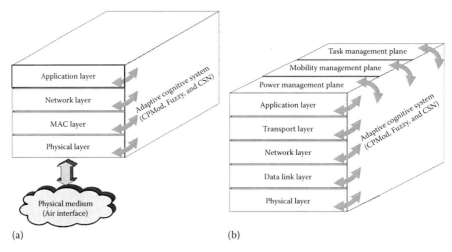

FIGURE 6.16
Layers stack with ACS multilayer approach. (a) 4 layers and (b) 5 layers.

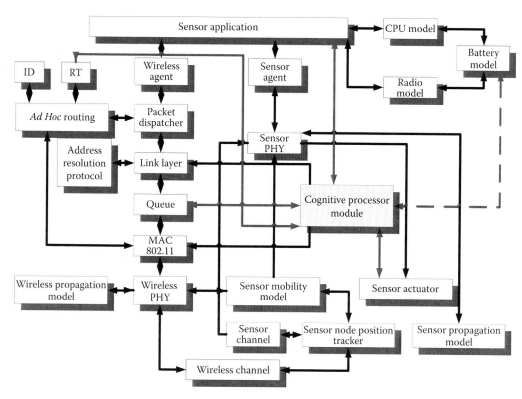

FIGURE 6.17
Common sensing node with routing capability and cognitive processor module.

the part connected to the energy which provides the generation of four vectors analysis for the rounds of analysis by the ACS.

From Wagner et al., (2015), the CPMod was created initially to operate according to the IEEE 802.15.4 standard for mobile *Ad Hoc* WSN networks in wireless local area network (WLAN), where ID corresponds to the identification of the node, the RT for the Routing Table (regarding routing protocol), and Address Resolution Protocol.

The implementation of CPMod occurs in the application layer, but it does not change the meaning and purpose of the process, that is, the realization of decision-making running on cognitive module will also affect internal parameters of other layers, such as adaptation in the processing queue of the packets, improving memory allocation, and modifications in the routing protocols adopted to optimize the packet forwarding routes, directly influencing the performance of the network. Here, it can be set up the use of TCNet protocol for the data routing in the network.

The CPMod perform analysis on the best condition according to the steps and run the data saving in another file as the values are defined in semantics block. The semantic block is associated with the CPMod analysis meaning part, namely as the progress of the test rounds on the system. CPMod uses fuzzy block to check the current network condition, comparing the data collected from WSN with values present in semantics. With the saving of the data on the best conditions for the events according to the adopted semantic, episodic memory is being formed and is the history of network building.

$$V_C = \{\text{Energy, Route, Queue, Memory}\} \tag{6.5}$$

From Equation 6.5, it can be seen that each detail is related with each possible metric pointed before, as vectors in binary 0_2 or 1_2. So, that is, if the node receives $V_C = \{0, 1, 0, 0\}$, it means that the routing protocol is subjected to be changed. In this approach, the communication protocol passes not to be the reference for the network, but it turns a powerful tool that can be changed and monitored to increase the WSN performance.

6.6 User Applications

A general vision about some cases of applying the WSN in the IoT environments and applications is done by the following:

- The use in the agriculture area with the utilization of drones: It implies in the range area to be covered and the use of localization scheme, the standard recognizing over the types of plantations, such as coffee, orange and soybeans, network connection between drones, and with a central node and also, big data regarding the data traffic to be in large scale.

- *Access and presence in educational environments*: The use of RFID tags for the control of the students presence in the class and to enable them for the access to specific environments, such as library, refectory, and gym room. For this application a great database for the information storage and possible consulting is necessary. It implies in security with high level, because the information about all students, professors, and employees are available in the database, so a cryptography system should be implemented here.

- *The use of tags for monitoring and control animals*: It implies in the use of RFID tags for the animal identification. These tags should be implemented in a way that the resistance to the climate would be higher than the normal situations, including the sensor to be water-proof and withstand high temperatures. Database is important here for the storage of animals' information and availability for the researchers consulting.

- *Smart grids*: They are in the real development area, because this issue directly influences the citizens' energy consumption. In this scenario, the use of a support for reading data (if possible in the real-time manner) and to control the decisions to be made over the stations, substations, and power lines is important. Big data surely is applied here. The ZigBee and LoRa technologies can be considered here for the nodes connection and monitoring.

- *Structural Health Monitoring (SHM)*: According to Akhras (1997), SHM defines the physical structures conditions of bridges, towers, cranes, and other large buildings. The measurements over a frame are used for computer models to analyze state data structure and make forecasts and warnings. One of the biggest challenges of SHM is that it is impossible to draw conclusions from just one measure about the use of only one sensor, which is well seen in the case of Figure 6.18 where a single sensor in a bridge collect data and no conclusions can be obtained from the measures, because this sensor not cover all area and possibilities of bridge disaster.

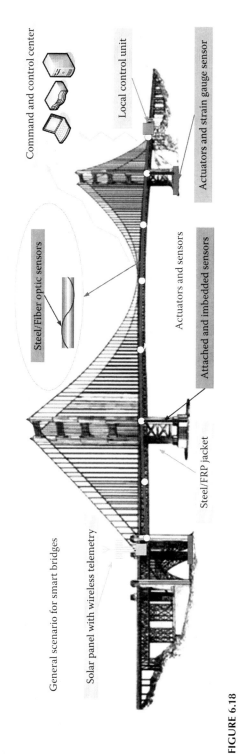

FIGURE 6.18
Scenario of the structure of a bridge with smart sensors.

The only feasible approach for SHM is to periodically measure a variety of physical quantities of acceleration, stress, pressure, temperature, wind speed, chemical reactions, and so on, which requires a considerable BW that can be solved taking into account frequency sampling and analysis methods online and offline.

6.7 Smart City with WSN

Micali (2016) at the Intel Developer Forum 2016 (an event to presents the trends and features of future of technology) has talked about virtual reality, robotics, visual intelligence, drones, as well as the power of autonomous vehicles, and the importance of the 5G on these scenarios. According to the event, over 50 billion things and devices must be connected by 2020, in addition with more than 200 billion connected sensors, all of them generating huge volumes of data. So, for this need, high processing capabilities, analysis, and storage in distributed network structure it will be necessary, in addition with permanent connectivity between things, network, and cloud.

Thinking in the presented scenario, this section will provide some cutting edge possible scenarios for the WSN use as the base of IoT in Smart Cities, based on the Brazil necessities, which can surely be applied in other places in the world. Below are described the authors' ideas concerning the IoT environments, taking into account the *all connected* philosophy:

- *Green line in the smart cities*: The idea here is to set the priority on the streets and avenues for ambulances, changing the aperture of the headlights, turning in this way the ambulances traffic more efficient, and decreasing deaths in the way to the hospitals. The principle applied here is that the ambulance contains a priority tag, that in contact with headlights as sensor nodes, the timing of traffic lights is changed. A control system can be applied for the monitoring and management of vehicles, including tags for the ambulance drivers. For instance, semaphores network will be created and can be clustered for connection between the lights and also with vehicles.

- *Interactivity of visually impaired or blind persons with traffic lights system*: The idea in this scenario is to enable visually impaired or blind people to more independent life in the urban centers. The main goal is to enable cell phone or smartphone with a SW for the user to receive audible information via the use of mobile devices. These messages are sent according with the environments, such as roads, bus stops, rail stations, airports, theatres, cinemas, galleries, events, shows, churches, schools, shopping, and a lot of other places where the user can interact.

  For the specific road scenario, semaphores can be connected as network nodes, executing an interaction between the lights and the blind person. The semaphores will interact with the users of the service receiving information about the traffic lights timing, how many time for the crossing of the road, about cars situation, and so on.

It is not only a beep or another type of code, but what is offered to the user is the information as it is, or in other worlds, as audible information. The control can be realized from the point of view of the network, determining if semaphores are out of service and other related things. This project is patented by Wagner et al. (2013) under the number PI 1105299-6 A2. Any queries about this issue can be treated with the patent authors.

- *Use of automobiles and motorcycles as network elements*: Cars, trucks, buses, and motorcycles, can represent majority traffic in the roads, avenues, and routes, so the use of them for the *Ad Hoc* M2M connection network is incredible. With this interaction between machines, a very large network can be implemented and it can offer an enormous range for the network and reach increasing for nodes. Not only vehicles can participate of the network, but semaphores (mobile elements in connection with fixed elements), smartphone, notebooks, tablets, and electronic devices in general.

  This solution promises to deliver the *all connect* concept to the smart city (the network coverage area extends to the edges of the city or between cities with the use of highways). The 5G will be the great responsible for turn this dreaming city in the real life possible.

- *The use of camera system interconnected with vehicles for the recognizing of a stolen vehicle and also to monitoring of the traffic in the roads, avenues, and routes*: This system works with a RFID in each camera point (or semaphore, traffic light, mast, or possibly trees) can aggregate the management of the vehicles (such as the cars, motorcycles, bicycles, buses, trucks, and other transport types), with tags inserted into the vehicle structure (for nonviolation of vehicle data), turning the smart city all connected for the government management (data can be collected from vehicles for verification, such as if the vehicle is regular with payment of taxes, and other emoluments). For this reason, is very important that the network and system offer very high security mechanism, including cryptography and tags checking for identifying tags RFID violations.

- *Intelligent traffic system (ITS) in urban clusters*: Several technologies are currently applied to what is being called as ITS, ranging from images pattern recognition until satellites, demanding complex infrastructures and large BW for data TX. The scenario depicted in Figure 6.19 shows an example of an application in which a vehicle can be tracked by transmitting its ID in broadcast mode to intelligent nodes. The network nodes are associated to the states of a FSM that implements a convolutional code using the TCNet protocol.

The sink or gateway nodes also need to be distributed along the city and the scenario presented in Figure 6.20 illustrating how vehicles can be tracked by trellis clusters of an ITS.

It is possible to change the configuration of the trellis changing the possible routes of the TCNet. The left side of Figure 6.21 shows two different trellis clusters with the same number of nodes but different number of routes. The first cluster has 8 direct paths between nodes and its complexity is of 3 registers and 2 XORs. The second has 16 direct paths and its complexity has increased only by 1 register and 1 XOR. This approach suggests that the model is scalable and quite robust in the presence of failures.

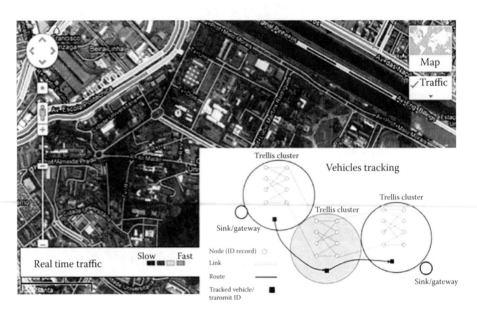

FIGURE 6.19
Scenario of vehicles tracking systems by means of trellis clusters.

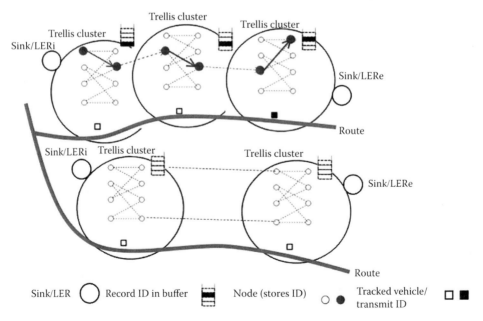

FIGURE 6.20
Illustration of vehicles tracked by trellis clusters.

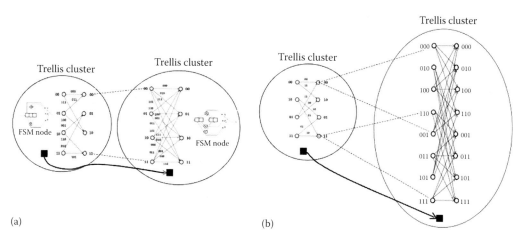

FIGURE 6.21
(a) Two different trellis clusters with the same number of nodes but different number of routes and (b) Illustration of how a coverage area can be changed by changing the capacity of a trellis.

6.8 Conclusion

Considering the application of TCNet and AODV in fixed structures, where nodes move slowly: for the TCNet performing a task, it is necessary to the node receive the obtained sequence $k_n(t)$ from a primary or using a graphical means for the simple cases using trellis inspection resources; in the case of AODV, the protocol performs the search of routing independently, with the disadvantage of increased latency. In the case of fixed structures where the nodes do not change the position, since the best routing table update, the AODV efficiency is almost equal to TCNet.

The ACS system is viable in the point of view of the network because it can actuates to improve the performance of the network related to the energy consumption of nodes, identifying the best network protocol to be used according to the network situation, the metric applied to the end-to-end delay and latency, was enhanced because the flow of packets by intermediate nodes is turned more controlled.

The cognitive approach can be inserted into an IoT device or for application in smart cities area, with the adaptation of the ACS/CPMod for operation in different nodes present in the WSN and not only as a traffic analysis based on application layer.

The universe of application possibilities in the IoT environments is only limited by our own thinking threshold and the connectivity is the key for the future development in the world where the Internet domain supports the addressing for IoT elements and the M2M, Machine-to-Living Beings and WSN-to-end/WSN are associated directly with the platform for communication, thus aggregating intelligence, reliability, security, and versatility with powerful smart devices usage, comprising a lot of variety of sensors and actuators to be employed into the control, monitoring, and management of networks.

References

Akhras, G. 1997. Smart structures and their applications in civil engineering. Civil Engineering. Report CE97-2, Royal Military College (RMC), Kingston, Ontario, Canada. Available: http://citeseerx.ist.psu.edu/viewdoc/download? doi=10.1.1.471.7279&rep=rep1&type=pdf.

Brazil. 2008. Presidência da República, Casa Civil, Subchefia para assuntos jurídicos. Law Decreto n. 6.654, 20/11/2008, that approve the Plano Geral de Outorgas de Serviço de Telecomunicações prestado no regime público. Brasília/DF, Brazil. Available: http://www.planalto.gov.br/ccivil_03/_Ato2007-2010/2008/Decreto/D6654.htm.

Comellas, F. and Miralles, A. 2011. Label-based routing for a family of scale-free, modular, planar and unclustered graphs. *Journal of Physics A: Mathematical and Theoretical.* 44(20): 205102. doi:10.1088/1751-8113/44/20/205102.

Dargie, W. and Poellabauer, C. 2010. *Fundamentals of Wireless Sensor Networks: Theory and Practice.* Chichester, UK: John Wiley & Sons.

Dias, R. R. F. 2016. *Internet das Coisas sem mistérios: uma nova inteligência para os negócios.* HP RFID Center of Excellence. São Paulo, Brazil: Netpress Books.

IEEE. 2011. Standard 802.15.4-standard for local and metropolitan area networks-Part 15.4: Low-rate wireless personal area networks (LR-WPANs). IEEE Computer Society. 5 September. Available: https://standards.ieee.org/about/get/802/802.15.html.

Krioukov, D. et al. 2010. Hyperbolic geometry of complex networks. *Physical Review E,* 82(3): 036106. doi:10.1103/PhysRevE.82.036106.

Lima Filho, D. F. and Amazonas, J. R. 2013. TCNet: Trellis Coded Network-implementation of QoS-aware routing protocols in WSNs. *IEEE Latin-America Conference on Communications (LATINCOM'2012).* Cuenca, Ecuador, April.

Lima Filho, D. F. and Amazonas, J. R. 2014. A Trellis Coded Networks-based approach to solve the hidden and exposed nodes problems in WSNs. *Segundo Seminario Taller Latinoamericano de Instrumentación Control y Telecomunicaciones (SICOTEL'2014).* Available: http://www.gituq.edu.co/memorias%20sicotel2014.pdf.

Lima Filho, D. F. 2015. Projeto e Implementação de um novo Algoritmo e Protocolo de Encaminhamento de Pacotes baseado em Códigos Convolucionais usando TCNet: Trellis Coded Network. 114 p. Doctoral Thesis. Sistemas Eletrônicos-Escola Politécnica, Universidade de São Paulo, São Paulo, Brazil.

Micali, B. 2016. Preparem-se: o futuro previsto pela Intel na tecnologia traz revoluções. Tecmundo in the Intel Developer Forum (IDF'2016). San Francisco, CA. Available: http://www.tecmundo.com.br/intel/108519-preparem-futuro-previsto-intel-tecnologia-traz-revolucoes.htm.

Mitola III, J. 2006. *Cognitive Radio Architecture: The Engineering Foundations of Radio XML.* Hoboken, NJ: John Wiley & Sons.

Papadimitriou, C. H. and Ratajczak, D. 2005. On a conjecture related to geometric routing. *Journal of Theoretical Computer Science-Algorithmic aspects of wireless sensor networks.* 344(1): 3–14. doi:10.1016/j.tcs.2005.06.022.

Silva, F. A. et al. 2014. Tecnologia de Nós sensores sem fio. *Revista Controle e Instrumentação.* UFMG. Available: http://homepages.dcc.ufmg.br/~linnyer/bianossensores.htm.

Sohraby, K., Minoli, D., and Znati, T. 2007. *Wireless Sensor Networks: Technology, Protocols, and Applications.* Hoboken, NJ: John Wiley & Sons.

Thomas, R. W. et al. 2006. Cognitive networks: Adaptation and learning to achieve end-to-end performance objectives. *Topics in Radio Communications-IEEE Communications Magazine.* 44(12): 51–57.

Wagner, M. S. 2009. Influência de protocolos de segurança sobre o desempenho de redes UMTS. 141 p. Master Thesis. Sistemas Eletrônicos-Escola Politécnica, Universidade de São Paulo, São Paulo, Brazil.

Wagner, M. S. 2016. Sistema cognitivo com tomada de decisão baseada em Lógica Fuzzy para aplicação em ambientes de redes de sensores sem fio com múltiplos saltos. 237 p. Doctoral Thesis. Sistemas Eletrônicos-Escola Politécnica, Universidade de São Paulo, São Paulo, Brazil.

Wagner, M. S., Ribas, C. A. M., and Oliveiras Junior, Z. D. 2013. PATENT number PI 1105299–6 A2. Processo e equipamento para transmissão e recepção de informações auditivas para portadores de deficiências visuais. Instituto Nacional da Propriedade Industrial (INPI). Revista da Propriedade Industrial (RPI) 2231 de 08/10/2013, Seção I, p. 65, São Paulo, Brazil. Available: http://revistas.inpi.gov.br/pdf/PATENTES2231.pdf.

Wagner, M. S., Ramírez, M. A., and Zucchi, W. L. 2015. Adaptive cognitive system applied on wireless sensor networks nodes decisions. *IEEE International Conference on Communication Systems and Network Technologies (CSNT'2015)*. Gwalior, India, April.

Wagner, M. S., Ramírez, M. A., and Zucchi, W. L. 2016. Adaptive cognitive system applied to WSN decisions at nodes with a fuzzy logic approach. *International Journal of Communication System and Network Technology*. 5(1). doi:10.18486/ijcsnt.2016.5.1.01.

7

Semantic Complex Service Composition within an IoT Ecosystem

Grigorios Tzortzis, Charilaos Akasiadis, and Evaggelos Spyrou

CONTENTS

7.1 Introduction

Recent technology advances in the fields of electronics, telecommunications, and informatics have allowed everyday physical objects to be enhanced with the embodiment of short-range and energy-efficient mobile transceivers. The extensive networking of everyday objects has led to the so-called Internet of Things (IoT) (Atzori et al., 2010), which has been considered by many to constitute the next industrial revolution (Gubbi et al., 2013). IoT is expected to find numerous applications in many diverse and heterogeneous areas, including, but not limited to, industry, logistics, building and home automation, smart cities and smart manufacturing. For further possible applications, the reader is encouraged to study the works of Guo et al. (2013) and Bandyopadhyay and Sen (2011).

It is typical for IoT ecosystems to incorporate a service-oriented architecture (SOA) (Al-Fuqaha et al., 2015). In other words, each *thing*, whether hardware (e.g., a device)

or software (e.g., an algorithm), is exposed to the outer world as a web service, a self-contained unit of functionality, which offers its services to other *things*. IoT services can be categorized into three distinctive, yet interdependent, types: (1) services that capture properties of the physical world and provide raw or slightly processed measurements (sensing services); (2) services that process the acquired measurements and provide the inferred results (processing services); (3) services that enable certain actions, based on the results of the processing services (actuating services). On the following, these three categories of services are referred to as S-, P-, and A-type services, respectively, or *SPA* services as a whole.

Commonly, applications designed on top of an IoT ecosystem exploit the SOA paradigm and involve distributed sensor networks at various scales that are interconnected with distributed processing modules (which autonomously process the measurements gathered by the sensors) and actuation elements that are triggered according to the produced results. Hence, the available SPA services can be combined to build complex applications that offer some new desired functionality that none of the existing services is able to provide. This procedure is referred to as *service composition* (Sheng et al., 2014). Key steps in composing a service (thus building an application) are the *discovery* of suitable services, and their appropriate *interconnection* using an IoT-ready platform, such that a functional composition, is ensured in which every service can be readily invoked (i.e., each service's inputs and pre-conditions are satisfied from the outputs and the effects of other services participating in the composition).

The main focus of this chapter is to describe the procedure of developing complex applications within SYNAISTHISI (Pierris et al., 2015), an IoT-ready platform in which available SPA services are semantically annotated using ontologies. In particular, a *smart meeting room* ontology is utilized, which introduces classes and relations (properties) to model the various service types encountered within the context of a meeting room that has been augmented with several S-, P-, and A-type services. Moreover, it is shown that a developer with knowledge of (1) the services required for the complex application and (2) their correct interconnection is able to exploit the semantic information in the service descriptions (via SPARQL queries [Harris & Seaborne, 2013]) to manually compose a service. Also, a service composition process is presented that utilizes semantics and allows developers to compose services in semiautomatic fashion, alleviating the assumptions made by the manual approach and considerably reducing the effort required by the developer and the risk of human error.

In this chapter, a developer's side perspective is adopted. Assume that there exists a set of already developed services, and the developer needs to reuse some of them so as to deliver new and more complex ones, as easily and as fast as possible. The following illustrate how a developer may search, discover, and interconnect available services using both the manual and the semiautomatic composition methodology. As a test case to demonstrate the potential of this framework, a *person counting* service is developed, implemented within the context of a smart meeting room pilot scenario that exploits a subset of the available sensing and processing services in the room.

The rest of this chapter is organized as follows: Section 7.2 provides an overview of related work concerning service composition and service ontologies. Then, in Section 7.3, the main characteristics of the SYNAISTHISI platform are outlined, along with the description of one of its pilot applications, namely the smart meeting room, in the context of which service composition is performed. The proposed smart meeting room ontology is analyzed in Section 7.4, whereas the two service composition approaches follow in Section 7.5.

Section 7.6 presents a use case of constructing a *person counting* service by combining existing services available in the smart meeting room pilot. Section 7.7 discusses various interesting features of the smart meeting room ontology and the two composition methods. Finally, conclusions are drawn in Section 7.8.

7.2 Related Work

It is well known that orchestrating a collective functionality using highly heterogeneous devices and modules exposed as web services is quite a difficult task. To obtain a comprehensive representation of web services and improve interoperability in the context of IoT so that tasks such as service discovery and composition can be performed in a more effective and efficient manner, *semantics* are employed to enhance the description of services. Several ontologies have been proposed for adding semantic content to web services, thus transforming them into *semantic web services* (McIlraith et al., 2001). OWL-S (Martin et al., 2004) and WSMO (Roman et al., 2005) are two renowned efforts, which provide highly expressive models for annotating web services in heterogeneous domains. Rich conceptual approaches, such as OWL-S and WSMO, demand considerable user effort to comprehend them and appropriately annotate the services, whereas reasoning becomes a computationally intensive process. Lately, a tendency has emerged to switch to lightweight semantic models that focus only on the core semantics of web services, trading expressivity for improved usability and complexity. SAWSDL (Kopecký et al., 2007) and WSMO-Lite (Vitvar et al., 2008), which builds upon SAWSDL, are two such examples that allow semantic annotations to be directly added to the WSDL description of services. Accordingly, for RESTful services, hRESTS (Kopecký et al., 2008) and MicroWSMO (Roman et al., 2015), which is based on hRESTS, have been developed, supporting the addition of semantics directly on the HTML description of services. The Minimal Service Model (MSM) (Pedrinaci et al., 2010) is another such approach that provides a common conceptual model for capturing semantics of services, whether they are WSDL based or REST based. As noted by Pedrinaci et al. (2010), MSM is largely compatible with OWL-S and WSMO service models. MSM has been incorporated in the iServe platform (Pedrinaci et al., 2010), a service publication and discovery platform.

The aforementioned ontologies constitute general purpose upper ontologies that include high-level concepts that can be utilized across diverse domains. However, low-level concepts specific to some particular domain are missing, and substantial effort is required to define them. To tackle this issue, web service ontologies that are domain specific have been exploited. Stavropoulos et al. (2012) proposed BOnSAI, a smart building ontology, to facilitate the development of a smart ambient setting that provides automation and energy savings. BOnSAI categorizes services/devices into sensors and actuators (i.e., S- and A-type services) and defines classes and properties related to a smart building (e.g., lighting, air-condition, smart plugs, communication protocols used by devices, location of devices, environmental parameters, such as temperature, measured [affected] by sensors [actuators]). Past work of Akasiadis et al. (2015) has presented a preliminary version of an ontology dedicated to smart meeting rooms that, in general, classifies services/devices into sensors, processors and actuators (i.e., S-, P-, and A-type services) and subsequently refines them by defining specific types of them in a class hierarchy, along with other important aspects such as location and the *IOPEs of services. This ontology is extended herein (Section 7.4) with additional concepts, to better describe the entities of smart meeting rooms and improve the execution of service composition tasks.

Web services may be perceived as self-contained units of functionality that can be readily published, discovered, and consumed using service-oriented architectures (SOA). To accomplish the vision of IoT it is of paramount importance that apart from simply using available services, users are able to easily combine them via service composition techniques and produce more complex ones that better suit their needs. Service composition approaches can be grouped into three main categories: manual, semiautomatic, and automatic (Sheng et al., 2014). In manual composition, the user is responsible for every step of the composition process. Appropriate services are manually discovered, and then their interconnection must be defined. It may even be the responsibility of the user to bind the web services according to the interconnection scheme. Although there is total freedom on how to perform the composition, this advantage is compensated by the need for strong programming skills (the user must be a developer) and the fact that the whole procedure is time-consuming and error-prone.

In the semiautomatic case, the underlying system assists users by automating some parts of the process (e.g., the discovery of suitable services, deciding whether two services can be connected or not) and, at various steps during composition, presents them with possible choices out of which they must select how to proceed. Liang et al. (2004) proposed an approach in which the user first interacts with a query composer to select service categories that suit his needs, thus forming a *service request*, then a service dependency graph that captures the input and output dependencies of the available services is automatically constructed and an AND-OR graph search algorithm is employed to extract multiple subgraphs that represent possible compositions. These compositions are shown to the user, who must acknowledge whether they are valid and satisfy his needs. Albreshne and Pasquier (2010) introduced a system in which developers build generic process templates that describe the control flow of possible composite services and end-users configure, via a graphical interface, those generic templates according to their preferences. Based on the configured template, a service discovery is executed to locate matching services and realize the composition.

Turning to automatic composition, the user typically sets his/her constraints and/or preferences in a service request and is not allowed to intervene at any other part of the process, which is executed in a fully automatic manner. Such methodologies are not demanding, nor require specialized skills, and are ideal for everyday users. However, as users have nocontrol on how the composition evolves, it may lead to a solution that does not entirely fulfill their original intentions. PORSCE II (Hatzi et al., 2012) transforms service composition to an AI planning problem that can be solved using off-the-shelf PDDL-compliant planners. The initial state consists of the data that are provided to the composition process (i.e., inputs), declared in a service request, whereas the goals are the desired outputs of the composite service, which are also specified in the service request. Actions for moving between states in the planning domain correspond to the available web services. The applicability of a service (i.e., action) at a certain state depends on its required inputs, whereas the resulting state is augmented with the outputs produced by the service. PORSCE II offers a mechanism to replace a service from those involved in the computed composite service if the user is unwilling to use it or the service has failed. Rodríguez-Mier et al. (2016) presented a graph-based method for automatic service composition and implement it as part of the ComposIT system. A service request similar to

---

* IOPEs: inputs, outputs, preconditions and effects.

that of PORSCE II must be declared, containing the inputs available to the composite service as well as the desired outputs it should produce, and a layered directed acyclic graph of services is incrementally constructed. Each layer contains all those services, inputs of which are provided by the outputs produced in previous layers and, therefore, are invokable at this layer. A heuristic search based on the A* algorithm is executed over the graph to find the composition (i.e., a subgraph) with the minimum number of services that fulfill the service request.

As a final note, all service composition approaches discussed here exploit semantic descriptions of the individual services in order to avoid syntactic barriers and boost their performance. However, this is performed in slightly different ways, according to the field of application (see, e.g., the work of Sheng et al. [2014] that surveys recent work regarding web service composition).

7.3 The SYNAISTHISI Platform

The platform presented herein has been developed within the context of the SYNAISTHISI project,[*] the goal of which was to deliver energy efficient, secure, and effective applications, as well as services to end users, aiming to minimize the environmental impact (services would run only when needed), monetary costs (unreasonable energy consumption should be prevented), user discomfort (room conditions should be comfortable for users), delays and utilization of resources (network traffic should be minimized). In this section, a brief description of the developed, IoT-ready, SYNAISTHISI platform (Pierris et al., 2015) is provided, along with an outline of how it enables the integration, interconnection, and coordination of a large number of heterogeneous devices and algorithms, which are *exposed as semantic web services*. The platform architecture is analytically discussed by Pierris et al. (2015), and going into the architectural details is out of the scope of this chapter. In brief, the platform supports communication channels that realize service interconnection, controls the available resources, permits the deployment of custom applications, and provides a set of tools to accommodate system administrators.

The available (semantic) services are registered into the service registry, implemented by an RDF triplestore equipped with a semantic reasoner, and annotated using an appropriate ontology, which shall be discussed in Section 7.4. Following the IoT paradigm discussed in Section 7.1, they are divided into three distinct categories: (1) S-type services corresponding to sensors that sense the physical world, (2) P-type services corresponding to processors (i.e., algorithms) that process the measurements of the S-type services and/or the processed results of other P-type services, and (3) A-type services corresponding to actuators that are used for the actuation of devices/signals based on the acquired results. Seen from the services' IOPEs perspective, S-type services produce outputs but do not have inputs (strictly speaking, they may only receive trigger signals for their (de)activation), P-type services have both inputs and outputs, whereas A-type services have inputs, and their output (actuation) is actually an *effect* that may be observed. The three service types will be collectively referred to as *SPA services* hereafter.

Within the platform, services communicate by publishing messages and/or by subscribing to topics managed by a Message-oriented Middleware (MoM), which is a message

---

[*] http://iot.synaisthisi.iit.demokritos.gr

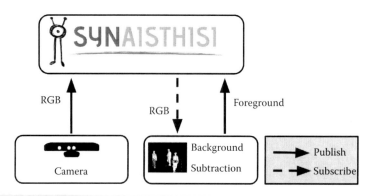

FIGURE 7.1
Two services interconnected with the SYNAISTHISI platform exchange messages. Solid arrows correspond to messages published to topics, whereas dashed arrows depict the messages arriving to the subscribers of those topics.

broker. Thus, all sensor measurements, processor results, and actuations are encapsulated into messages and communicated via these topics. Once a message is published to a topic, the MoM informs and delivers it to all clients that are subscribed to that topic. The MoM, among others, supports the MQTT protocol (Banks & Gupta, 2014), main role of which is to orchestrate the intercommunication among services in the aforementioned way. Each service incorporates an MQTT client to be connected to the MoM. The MoM is responsible to deliver the messages to subscribed clients. Scalability is facilitated, as new services may be added with almost zero configuration. Moreover, the MQTT is an extremely lightweight protocol, designed to require small code footprint and network bandwidth, while also being available for many heterogeneous devices and programming languages. An example of the interconnection of services using a MoM is depicted in Figure 7.1. A camera is an S-type service that captures RGB video data from the physical world and publishes them to a topic. A Background Subtraction module is a P-type service, which receives the camera-generated data by subscribing at the same topic. Upon processing, it publishes its output using another topic.

The SYNAISTHISI smart meeting room (Sfikas et al., 2015) is one of the pilot applications of the SYNAISTHISI project and resulted upon the enhancement of a typical meeting room with a set of SPA services that are integrated into the SYNAISTHISI platform. More specifically, it incorporates sensors and actuators, physically installed in the room, as well as processing units running in the cloud. A Decision Maker (DM) service performs the main actuations (implemented as A-type services), using measurements and results from S- and P-type services. Upon recognizing specific simple events occurring within the room (e.g., temperature, humidity, luminosity, electricity consumption levels and occupancy), it triggers actuations (e.g., controlling HVAC,* lights, the projector) aiming to minimize energy losses, while maintaining certain comfort levels to improve the overall working environment. The smart meeting room scenario provided the motivation behind the ontology and the service composition algorithms developed herein, although it must be made clear that the service composition techniques are general and by no means limited to this scenario.

---

* HVAC: Heating, Ventilation, Air-Conditioning.

7.4 An Ontology for Smart Meeting Rooms

To enhance the efficacy of service discovery and composition toward realizing complex applications, the SPA services of the SYNAISTHISI platform service registry are *semantically enriched* using ontologies. Specifically, a *domain-specific* ontology has been developed that models the concepts related to smart meeting rooms (Section 7.3), in which all sensors, processors, and actuators are exposed as web services.

The developed semantic model imports existing ontologies and reuses their knowledge. The Internet of Things Architecture (IoT-A) ontology (De et al., 2012) includes high-level concepts for describing key aspects of the IoT domain and forms its basis. Also, the popular Semantic Sensor Network (SSN) ontology is utilized to represent various features of sensors and actuators (e.g., the measurement capabilities of a sensor), whereas the QU and QUDT ontologies provide descriptions for physical quantities (e.g., temperature) and their measuring units.

In what follows, the main classes and properties that make up the proposed ontology are outlined, omitting secondary elements to avoid cluttering the presentation. A preliminary version of the ontology can be found in the work of Akasiadis et al. (2015), which is here refined and extended with additional concepts.

The resource and service model of the IoT-A ontology are specialized by introducing classes, subclasses, and properties related to the smart meeting room domain. In general, the resource model is responsible for describing the characteristics of the actual device (i.e., sensor or actuator) or processor, referred collectively to as *resources* hereafter, that is hidden behind a web service, whereas the service model is responsible for describing the characteristics of the web service (e.g., service input/output, service endpoint) that is used to access a resource and expose its functionality to the outer world. Note that in the context of this chapter, the notion of SPA services encapsulates both resource and service model concepts.

7.4.1 Resource Model

The resource model, illustrated in Figure 7.2, consists of a core class, namely `Resource`, that captures the notion of a resource. A hierarchy is defined to represent various types of resources that may exist in a smart meeting room, many of which were not part of the ontology of Akasiadis et al. (2015). The hierarchy includes specific types of sensors, for example, temperature sensors, cameras and microphones; processors, for example, modules for subtracting the background of a scene; and actuators, for example, plug switches. Class names are quite indicative of the resources' functionality. A resource is equipped with properties specifying its name (`hasName`), an ID (`hasResourceID`), some keywords describing it (`hasTag`) and, importantly, the web service that exposes its functionality to the outer world (`isExposedThroughService`). The `Location` class defines the location of a resource, while the `Network` class is used to describe the network interface which makes the resource accessible through the web. Product-related information, for example, the manufacturer's name, can be provided using the `Product` concept. The ontology also allows to associate a resource with its owner (`ResourceOwner`). Finally, for sensing and actuating devices the SSN ontology concepts can be exploited using the link to the SSN `Device` class.

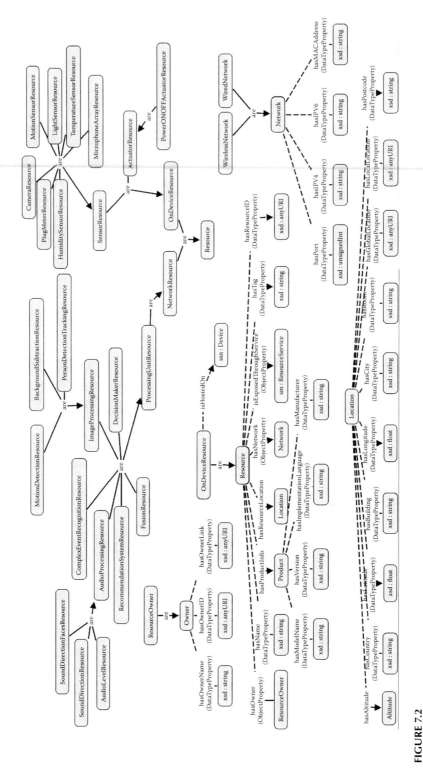

FIGURE 7.2
Core classes and properties of the resource model. The sm and ssn namespaces refer to the service model and the SSN ontology, respectively.

7.4.2 Service Model

The service model, depicted in Figure 7.3, consists of a core class, namely `Service`, that captures the notion of a web service and contains the necessary information for discovering and invoking the service. Various service types are defined in a hierarchy, corresponding to the aforementioned resource types. A service has a name (`hasName`), an ID (`hasServiceID`), an owner (`ServiceOwner`), a link to the resource that is accessed using the service (`exposes`) and an endpoint from where client applications can access the service (`ServiceEndpoint`). The (physical) area that is affected when the service is invoked can be determined via the `hasServiceArea` property. This property is particularly useful for S-type services that declare the area that is observed by the sensor, and for A-type services that declare the area that is affected by an actuation. Possible time constraints on the availability of a service can be defined using the `ServiceSchedule` class. A very important element in the semantic description of a service, especially for service composition tasks, is to accurately model its IOPEs. Here, this is accomplished through the subclasses of the Parameter class and by using the `hasParameterType` property to annotate the IOPEs by providing the URI of the concepts (defined in some appropriate ontology) that capture their meaning. Note that effects are considered as a special type of output intended to describe the actuation of A-type services. Finally, it is possible to declare a price for trading a service in a service marketplace (`ServicePrice`).

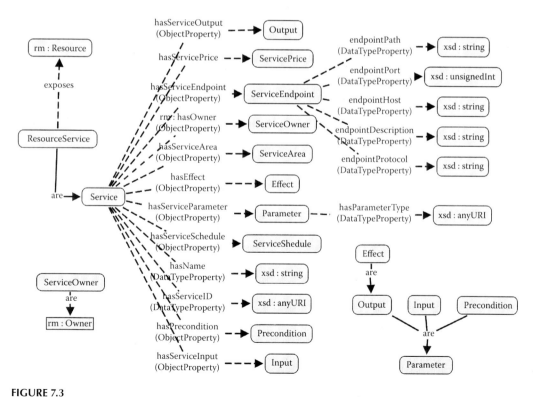

FIGURE 7.3

Core classes and properties of the service model. The `rm` namespace refers to the resource model. The hierarchy of service types is not shown here, as it has a one-to-one correspondence to the resource type hierarchy.

7.5 Service Composition

The proliferation of the IoT relies heavily on the ability of users to readily combine the available services within an IoT ecosystem and build complex applications that satisfy their needs. On the following, *two semantically-aware service composition methods* that are integrated into the SYNAISTHISI platform are presented; one falls under the manual composition paradigm while the other under the semi-automatic composition paradigm. In its core, similar to Hatzi et al. (2012), Liang et al. (2004), and Rodríguez-Mier et al. (2016), service composition is performed by *matching*-connecting the inputs of services to the outputs produced by other (preceding) services in the composition, so that all participating services can be invoked, thus ensuring a functional composition. Note that preconditions of services are not considered in this framework (similar to Liang et al. [2004]; Rodríguez-Mier et al. [2016]) and effects are treated as a special case (i.e., subclass) of output produced by A-type services, therefore, hereafter, focus is entirely on the inputs and outputs of services. On the following, the smart meeting room ontology (Section 7.4) is exploited to obtain the semantic descriptions of SPA services, however the discussed approaches are general, in the sense that they can be applied over any SPA services ontology.

7.5.1 Semantic Input/Output Matchmaking

In service composition an output of a service is regarded to *match* the input of another service when the outputted data are compatible with the required input data and thus can be readily consumed when inputted to the other service. To decide whether a match between an input and an output exists, usually, the corresponding service descriptions are checked to establish whether the input and the output reflect the *same* concept. In the most simple case, two concepts are considered to be the same if they carry identical names, hence the matching is done at a syntactical level (Sheng et al., 2014). However, as it is common in real life, different names may be used to represent the same thing. Unavoidably in this case, syntactic matchmaking is destined to fail. Semantics can greatly enhance the matchmaking process by eliminating syntactic barriers through the identification of semantically relevant (i.e., matching) concepts defined in some ontology.

An output concept A is considered to be semantically relevant to an input concept B when one of the following *hierarchical relationships* (Paolucci et al., 2002) exists between them: (1) exact(A, B) when the two concepts have the same URI or are equivalent in terms of OWL equivalence, (2) plugin(A, B) when A is subsumed by (is a subclass of) B, and (3) subsume(A, B) when A subsumes (is a superclass of) B. These hierarchical relationships represent different degrees of *semantic relaxation*, in increasing order, and are typical in the service composition literature (Hatzi et al., 2012; RodríguezMier et al., 2016). The exact relationship implies that two concepts are the same and can be used interchangeably during composition. Plugin and subsume imply an approximate relationship between two concepts and can be utilized to approximately match inputs to outputs, a useful feature, especially when exact matches do not exist. Specifically, the plugin relationship means that the output providing service produces a more specific type of data, than the input receiving service consumes. The subsume relationship suggests that the provided output data are of a more general type than the one that is consumed in the input. Hence, a composition containing a subsume match is not guaranteed to execute smoothly during runtime, as there is a risk that a data-flow incompatibility may arise. Exact and plugin matches avoid this pitfall and are thus a much safer option. In general, it is anticipated that exact matches will be preferred over plugin matches, and plugin matches over subsume matches. Hierarchical relationships are calculated by applying a semantic reasoner over the ontology defining the concepts used to annotate the IOPEs of the SPA services (not the smart meeting room ontology).

7.5.2 Manual Service Composition

The SYNAISTHISI platform is equipped with a manual service composition mechanism (Akasiadis et al., 2015), in which the main steps for building a complex application are manually performed. Specifically, the necessary SPA services must be first discovered, and then their interconnection (i.e., matching outputs to inputs) must be determined. The platform provides the tools to support the composition process (e.g., the service registry, a SPARQL query [Harris & Seaborne, 2013] endpoint, the mechanism to state the interconnection scheme), but it is not actively involved in the process per se. Obviously, manual composition is a quite daunting task (see also Section 7.2); thus, a developer is needed to execute it. From the developer's side, assume that there exists some *prior knowledge regarding service composition*; that is, what kind of SPA services are required, and how they should be meaningfully interconnected.

To discover the services, the developer writes and issues through the platform SPARQL queries conforming to the smart meeting room ontology, which describes the characteristics of the desired services and subsequently recovers the matching services (retaining one service for each part of the composition). By appropriately formulating the SPARQL queries, the different hierarchical relationships between input/output concepts outlined in the previous section can be exploited. As an example, assume that a developer has decided to use in a composition a service that requires as input a video stream (declared using the Video concept from some suitable external ontology where services' IOPE-related semantics are captured). The developer knows that such output is provided by camera services. Therefore, he writes an appropriate SPARQL query adhering to the smart meeting room ontology, shown in Chart 1.1, to discover these cameras. This query will return all camera services that their output is either Video (exact match) or a subclass of Video (plugin match), but not a super class of Video (subsume match), and the developer has to choose one of them to include in the composite service.

Chart 1.1: Example developer-composed SPARQL query for discovering camera services that output Video data. The sm and iope namespaces refer to the service model from Section 7.4.2 and the external ontology where the Video concept is defined.

```
SELECT ?cam serv WHERE {
    ?cam serv a sm:CameraService.
    ?cam servsm:hasServiceOutput ?output.
    {?outputsm:hasParameterTypeiope:Video.}
    UNION
    {?outputsm:hasParameterType ?type.
     ?type rdfs:subClassOfiope:Video.}
}
```

Having discovered the services to use, the next step is to formulate their interconnection scheme. In particular, the inputs of each service participating in the composition must be connected to compatible (according to the semantic hierarchical relationships) outputs produced by some of the other participating services, to ensure all the involved services can be invoked. Implicitly, this defines a *partial ordering* of the services, in which an output providing service precedes those consuming the generated data through their inputs. Remember that S-type services can be readily invoked as they do not have inputs (Section 7.3). Hence, by utilizing one or more such services (recovered in the service discovery phase), it is possible to establish a partial ordering in which all inputs are satisfied (S-type services will be at the very beginning of this ordering). To describe the interconnection scheme the developer assigns *a unique identifier* (a nonnegative integer number) to each service as a whole, as well as to each of the service's inputs/outputs. Then, a combined identifier consisting of a service identifier and an output identifier is paired with a combined identifier that consists of a service identifier and an input identifier; this reflects that this particular output of the first service provides the data for the specific input of the second service. The interconnection scheme is completed when such pairs of combined identifiers are declared for all services' inputs involved in the composition. The resulting scheme is submitted to the platform that undertakes to enable the participating services and bring the required communication channels between them into realization so that the composite service is executed. Note that the developer is responsible for guaranteeing compatibility when pairing inputs and outputs by respecting the semantic hierarchical relationships. The main steps for manual service composition are summarized in Algorithm 1.

Algorithm 1   Manual Service Composition Steps

1: Decide what kind of SPA services are needed
2: **for each** kind of SPA service **do** // Service discovery.
3:      Write a SPARQL query to retrieve matching services from the service registry
4:      Select one of those services to be used in the composition
5: **end for**
6: **for each** selected SPA service **do** // Interconnection scheme.
7:      **for each** input of the service **do**
8:          Define a service-input combined identifier
9:          Pair the combined identifier with a service-output combined identifier to designate the origin of the data consumed in the input
10:      **end for**
11: **end for**
12: Submit the interconnection scheme to the platform

7.5.3 Semiautomatic Service Composition

The manual service composition approach imposes some rather stringent assumptions that require a considerable amount of effort on the developer's side, high-level programming skills, and a priori knowledge of the IoT ecosystem he operates in. Here, part of this effort is transferred to the SYNAISTHISI platform that will guide the developer in building a complex application, by proposing a semiautomatic service composition method that makes extensive use of the semantic input/output matchmaking mechanism presented in Section 7.5.1.

Initially, the developer must specify a *service request* describing the desired functionality of the composite service in terms of outputs that should be generated when executed. Each output is declared using a suitable concept from an external ontology in which services' IOPE-related semantics are captured. For each concept in the service request the platform employs a distinct service discovery procedure that searches over the SPA services available in the service registry to locate those that produce an output (output concept) that matches, hence satisfies, the particular service request concept and organizes them in a *matching services' list.* To populate this list, let us consider that two concepts match when any of the three hierarchical relationships (exact, plugin, subsume) exists between them, and a platform-generated SPARQL query conforming to the smart meeting room ontology is issued to retrieve the corresponding services. An indicative SPARQL query formed by a service discovery procedure that searches for services, the output of which matches, in terms of either of the three hierarchical relationships; the Video concept determined in a service request is illustrated in Chart 1.2. Services included in a matching list represent alternative choices for satisfying a concept of the service request and are presented to the developer who must select one to be incorporated into the composite service. Apparently, a selection must be made for each service request concept. To ease selection, services in a matching list are ordered according to the hierarchical relationship their output exhibits *w.r.t.* the service request concept considered by the particular discovery process. Exact matches precede plugin matches, which precede subsume matches, that is, the ordering is done in increasing degree of semantic relaxation.

Chart 1.2: Example platform-generated SPARQL query for discovering SPA services that output Video data. The sm and iope namespaces refer to the service model from Section 7.4.2 and the external ontology where the Video concept is defined.

```
SELECT ?serv WHERE {
    ?serv a sm:Service.
    ?servsm:hasServiceOutput ?output.
    {?outputsm:hasParameterTypeiope:Video.}
    UNION
    {?outputsm:hasParameterType ?type.
     ?type rdfs:subClassOfiope:Video.}
UNION
    {?outputsm:hasParameterType ?type.
    iope:Videordfs:subClassOf ?type.}
}
```

After selecting a service, it must be ensured that the service can be invoked, that is, all the inputs of the service can be supplied with appropriate data, otherwise the composite service will not execute. This is accomplished by retrieving the inputs (input concepts) of the service and launching a separate service discovery procedure (analogous to the one described above) for each input to recover those SPA services those SPA services that are available in the service registry that produce an output (output concept) that matches the input; hence, they provide suitable data for the input. Subsequently, for each input, the developer is prompted to choose a service from the corresponding matching list that will be added to the composite service and a new round of service discovery and selection is initialized. This process is repeated until there are no more SPA services participating in the composite service that cannot yet be invoked. This way an interconnection scheme is semiautomatically formed in the background and a partial ordering of the services is established. Note that S-type services can be readily invoked as it do not have inputs (Section 7.3); thus, there is no need to carry out service discovery if such a service is chosen. By progressively selecting S-type services, it is possible to reach a state, in which all services are invokable and service composition terminates. After termination, the platform proceeds with realizing and executing the composite service. Note that if, at any point, the service discovery procedure for an input returns an empty matching list, the composition process fails, and no composite service is created, as all available SPA services are deemed unsuitable.

Algorithm 2   Semi-Automatic Service Composition Procedure

Input: Dummy service *SR*, representing the service request
Output: Composite service *CS*, or
fail

1: Set $Q = \emptyset$ // Queue structure with services as elements.
2: ENQUEUE(*Q*, *SR*)
3: **repeat**
4:      *S* = DEQUEUE(*Q*)
5:      **for each** input I_S of service *S* **do**
6:              *L* = SERVICE DISCOVERY(I_S) // Matching services list.
7:              **if** $L == \emptyset$ **then** // If the list is empty.
8:                      **return fail**
9:              **end if**
10:             Prompt the developer to select a service S_L from *L*
11:             Add S_L to *CS* to supply the data for input I_S
12:             **if** S_L **not** an S-type service **then**
13:                     ENQUEUE(*Q*, S_L)
14:             **end if**
15:     **end for**
16: **until** $Q == \emptyset$ // Until all services can be invoked.
17: **return** *CS*

The semiautomatic service composition process is summarized in Algorithm 2, in which the service request is represented as a dummy service without outputs, having the service request concepts as inputs. A graphic example is depicted in Figure 7.4.

7.6 Composition of Complex Applications in an IoT Ecosystem: A Use Case

As has already been mentioned in Section 7.3, in the context of the SYNAISTHISI project, a typical meeting room has been equipped with several sensors, interconnected through the SYNAISTHISI platform, to provide S-type services. Moreover, a number of services performing audiovisual analysis have been incorporated in order to provide cloud-based P-type services. All these services have been semantically annotated using the proposed smart meeting room ontology and form its instances. Following is a brief presentation of some of these simple services, which are then utilized to compose a complex Person _ Counting service using both service composition approaches.

7.6.1 Available Services

The Person _ Tracking service is a P-type service that aims to continuously and in real time detect and track people present in the smart meeting room. The algorithm applied (Sgouropoulos et al., 2015) uses a blob-detection approach and is functioning under the assumption that a moving blob at the foreground should belong to a human,

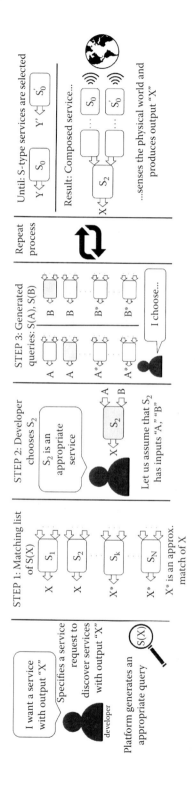

FIGURE 7.4

The process of semiautomatic service composition.

TABLE 7.1

S- and P-Type Services Utilized in Composing the `Person _ Counting` Service

Service	Type	Input(s)	Output
Static_Camera	S	–	RGB Video (RGB)
Stereo_Camera	S	–	Depth Information (depth)
Dynamic_BG	P	RGB Video (RGB)	Foreground (fg)
Person_Tracking	P	Foreground (fg), Depth Information (depth)	Number of Persons (nop)
Person_Fusion	P	Number of Persons (nop)	Fused Number of Persons (fnop)

as another moving entity is not expected to enter the meeting room. When a new person enters the room, he/she is continuously tracked during his/her stay in the room. The output of this service consists of the coordinates of the set of bounding boxes that enclose the detected blobs, which are also enumerated; thus, their total number is trivially calculated and outputted by the service. Accordingly, the number of tracked objects is assumed to be the number of people present within the room.

However, in order for this P-type service to work, several other S- and P-type services should interconnect and exchange information. More specifically, there exist several RGB-D cameras that continuously capture video of the room's interior. Each RGB-D camera provides two services, one for RGB video and the other for depth. Thus, the available S-type services are `Static _ Camera` and `Stereo _ Camera`, respectively.

The output of the `Static _ Camera` service is processed (i.e., is fed as input) by the `Dynamic _ BG` service, a background subtraction module, which *divides* video content into two parts based on motion: the background, which is discarded, and the foreground, which consists the input (along with the depth dimension of the `Stereo _ Camera` service) for the `Person _ Tracking` service that outputs the number of people present in the foreground to the platform. The aforementioned services are summarized in Table 7.1.

As in general more than one camera may be used to surveil different parts of the room, their outputs should be appropriately fused in order for the composed `Person _ Counting` service to provide an output that reflects the total number of people in the room. Data fusion, in general, denotes the act of combining data from disparate and even heterogeneous sources, in order to obtain improved information compared to what is possible when each source is used individually. To this goal, the implemented `Person _ Fusion` service aims to combine measurements or raw data from heterogeneous or homogeneous sensors. Specifically, `Person _ Fusion` is a P-type service, fed with the output of the `Person _ Tracking` module produced by using two different cameras as input, that is, from two different parts of the smart meeting room.

7.6.2 Manual Service Composition Example

Consider that the goal of the developer is to compose a `Person _ Counting` service, that is, a service that estimates the number of the people present within the SYNAISTHISI smart meeting room, by utilizing the two overlooking cameras. It is fair to assume that the developer is familiar with basic computer vision tasks, thus is able to appropriately interconnect two services upon observation of their inputs/outputs.

As has been already discussed, the room is fully covered by two RGB-D cameras (each comprising of a `Static_Camera` and a `Stereo_Camera` service). Thus, the developer first needs to discover these camera services. The developer is assumed to know that a

`Person_Tracking` service is needed and that it operates on the foreground of the scene captured by a `Static_Camera`, and extracted by the `Dynamic_BG` service, whereas it also requires, as input, the depth information of the scene produced by the corresponding `Stereo_Camera`. He/she then interconnects the outputs of `Dynamic_BG` and `Stereo_Camera` with the input of the `Person_Tracking` service. This process should be repeated for both `Stereo_Camera` services. Also each `Static_Camera` must be interconnected with the `Dynamic_BG` service to supply its input. Notice that the `Person_Tracking` service is executed twice, once for each of the two cameras. The developer is aware that the outputs of the two executions should be fused by the `Person_Fusion` service; therefore, this service must be discovered and interconnected with the `Person_Tracking` service. The `Person_Fusion` output corresponds to the desired output of the complex `Person_Counting` service, that is, the total number of people in the room.

Service discovery is performed by issuing appropriate SPARQL queries to the smart meeting room ontology via the SYNAISTHISI platform, as shown in Section 7.5.2. The composed `Person _ Counting` service is illustrated in Figure 7.5. Note that the flow of information between services should be considered before commencing the composition process and that the developer is the only responsible for the correct interconnection of the services.

7.6.3 Semiautomatic Service Composition Example

To present a use case of the semiautomatic service composition methodology, let us revisit the previously described manual service composition example toward a `Person_Counting` service. The goal of the developer is once again assumed to be the composition of a service that counts the persons in the smart meeting room by utilizing the two cameras. Here, the developer is assumed to be familiar with basic computer vision tasks (not necessary as advanced as in the manual composition case), thus is able to understand the flow of services suggested by the platform at each step and proceed to meaningful choices.

In Figure 7.6, the steps of the service composition process are illustrated. The process begins with the developer's intention to create a service with a specific functionality, that

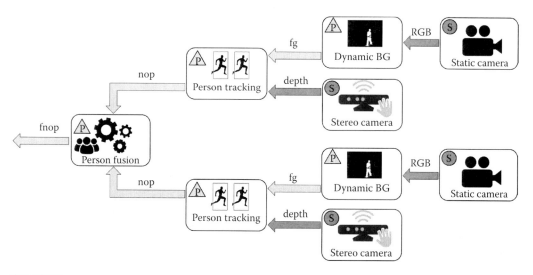

FIGURE 7.5
The composed `Person _ Counting` service using either manual or semiautomatic composition.

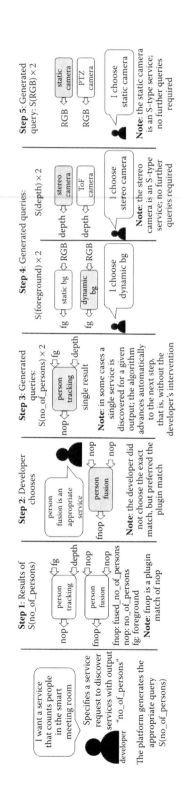

FIGURE 7.6
Semiautomatic service composition use case: Creating a person counting service. The depicted services' IOPE concepts are assumed to be defined in an appropriate external ontology.

is, he/she knows the service's output. He/she expresses this intention via a service request, and then two services are returned in the matching list. Even though the output of the second service (namely `Person _ Fusion`) is not an exact match, but a plugin match,[*] his/her intention to interconnect more than one cameras, along with his/her domain knowledge, constitute the second service a preferable option over the first. The platform then autoqueries, by generating suitable SPARQL queries of the form shown in Section 7.5.3, the ontology and a single matching service, namely `Person _ Tracking`, that matches both queries (one query for each input of the `Person _ Fusion` service) is returned. A further autoquery allows the developer to select the next services, namely `Dynamic _ BG` and `Stereo _ Camera`, thus satisfy the inputs of `Person _ Tracking`. `Stereo _ Camera` is an S-type service, thus does not trigger another autoquery. Finally the process is terminated when the developer selects another S-type service, namely a `Static _ Camera` to supply the input of `Dynamic _ BG`. The composed service is the same as in the manual composition example and is illustrated in Figure 7.5.

7.7 Discussion

This section discusses various aspects concerning the smart meeting room ontology and the two service composition methods. Regarding the ontology, its most prominent characteristics are (1) the utilization of existing and well-established semantic models; (2) the ease of use in instantiating a smart meeting room, as the domain-specific nature of the ontology means that not only high-level, but also low-level concepts are provided; and (3) the ability to readily extend the ontology to include a new kind of SPA service, by adding a new class in the resource-service type hierarchy. On the downside, as the ontology is oriented toward modeling SPA services of smart meeting rooms, adaptation is necessary so as to apply it on another domain.

Regarding the service composition, the exploitation of semantics allows one to overcome syntactic barriers, which, otherwise, could lead to failures in producing the requested services—for example, when exact matches do not exist, and the inputs and outputs of services are approximately matched through plugin and subsume relationships. Also note that both approaches currently operate over the smart meeting room ontology; however, they are not limited to it. If SPARQL queries are appropriately rephrased, they can be readily applied over any other ontology. The manual composition is targeted toward developers who possess the necessary programming skills in order to manually discover services and interconnect them. A severely limiting factor in this kind of composition is the need to know in advance which services are available in the registry so they can be discovered. Such an assumption is not realistic in IoT ecosystems with millions of services that are dynamically added and removed. Also the whole process is error-prone, as every step is manually executed by the developer. In the semiautomatic framework the involvement of the developer is limited to the basic tasks of defining a service request and selecting services from platform-generated matching lists; hence, the whole process evolves with minimum human intervention. Service discovery and interconnection is performed by the platform, alleviating this way the need of a priori knowledge about the available services,

---

[*] This is just an assumption in order to showcase the semantic matchmaking mechanism.

and also acting as safeguard against human errors. Composing a service now becomes a quite straightforward task, which can be even accomplished by an experienced user who is not a developer but is skillful enough to comprehend the semantics surrounding the service descriptions.

The semiautomatic composition algorithm guarantees that if a composite service is created, all concepts included in the service request will be satisfied, and all services combined in the composite service can be invoked. A drawback of the algorithm described in Section 7.5.3 is the possibility of failing to produce a composite service, although the necessary services are available. This is attributed to the fact that the composition process immediately stops when an empty matching list is encountered after running the service discovery procedure for one of the inputs of an already selected service. However, this limitation is easily circumvented if, instead of stopping, the composition process prompts the developer to replace the service previously selected with an alternative one contained in the corresponding matching list. If no alternatives are available, it is possible to revert to an earlier step of the composition process, replace the service at that step, and resume the composition process from there.

7.8 Conclusion

Among the main visions of IoT is to seamlessly interconnect heterogeneous devices. This chapter has built upon the SYNAISTHISI platform (Pierris et al., 2015) that follows the service-oriented architecture (SOA) of IoT and exposes available resources as SPA semantic web services. This chapter primarily focused on service composition, which constitutes one of the most important and demanding tasks in unlocking the full potential of the service-oriented architecture of IoT. The objective of service composition is to combine a number of services, available in a service registry, in order to satisfy some user need when none of the available services in the registry can do so. Key steps in composing a complex service are the discovery of suitable services from the registry and their appropriate interconnection so as to ensure a functional composition, that is, one that every service in the composition can be invoked, which, in general, translates to that all inputs and preconditions of a service are provided/satisfied from the outputs and the effects of some other service(s) that is (are) part of the composite service.

The way service composition can be performed in both manual and semiautomatic manner within the SYNAISTHISI platform was demonstrated. The manual approach requires considerable user intervention in every step of the composition process, both to discover the services and to interconnect them, whereas in the semiautomated, the user only specifies a service request and chooses services out of automatically populated matching services' lists presented to him, thus implicitly defining the service interconnection scheme. To achieve this, an ontology was utilized in order to add semantic content to the SPA services descriptions so that complete and accurate information regarding the functionality offered by each service can be easily extracted and to permit the semantic matchmaking between the inputs and the outputs of services. The ontology is domain-specific, providing all necessary high- and low-level concepts to model a service-oriented smart meeting room, similar to the one considered in the SYNAISTHISI project use cases. However, the proposed service composition methods can operate over any ontology if SPARQL queries are adapted accordingly.

Acknowledgment

This work was part of the "SYNAISTHISI" project results. The project was co-financed by the Greek General Secretariat for R&T, Ministry of Culture, Education & RA and the European RDF of the EC under the Operational Program "Competitiveness and Entrepreneurship" (OPCE II), in the action of Development Grants For Research Institutions (KRIPIS).

References

Akasiadis, C., Tzortzis, G., Spyrou, E., & Spyropoulos, C. (2015). Developing complex services in an IoT ecosystem. In *IEEE World Forum on Internet of Things (WF-IoT)* (pp. 52–56).

Al-Fuqaha, A. I., Guizani, M., Mohammadi, M., Aledhari, M., & Ayyash, M. (2015). Internet of Things: A survey on enabling technologies, protocols, and applications. *IEEE Communications Surveys and Tutorials, 17*(4), 2347–2376.

Albreshne, A., & Pasquier, J. (2010). Semantic-based semi-automatic web service composition. Computer Department, Switzerland.

Atzori, L., Iera, A., & Morabito, G. (2010). The Internet of Things: A survey. *Computer Networks, 54*(15), 2787–2805.

Bandyopadhyay, D., & Sen, J. (2011). Internet of things: Applications and challenges in technology and standardization. *Wireless Personal Communications, 58*(1), 49–69.

Banks, A., & Gupta, R. (2014). *MQTT protocol specification.* Retrieved from http://docs.oasis-open.org/mqtt/mqtt/v3.1.1/mqtt-v3.1.1.html.

De, S., Elsaleh, T., Barnaghi, P. M., & Meissner, S. (2012). An internet of things platform for real-world and digital objects. *Scalable Computing: Practice and Experience, 13*(1), 45–58.

Gubbi, J., Buyya, R., Marusic, S., & Palaniswami, M. (2013). Internet of things (IoT): A vision, architectural elements, and future directions. *Future Generation Computer Systems, 29*(7), 1645–1660.

Guo, B., Zhang, D., Wang, Z., Yu, Z., & Zhou, X. (2013). Opportunistic IoT: Exploring the harmonious interaction between human and the internet of things. *Journal of Network and Computer Applications, 36*(6), 1531–1539.

Harris, S., & Seaborne, A. (2013). *SPARQL 1.1 query language, W3C Recommendation.* Retrieved from http://www.w3.org/TR/sparql11-query/.

Hatzi, O., Vrakas, D., Nikolaidou, M., Bassiliades, N., Anagnostopoulos, D., & Vlahavas, I. (2012). An integrated approach to automated semantic web service composition through planning. *IEEE Transactions on Services Computing, 5*(3), 319–332.

Kopecký, J., Gomadam, K., & Vitvar, T. (2008). hRESTS: An HTML microformat for describing restful web services. In *IEEE/WIC/ACM International Conference on Web Intelligence (WI)* (pp. 619–625).

Kopecký, J., Vitvar, T., Bournez, C., & Farrell, J. (2007). SAWSDL: Semantic annotations for WSDL and XML schema. *IEEE Internet Computing, 11*(6), 60–67.

Liang, Q., Chakarapani, L. N., Su, S. Y., Chikkamagalur, R. N., & Lam, H. (2004). A semiautomatic approach to composite web services discovery, description and invocation. *International Journal of Web Services Research (IJWSR), 1*(4), 64–89.

Martin, D., Burstein, M., Hobbs, J., Lassila, O., McDermott, D., McIlraith, S., … Sycara, K. (2004). *OWL-S: Semantic markup for web services, W3C Member Submission.* http://www.w3.org/Submission/OWL-S.

McIlraith, S. A., Son, T. C., & Zeng, H. (2001). Semantic web services. *IEEE Intelligent Systems, 16*(2), 46–53.

Paolucci, M., Kawamura, T., Payne, T., & Sycara, K. (2002). Semantic matching of web services capabilities. In *International Semantic Web Conference (ISWC)* (pp. 333–347). Berlin, Germany: Springer.

Pedrinaci, C., Liu, D., Maleshkova, M., Lambert, D., Kopecky, J., & Domingue, J. (2010). iServe: A linked services publishing platform. In *Ontology Repositories and Editors for the Semantic Web Workshop at the Extended Semantic Web Conference (ESWC)* (pp. 71–82). Hersonissos, Greece.

Pierris, G., Kothris, D., Spyrou, E., & Spyropoulos, C. (2015). SYNAISTHISI: An enabling platform for the current internet of things ecosystem. In *Panhellenic Conference on Informatics (PCI)* (pp. 439–444). Athens, Greece: ACM.

Rodríguez-Mier, P., Pedrinaci, C., Lama, M., & Mucientes, M. (2016). An integrated semantic web service discovery and composition framework. *IEEE Transactions on Services Computing, 9*(4), 537–550.

Roman, D., Keller, U., Lausen, H., de Bruijn, J., Lara, R., Stollberg, M., ... Fensel, D. (2005). Web service modeling ontology. *Applied Ontology, 1*(1), 77–106.

Roman, D., Kopecký, J., Vitvar, T., Domingue, J., & Fensel, D. (2015). WSMO-Lite and hRESTS: Lightweight semantic annotations for web services and RESTful APIs. *Journal of Web Semantics, 31*, 39–58.

Sfikas, G., Akasiadis, C., & Spyrou, E. (2015). Creating a smart room using an IoT approach. In *Artificial Intelligence and Internet of Things (AI-IoT) Workshop, in conjunction with the 9th Hellenic Conference on Artificial Intelligence (SETN)*, Thessaloniki, Greece.

Sgouropoulos, D., Spyrou, E., Siantikos, G., & Giannakopoulos, T. (2015). Counting and tracking people in a smart room: An IoT approach. In *IEEE International Workshop on Semantic and Social Media Adaptation and Personalization (SMAP)* (pp. 1–5).

Sheng, Q. Z., Qiao, X., Vasilakos, A. V., Szabo, C., Bourne, S., & Xu, X. (2014). Web services composition: A decade's overview. *Information Sciences, 280*, 218–238.

Stavropoulos, T. G., Vrakas, D., Vlachava, D., & Bassiliades, N. (2012). BOnSAI: A smart building ontology for ambient intelligence. In *International Conference on Web Intelligence, Mining and Semantics (WIMS)*. Craiova, Romania: ACM.

Vitvar, T., Kopecký, J., Viskova, J., & Fensel, D. (2008). WSMO-Lite Annotations for Web Services. In S. Bechhofer, M. Hauswirth, J. Hoffmann, & M. Koubarakis (Eds.), *The Semantic Web: Research and Applications. ESWC 2008.* Lecture Notes in Computer Science, Vol. 5021. Berlin, Germany: Springer.

8

Enabling Smart Cities through IoT:
The ALMANAC Way

Dario Bonino, Maria Teresa Delgado, Claudio Pastrone, Maurizio Spirito,
Alexandre Alapetite, Thomas Gilbert, Mathias Axling, Matts Ahlsen,
Peter Rosengren, Marco Jahn, Raphael Ahrens, Otilia Werner-Kytölä,
and Jose Angel Carvajal Soto

CONTENTS

8.1 Introduction

Smart cities advocate future environments where sensor pervasiveness, data delivery and exchange, and information mash-up enable better support of a variety of aspects in everyday life. As this vision matures, it addresses several application scenarios and adoption perspectives, which call for scalable, pervasive, flexible, and replicable infrastructures.

This need is currently fostering new design efforts to deliver architectures and platforms that cover the requirements on performance, reuse, and device and system interoperability capabilities. The adoption of Internet of Things (IoT) as smart city enabler is challenging. On one hand, IoT technologies are widely recognized as key components of successful smart cities deployment. They, in fact, allow one to effectively tackle scenarios involving multitudes of heterogeneous devices and subsystems interconnected through a variety of networks, with different reliability, connectivity, and performance. On the other hand, city life involves humans, businesses, and administrative entities that shall effectively cooperate by means of the IoT technological infrastructure.

This chapter describes the smart city-enabling path pursued by the ALMANAC project to develop a service delivery platform specifically targeted to smart cities, by integrating IoT and capillary networks with telecommunication networks. According to the evidences emerging from the project, successful IoT adoption in smart city scenarios is better achieved through the following:

1. Sound and informed design of an IoT technological platform tailored to the specific needs of smart cities.
2. Early involvement of smart city actors, for example, public administrations, social communities, and utility services.
3. Explicit account of data ownership and access policies for data exchanged within entities.
4. Direct involvement of citizens.

In this chapter, technological choices, their underlying rationale, and the resulting ALMANAC platform architecture are described in depth. Moreover, application scenarios targeted by the project are discussed, and evidence from a real-world pilot is reported. For each professional domain targeted by the project, namely water and waste management, requirements gathered are presented and related to the technological solutions developed by ALMANAC. Particular emphasis will be devoted to the pilot installed in Turin, in which the ALMANAC platform has been deployed and tested. Finally, insights gained from the pilot, outcomes of the real-world experimentation, and citizen feedback will be discussed in relation to IoT technologies adoption in smart cities.

8.2 Related Works

Over the last years, the research community has reached a quite clear consensus about the necessity and usefulness of making cities smarter through the use of innovative technologies. Many studies have been carried out on market potential (Sma, 2015), possible architectures (e.g., I-architecture [Sma, 2012], IoT-A [Bauer et al., 2013]), deployments, and roadmaps for real-world exploitation (Vermesan, 2009). Nevertheless, smart city is still a very active research field, with solutions that are still experiencing difficulties demonstrating their full potential in a smart city context.

To address the challenges emerging in this domain, researchers from industry and academia propose approaches and platforms supporting the *smartification* of existing cities. On the industrial side, big firms such as IBM (Fritz et al., 2012, Hogan et al., 2011), HP, BOSCH, Siemens, and Cisco are pursuing smart city solutions to enable cities to become *smarter* by connecting city events and data, possibly exploiting integrated cloud solutions.

On the research side, several approaches address different aspects of smart city design and exploitation, from sustainable deployment and growth, to next-generation semantics powered smart city platforms (SCPs). In their paper, Vilajosana et al. (2013) investigate the underlying reasons for the slow adoption of smart cities and propose a procedure based on big data exploitation through the so-called application programming interface (API) stores concept. A viable approach for scaling business within city ecosystems is discussed and currently available ICT technologies are described, exemplifying all findings by means of a sustainable smart city application. Although this approach describes the motivations for smart cities and establishes a general framework for their adoption, the proposed ALMANAC framework is more oriented to the actual implementation of a federated SCP.

Bellini et al. (2014) propose a system for data ingestion and reconciliation of smart city aspects. The system allows managing a large volume of static and dynamic data coming from a variety of sources. These data are mapped to a smart-city ontology, called KM4City,[*] and stored into an RDF[†] store in which it is available for applications via SPARQL[‡] queries, providing new services to the users via specific applications for public administration and/or enterprises. Examples of potential uses of the knowledge base produced by the Bellini et al. (2014) system are also offered, and they are accessible from the resource description framework (RDF) store and related services. The framework and platform architecture presented in this paper share several design principles with the approach of KM4City, that is, semantic modeling of smart-city entities and big data approach to data capture. However, while the former is more focused on ontology and dataset population, ALMANAC is more concerned with federation of several instances and direct exploitation of platform services by third-party application developers.

Several research projects are currently investigating aspects related to smart city design and adoption both in Europe and the United States, among them:

- The Smart Santander project[§] provides a city-scale experimental research facility in support of applications and services for a smart city. ALMANAC solutions already integrate data managed by Smart Santander thanks to a strong collaboration between the two projects.
- The Urban Water project,[¶] installing smart meters in Almería (Spain) to achieve greater efficiencies in water management.
- The Open IoT project[**] defines an open-source cloud solution for the IoT, which can be seen as a technological enabler for SCPs such as the one presented in this chapter.
- The Mobosens U.S. research project[††] provides citizens with a platform for collecting and sharing environmental data, from stream quality to drinking water safety.

---

[*] Knowledge Model for City.
[†] Resource Description Framework https://www.w3.org/RDF/, last visited September, 2016.
[‡] Query Language for RDF https://www.w3.org/TR/rdf-sparql-query/, last visited September, 2016.
[§] http://www.smartsantander.eu/, last visited March, 2015.
[¶] http://urbanwater-ict.eu/, last visited March, 2015.
[**] http://openiot.eu/, last visited March, 2015.
[††] http://nanobionics.mntl.illinois.edu/mobosens/, last visited March, 2015.

8.3 The ALMANAC Platform

The ALMANAC SCP collects, aggregates, and analyses real-time data from resources (appliances, sensors and actuators, smart meters, etc.) deployed to implement smart cities. Three specific applications, waste management, water supply, and citizens' engagement, have been selected for proof-of-concept implementation and evaluation.

8.3.1 Architecture Overview

Figure 8.1 depicts the logic view of the final ALMANAC architecture. It can be described along two main axes, by vertical and horizontal divisions: the functional axis and the sub-system axis. Although the functional axis is rendered as a stacked set of platform modules each deserving specific documentation,[*] the subsystem axis (mainly horizontal) can be exploited to gain a quick understanding of the main platform functions.

Subsystems composing the platform encompass from top to bottom:

- The *cloud-based APIs*, including all modules acting as service access point for external, possibly third-party applications exploiting the ALMANAC platform. They mostly include REST end points mapping to the offered platform functionalities.

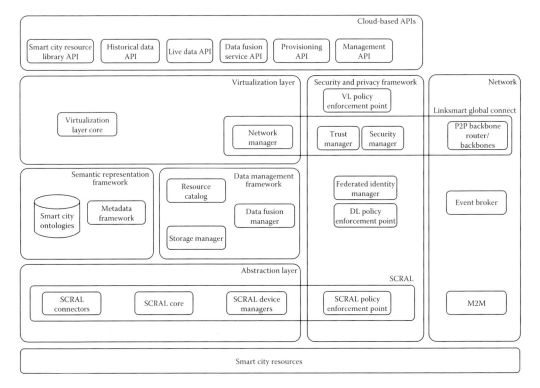

FIGURE 8.1
The ALMANAC logic architecture overview.

---

[*] Interested readers may refer to Bonino et al. (2015).

- The *virtualization layer,* which orchestrates the platform modules to fulfill Cloud APIs requests, both within a single platform instance (PI) and among different instances (federation). Moreover, it supports search, lookup, and addressing of smart city resources and services.

- The *semantic representation framework*, which encompasses updated smart city models based on ontologies as well as a metadata management framework enabling effective handling of machine understandable metadata on smart city resources.

- The *data management framework,* grouping all data-management modules that handle effective storage, elaboration and fusion of live data, and that offer directory services for resources registered in the platform.

- The *abstraction layer,* which provides technology independent, uniform access to physical resources and maps both live and off-line metadata into shared, standard representations that can be easily handled by upper platform modules.

- The *security framework,* which crosses all layers and provides services and utilities to secure communication between platform modules, to check access permissions and verify information disclosure on the basis of a policy mechanism. The framework encompasses several enforcement end points, where application/user permissions are checked and accept/deny actions applied, and one federated identity manager handling roles, permissions, and supporting end-to-end control on transferred data, both inside and outside of the platform.

- The *networking subsystem,* supporting effective communication between platform modules, among different platforms (federation) and between the platform and the outside world (e.g., through M2M communication).

8.3.2 Data Flow

The main information source in ALMANAC is the heterogeneous sensing network constituting the smart city *nervous* system. Similar to capillaries in living beings, sensors deployed in the city *capture* vital information on city processes and states (e.g., pollution level, traffic, waste levels) and deliver it to the core elaboration center represented by the ALMANAC platform. In this path, every single bit of information traverses a complex set of modules, each performing a different operation, either on the data itself or on the so-called metadata (i.e., information about delivered data). Raw values generated by sensors are gathered by a capillary communication network, or directly by the smart city resource abstraction layer (SCRAL) and undergo several checks and modifications to transform each measurement point into meaningful and exploitable information. For each sensor-generating data stream, a corresponding metadata description is generated, and kept up to date, exploiting standard representations (OGC SensorThings API[*]), open data, and semantic models. The resulting metadata is attached to the data stream and exploited to better characterize every single measure. For example, the raw temperature value of 20°C measured by a DIY[†] sensor located in the center of Copenhagen is enriched at SCRAL level with metadata about the source. This means that at SCRAL level, the raw temperature value becomes an OGC Sensor Things API observation

---

[*] https://ogc-iot.github.io/ogc-iot-api/, last visited September, 2016.
[†] Do It Yourself.

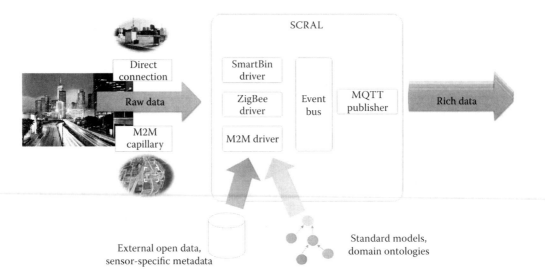

FIGURE 8.2
Data flow from the city to the ALMANAC lower layers.

referred to a well-defined property (e.g., the outside temperature), with a precise time stamp, a standard unit of measure, and a geographical positioning in terms of latitude and longitude. Attached to this bit of information, additional (meta-)data permits to identify the sensor owner, for example, a private user, with medium access level, and possibly data-quality indicators, for example, amateur-level sensor versus professional measure system.

At the upper boundary of the SCRAL, data injected in the platform is no longer raw and uncategorized. Instead, it is rich and expressed in a shared, standard, uniform, and machine-understandable format, which can easily be handled by higher platform levels where intelligent elaboration takes place (see Figure 8.2, which summarizes this first step).

Enriched data flow into the platform along several parallel paths (Figure 8.3). First, data are captured by the platform storage manager (SM) which provides persistence to gathered data. In parallel, the same data flow (selectively) into the data fusion manager, which applies complex event-processing operators to the incoming data streams and generates new (rich) data, either referred to same originating sensors (e.g., temporal aggregations, statistical properties) or pertaining to completely new sources, that is, virtual sensors generated by combining together different data streams. New data are in turn routed toward the SM for persistence. Although both the SM and the data fusion manager handle *live* data, that is, measures plus metadata, the metadata information flow is also routed to the resource catalog component, which is responsible for maintaining a directory of currently connected data sources with its corresponding metadata. Metadata can be further enriched by exploiting the inference mechanisms provided by the metadata framework, which hosts domain ontologies exploited by the ALMANAC platform.

At the highest architectural layer of the ALMANAC platform, rich data and metadata are made available to end users, that is, applications using the ALMANAC Cloud APIs and to other federated platforms. Data are either delivered through REST or Web Socket. This layer can be seen as the *end point* of the field-to-platform information flow, and the boundary at which the user-to-platform information flow begins.

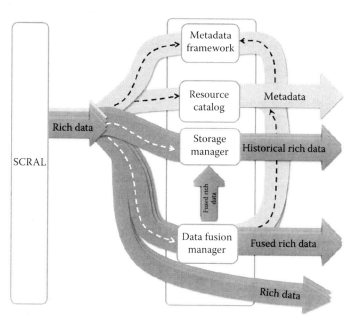

FIGURE 8.3
Data flow, from SCRAL to data management.

8.3.3 Federation

Every smart city is peculiar, with its own metering infrastructure, its own administrative constraints, and its variegated set of actors and entities interacting together in everyday city processes.

This scenario often leads to tight requirements on ICT systems, which are required to support different—sometimes collaborative—entities, while having the ability to enable independent operations of connected entities. Differently from several existing state-of-the-art approaches, which concentrate on technical requirements and only consider social and legal interactions at a later design stage, ALMANAC tackles these issues from the very early design phases. Interaction of entities involved in smart cities is addressed by means of the federation concept, an extension of the well-known *cloud federation* idea (Rochwerger et al., 2009) to real-world city requirements and use cases. Broadly speaking, the term *cloud federation* refers to interoperability of different cloud systems as if they were one, with each single system maintaining a firm control on shared information and functions. This concept can be extended quite well to smart cities where different stakeholders need to work together as a *single entity* depending on the current context and needs. The main challenge is addressing the fact that context and stakeholders being part of a federation might change in time, and that their exchanges are shaped and constrained by nontechnical boundaries set by policies and regulations, which may also vary in time.

ALMANAC defines a framework in which multiple PIs can participate in federations (Figure 8.4) offering cross-city, cross-nation, and cross-entity services. Such federations are typically built around service-exchange agreements that consider the involved political and legal issues. From a technical standpoint, they are deployed by establishing sets

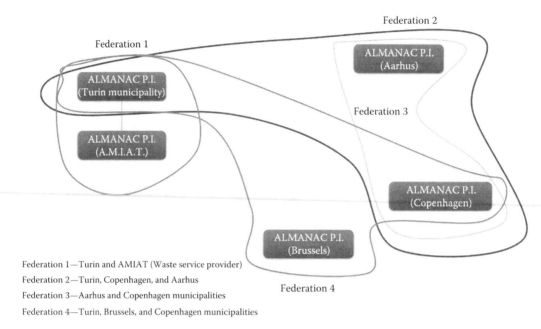

FIGURE 8.4
Multiple federations between ALMANAC platform instances.

of trust domains including two or more instances. PIs exchange data and distribute tasks according to roles and privileges defined in the federation agreement (e.g., in a simplistic on/off case, all users/platforms belonging to the federation are allowed to perform tasks, whereas external platforms/users cannot exploit the federated services), and managed through state-of-the-art solutions for federated identity and role management, for example, based on the SAML* specification.

8.4 Application Scenarios and Real-World Pilots

Since the beginning of the ALMANAC project, a scenario-thinking approach has been implemented to analyze and cocreate the selected application scenarios with relevant stakeholders. Although the development of the ALMANAC SCP has been tailored to be easily scalable to other cities and domains, the validation of ALMANAC results in the waste and citizen-centric domains has been performed in the city of Turin, whereas activities regarding application of the ALMANAC platform in water management scenarios are referred to a small pilot in Denmark.

---

* https://www.oasis-open.org/committees/tc_home.php?wg_abbrev=security, last visited September, 2016.

8.4.1 Waste Management Domain

In order to get a deeper insight into the waste domain in the Turin context, Turin's Municipality—which is one of the partners of the ALMANAC consortium—played an important role in providing domain knowledge.

Thanks to the knowledge gained from the Environmental Department of the City of Turin, the project learned that the waste management domain has mainly three protagonists: the citizens, the utility service operator (Amiat[*]), and the public administration. The latter acts as both client and controller of the services provided by Amiat, while also behaving as the interface between this and the general public, for example, by gathering suggestions and complaints from the citizens concerning the services.

To include all relevant perspectives, these three main actors were engaged in a cocreation process that puts in place ordinary workshop tools together with a set of ethnographic tools, including field visits, scenario observations, and interviews with people from the city environmental office and from the utility company.

From these activities, information related to the three main types of waste collection methods in Turin was gathered: (a) street-based collection, based on waste containers placed on the streets; (b) curbside collection, based on small waste bins located directly inside households; and (c) underground ecological islands (UEIs), which are underground containers with a large storage capacity (5 m^3).

Regarding the quality of collected waste, street-based collection shows the lowest: recyclable waste is often of such poor quality it cannot be reused and ends up as unsorted waste that is taken to an incineration plant or landfill. On the contrary, the quality of waste collected through curbside collection is the highest. Citizens might feel a higher degree of ownerships and pay more attention to the way waste is sorted, whereas it is easier for the city to apply individual fines based on the quality of the collected waste. Nonetheless, curbside collection is not always feasible due to space, aesthetic, and health reasons. UEIs are not deemed to be as effective as the curbside collection (e.g., frequent abandoned waste issues around the UEI); however, they are a good alternative to curbside collection both in terms of the citizens feeling of ownership and in terms of collection optimization.

8.4.1.1 Smart Waste Pilot in Turin

From a technological point of view, the waste domain represents an interesting application for the solutions developed by ALMANAC, especially for those considering federation of IoT platforms operated by different owners—enabling data sharing and reuse of management functionalities in a safe and controlled way—to optimize a service of common interest. For these reasons, two UEIs were selected to deploy technology solutions considered in ALMANAC. Data referred to waste collection in selected locations encompass both fill level and weight measurements from collection trucks serving the UEIs.

The two locations selected for the pilot deployment are located in Turin city center, see Figure 8.5, and they serve around 300 families and 90 businesses.

8.4.1.2 Deployment Activities

Deployment activities in the two selected UEIs started during August 2015. Fill-level sensors and solar panels were installed in each waste container, and the two nonrecyclable waste containers were also equipped with radio-frequency identification (RFID)

---

[*] Azienda Multiservizi Igiene Ambientale Torino.

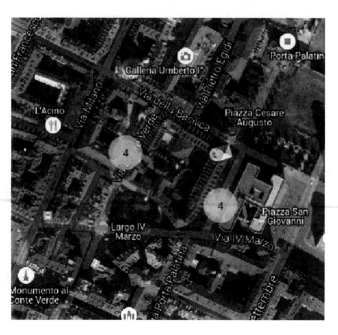

FIGURE 8.5
The sites selected for the ALMANAC project pilot.

access-control modules to grant exclusive access to the citizens involved in the experimentation. During September 2015, the municipality proceeded to communicate to the residents and businesses involved all relevant information. Users were informed about the starting date of the trial and briefed on the new conferral mode for the nonrecyclable waste that would include the use of a magnetic key to access the respective container. During this time, Amiat validated the correct installation and corroborated the normal functioning of fill-level sensors, solar panels, radio modules, and RFID-controlled access modules, whereas the ALMANAC staff started the integration of the data generated by these devices into the ALMANAC platform and reported any anomalies founded in the collected data. On September 28th 2015, the experimentation officially started which is foreseen to last until the end of 2016.

To assess the citizens' perception of the experimentation activities, ALMANAC distributed a short questionnaire aimed at evaluating the perceived benefits after the end of the trial period. The questionnaire was given to users involved in the experimentation and was also available online.[*] A summary of the deployment-related activities described so far is presented in Table 8.1.

8.4.1.3 Technical Details

Each site hosts UEI equipped with four waste containers[†] collecting the organic, glass, paper, and not-recyclable fractions (Figure 8.6).

_____

[*] https://it.surveymonkey.com/r/B2WJD3H.
[†] Funded by Amiat, the Turin Waste utility, and partly supported by Nord Engineering, a third party provider, see http: //www.nordengineering.com/en/, last visited April 20, 2016.

TABLE 8.1

Summary of the ALMANAC Pilot Deployment Activities

Activity	Date
UEIs equipment upgrade (sensors, solar-panels, and controlled access modules installation)	August 15–31, 2015
Informative letters sent to residents and businesses involved	September 1–14, 2015
Key delivery to involved businesses	September 1–26, 2015
Key delivery in city offices for involved residents	September 14–26, 2015
Questionnaire to assess the perceived benefits of the deployment activities	May 1, 2016

FIGURE 8.6
One of the UEIs involved in the ALMANAC pilot, located in Via Porta Palatina.

All containers were equipped with commercial-grade wireless fill-level sensors provided by EcoTec* (specifications reported in Table 8.2) and kindly funded by Amiat. Nonrecyclable containers carry an RFID-based access control system that allows waste disposal from authorized citizens only. All sensors communicate with a central, legacy, data collection service also provided by EcoTec. This setting reflects quite well the typical smart city scenario, where administrative constraints and commercial aspects are strictly intertwined, often leading to deployment of heterogeneous data collection infrastructures that shall be coordinated and operated as a single complex SCP.

The legacy data collection system gathering data from the pilot UEIs was integrated into the ALMANAC platform thanks to a dedicated device manager, part of the platform resource adaptation layer. Data collected every hour and at every waste disposal (for the nonrecyclable fraction) are fed into the platform and stored for further processing and

---

* http://www.ecotecsolution.com/, last visited April 20, 2016.

TABLE 8.2

Specifications of Smart Waste Bins Deployed in the Pilot

Ecotec Smart Waste Bin	
Power supply	12 V
#loggable disposal operation	250,000
#rfid keys	16,000
RFID reader	125 kHz or 13.56 MHz
Data transmission	GSM/GPRS
Fill Level Sensor	
Current rating	<50 mA
Distance range	200–400 mm
Output range	4–20 mA
Distance resolution	4 mm
Operative temperature range	−25°C + 70°C

elaboration. Collected data are available through the ALMANAC Cloud APIs, which fully subsume the OGC SensorThings API SensingProfile,[*] thus offering standard means for querying/retrieving specific data ranges.

8.4.1.4 Citizen Engagement

Citizen engagement in city processes is one of the crucial factors for successful deployment of smart cities. ALMANAC sought citizen involvement from the beginning of the project. Two main activities were carried respectively targeted at (a) attracting, engaging, and keeping the *pace* with citizens living in the city quarter hosting the UEI pilot and (b) supporting the whole citizenship in the tasks of duly disposing recycled waste.

8.4.1.4.1 Involvement in Pilot Activities

As part of the pilot deployment activities in Turin's city center UEIs, the municipality together with ALMANAC puts in place a campaign to engage citizens and promote more sustainable behaviors. General information regarding the ALMANAC project, the scope of the pilot activities, and recycling educational material were distributed to the households and businesses involved in the experimentation. During the final phase of the experimentation, a questionnaire was distributed to collect the perceived benefits and disadvantages of the UEIs experimentation. Questionnaires were analyzed to assess the success of the experimentation and to provide recommendations for a possible extension of the pilot to other UEIs in the city in a near future. Involved citizens were invited to a final meeting where they were given the opportunity to express themselves and provide further insights on the subject.

8.4.1.4.2 Recycling Support

In parallel with these activities, ALMANAC implemented a cocreation approach (Figure 8.7) to design and develop a recycling support tool engaging the citizens of Turin. The codesign activities have been carried out in collaboration with a selected community from a temporary social housing project in the city of Turin[†] to mutually produce the desired value outcome.

---

[*] http://ogc-iot.github.io/ogc-iot-api/, last visited April 29, 2016.
[†] http://www.sharing.to.it/index.php/en.html, last visited April, 2016.

FIGURE 8.7
Codesign workshop activities with a group of Turin citizens.

Interaction with the community consisted of a set of interviews and a one-off workshop, organized over a six-month timespan and aimed at designing an effectively engaging recycling-support application. Although the heterogeneous and dynamic nature of the sample did not allow for participation over a longer period of time, the quality of the interactions compensated for fewer encounters and kept the community willingly and happy to engage in the process.

Main findings from the codesign workshop included people facing difficulties in the recycling process (e.g., *what goes where*?), thus becoming frustrated and developing bad recycling habits. Also, the need to find a specific recycling bin emerged during the workshop, due to overfilled bins in the area or simply to dispose a specific packaging, for example, a plastic bottle of water, while walking around in the city.

Needs and desires of the citizens concerning recycling and waste disposal were collected and translated into high-level requirements for a recycling supporting mobile application *The WasteApp*. The developed application has four (4) basic functions (Figure 8.8): (1) waste collection calendar, (2) waste recycling guide extended with crowd-sourcing techniques for harvesting waste classification from users and enabling direct bar code scanning, (3) waste bin location map, and (4) drop-off centers map. The process of cocreation, the workshop results, and the high-level requirements and related features of the mobile application developed within ALMANAC are described in Bonino et al. (2016).

8.4.2 Water Management Domain

Water management is another interesting domain that can be exploited for benefiting the whole smart city ecosystem. For the water utilities, the deployment of smart metering devices (with flow meters) can bring huge economical and technical benefits, saving on

FIGURE 8.8
The WasteApp Home showing the four main features.

metering costs, and enabling advanced features such as real-time detection of the reverse flow and tampering. Once the water management infrastructure is in place, smart metering devices can be used to provide citizens immediate feedback on consumption, alert them about leakages in private water pipes, and so on. As a result, water management applications can increase citizens' awareness of smart city processes and promote motivational schemes for sustainable behaviors through active engagement in individual and community activities supported by new business models.

Water management applications developed in ALMANAC are linked to the requirements related to the consumption awareness scenario analyzed in the process. The areas of interest are as follows:

- Showing user information about their own consumption, both historical and current
- Aggregating results and showing consumption trends
- Sharing consumption data on social media
- Notification of leaks and other events.

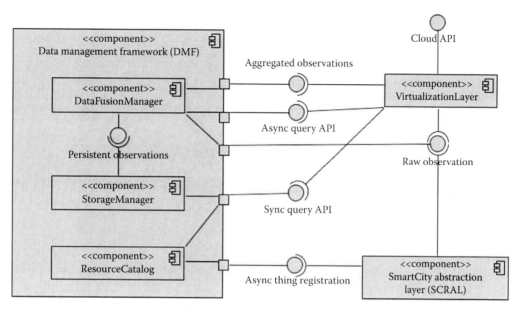

FIGURE 8.9

ALMANAC Cloud APIs exposed by the virtualization layer and exploited to build the water management applications.

Data flow from a WM Bus smart meter deployed in Aarhus to the ALMANAC SCP, and then into the end-user mobile application. The applications show integrations with both the SCP and external providers like Microsoft Azure and Apple apple push notification service (APNS). Just as important as the actual applications are the libraries that have been developed. These are now available not only as source code on GitHub but also as easy-to-consume libraries that can be included in other applications.

The water management applications communicate with the open end points of the ALMANAC SCP. This means that all communication is done through the virtualization layer (Figure 8.9), which forward requests to all the relevant platform modules—and enforces the security and policy frameworks configured within the ALMANAC SCP.

The water management application suite is composed of two separate software components: (a) a mobile iOS application for citizens and water consumers, connected to the ALMANAC SCP providing information about water consumption pertaining to the smart meter installed in the user's home; and (b) a city management and communication application where city officials can push notifications to their citizens through the mobile app, which is a desktop application designed to allow employees of the water utility or the city to transmit notifications to users who live in the city.

8.4.2.1 Mobile iOS Application: My Water

My Water is an end-user application, which provides multiple views on the user's water consumption: it shows current consumption and compares it to the average from the last 30 days, and shows alerts and notifications. For normal use, the end user has to log-in using the ALMANAC security and policy framework. This is done by loading an

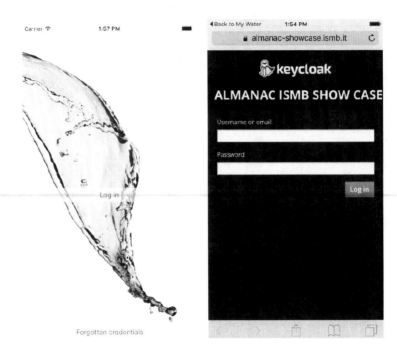

FIGURE 8.10
Security and log-in in My Water.

external log-in page from within the application, which returns the necessary security tokens to the application logic. The application can afterward communicate freely with the ALMANAC SCP.

The security framework (Figure 8.10) is completely decoupled from the rest of the application and is based on the techniques used by Google and Facebook. This allows the application to seamlessly switch between security frameworks, without having to modify core application logic.

Once the user is logged in, there are three tabs available to the user. One showing current consumption, or for a specific day; another with graphs showing consumption trends and more historical data; and finally one page showing historical alerts (Figure 8.11).

The application also supports sharing personal consumption using the Facebook social media. The user can choose to share today's consumption with the public or just friends—by clicking the *thumbs up* icon (Figure 8.12).

8.4.2.2 City Management and Communication Application

Through this desktop application, it is possible for a utility company to notify citizens of water quality issues, for example, if a pipe has burst or if contaminants have been detected somewhere in the water network infrastructure.

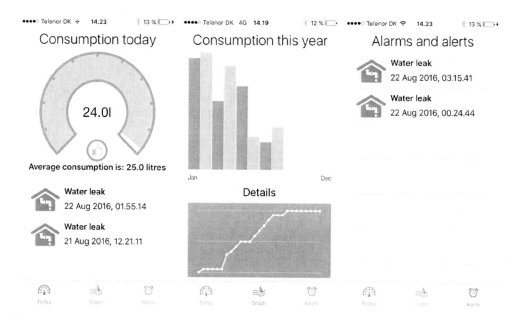

FIGURE 8.11
The My Water app.

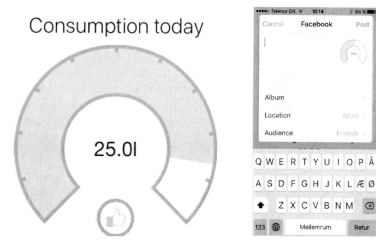

FIGURE 8.12
Facebook-like button and feedback sharing page.

FIGURE 8.13
City notification service application.

The application can deliver notifications (Figure 8.13) to residents of specific target areas or to the entire city, if needed.

8.5 Results

To better understand the impact of the *ALMANAC Way* of implementing a smart city ICT infrastructure, the project consortium pursued several evaluation activities including: (a) evaluation of the overall platform scalability, (b) evaluation of developed sample applications in terms of usability and willingness to adopt by end users, and (c) evaluation of real-world deployment of the platform in terms of user satisfaction and perceived impact on the user behavior.

Although full reporting of evidence and outcomes from experimentation activities is slightly out of scope for this chapter, this section summarizes main results and relevant insights gained during the experimentation of the ALMANAC approach.

8.5.1 Waste Management Domain

8.5.1.1 Smart Waste Pilot in Turin

To start understanding how citizens perceived pilot activities, a brief questionnaire (seven questions) was designed and distributed among residents who were involved in the pilot. The questionnaire was available in both paper form and online, focused on the waste disposal habits, including assessment of collection sites neatness and organization, and accessibility of waste containers. Questions are reported in Table 8.3. Readers must be aware that translation from Italian to English might not retain all the nuances and careful formulation of the survey questions.

TABLE 8.3

Turin Pilot Evaluation Questionnaire

Question Number	Question Text
1	Do you think that using the electronic keys is easy?
2	Do you think controlled access to unsorted waste bins provides emptier containers?
3	Do you think that access control has contributed to a better waste differentiation?
4	During the trial, have you changed the way you sort waste at home?
5	Do you think the amount of abandoned waste has increased close to the underground ecological islands?
6	Do you think that the installation of underground ecological islands with controlled access would be an efficient way to improve the quality of waste collection?
7	Overall, is your opinion of the trial with the underground ecological islands positive?

The amount of questions in the questionnaire was purposely kept low with the goal of reducing its complexity and obtrusiveness. Each question consisted of a statement for which citizens were required to express their relative agreement using a 5-value Likert scale ranging from *strongly disagree* to *strongly agree*.

8.5.1.1.1 Survey Outcomes

Out of 400 citizens and businesses who were engaged in the UEIs experimentation, only a group of 30 people provided their opinions and perceptions through the questionnaire. This could be related to the fact that no follow-up activities were taken by the partners deputed to administer the survey (Turin municipality and Amiat) to promote a stronger citizen engagement. Although results have low statistical significance due to poor coverage of the pilot user base, they provide interesting insights on the user perception of pilot activities. Several unexpected findings emerged from the questionnaire answers also fostered by the availability of open comment spaces in all questions. Most participants took advantage of this space to clarify their answers, often explaining motivations underlying their responses. Only 2 out 30 persons submitted partially filled questionnaires, skipping Question 2 and 3, respectively. In the following paragraphs, answers to the questions are discussed in detail summarizing survey findings and discussing the corresponding outcomes.

Question 1: Question 1 aimed at understanding to which degree access control is perceived as negative and annoying as feared by some of the most prudent administrators. Survey answers demonstrate that there is no negative connotation associated to access control and no particular complaints were registered about the access key adoption. Overall, positive judgments covered 86% of the available answers.

Question 2: Questions 2 and 3 aimed at verifying the assumption that forcing access control on the nonrecyclable waste would trigger better recycling behaviors by increasing the share of recycled materials. This design choice however did not account for possible counter effects such as reduced quality of the other fractions collected due to improper recycling or for possible increases of abandoned waste. The results for this question show that 75% of the citizens found that the containers for unsorted waste in UEI were emptier than traditional containers. The remaining 24% did not find any difference, possibly because visual inspection is physically impossible in the UEI container, thus reducing the answers to pure *perceptions*.

Question 3: Due to the impossibility of performing visual inspection of the actual container fill-level, answers to Question 3 are again purely perceptual. Nevertheless, by looking at the question outcomes, citizens involved in the pilot feel that the overall recycling efficiency has been improved by the access control measures: 70% of them, in fact, provided a positive answer to the question, whereas only 30% reported no changes.

Question 4: Although Questions 2 and 3 invited people to comment on other people's behavior and usage of the UEIs, Question 4 was related to changes in personal behaviors. In this case, the pilot seems to have forced some slight change in the recycling behavior of interviewed people. Around 53% of answers reflected changes in the way waste is sorted at home, which range from slight to significant (only 13% reported disruptive changes). Nearly half of the respondents (47%) said that they had not changed their behavior.

Question 5: Question 5 aimed at detecting if restraining the access to the nonrecyclable container had a negative effect, for example, increase of the waste abandoned in the proximity of the UEIs. Evidence from the survey demonstrates, in contrast with many user comments, that there was no counter effect in terms of abandoned waste. Quality of recycled materials could have decreased, but this was not captured by the questionnaire. Out of the 30 respondents, 57% judged that abandoned waste actually decreased during the experimentation, whereas 43% did not notice any change.

Question 6: Question 6 was aimed at understanding whether people think that the installation of the UEIs with controlled access would be an efficient way to improve the quality of waste collection. Almost 43% of the respondents think that controlled access brings little or no changes in recycling behaviors, whereas the majority (57%) thinks that it would actually be beneficial.

Question 7: Question 7 wanted to sum up citizens' overall experience achieving overall positive results (80%), thus confirming the acceptance and viability of the pilot.

8.5.1.1.2 Outcomes from the Meeting with Public Authorities

As expected, citizens who attended the public meeting organized by Torino, with the participation of representatives from istituto superiore mario boella (ISMB) and Amiat provided feedback in line with the comments written in the questionnaires. Additionally, they also provided some unrelated, yet interesting context information regarding noise pollution due to glass disposal late at night, mainly from restaurants and other businesses around. Among suggestions, the most recurring was enforcing by low glass disposal during the daytime, to reduce noise annoyances during night hours. In line with this, citizens proposed to extend access control to all waste containers, to reduce, or at least trace, behaviors that might hamper the quietness of the neighborhood.

8.5.1.1.3 Discussion

Overall, the pilot in Turin was positively perceived by users and provided precious insights to both city administrators and waste utility operators. A more structured approach at surveying user perception together with follow-up techniques is needed to improve the base of *collaborative users* willing to aid the city in identifying best ways to improve waste management. Further investigations in this sense are needed, both to cover the several *dark* points emerging from this survey and to better discern actual facts from feelings. Extending the pilot to more locations in different city quarters would help in spotting and compensating geographical and social biases. Moreover, it would provide a larger user set, which together with proper follow-up techniques should provide statistical significance to survey outcomes that at the time of writing are purely qualitative.

8.5.1.2 Citizen-Centric Application Usability

The citizen-centric application (namely WasteApp) was tested for usability, with a selected group of users from the SHARING community. This preliminary test was aimed at spotting the main usability issues in the prototype, thus allowing corrective actions to be taken before final, long-term trials with users from SHARING and other connected communities in Turin.

To evaluate usability of the interactive prototype, four participants used the Citizen-centric application in a controlled environment, performing six tasks each (Table 8.4).

Tasks were planned taking into account requirements and needs emerging from the codesign workshop and were tailored to fit the ALMANAC waste scenario. The set of tasks performed by the participants involved in the usability study reflect the typical interactions we envision for the proposed application. Participants never met during the evaluation. Their observations, together with data collected about their sessions with the citizen-centric app, allowed us to perform a qualitative analysis of the application usability, and helped to identify strong and weak points of the interface as well as to direct future works and research activities.

Our analysis focuses on five basic questions about the usability of the citizen-centric application:

- How easily do users understand the meaning and intended use of the four main sections of the app?
- How quickly can users identify where to dispose of a specific item, that is, the type of waste?
- How easily can users understand where to go for disposing waste of a given type?
- How easily users identify and understand the curbside collection calendar?
- How easily can users configure the app to send collection notifications?

8.5.1.2.1 User Sample

We recruited four participants for our user study (Table 8.5): two females and two males, aged 31–55, residents of the SHARING community, who freely volunteered to participate in the study. Three out of four participants are Italian while one is originally from a

TABLE 8.4

Tasks Numbering and Its Descriptions

Task Number	Task Description
T1	Where is the nearest *paper/cardboard* container located?
T2	Given a specific object that needs to be disposed of, identify the right waste type using the citizen-centric application.
T3	When is *glass and can* waste being collected?
T4	Where should you dispose of your Microwave Oven?
T5	Which waste category does *used tires* belong to?
T6	Configure the citizen-centric application to generate a waste collection notification at 7:00 a.m.

TABLE 8.5

Tasks Used for the Small-Scale Usability Study

Participant	Age Range
P1	31–35
P2	51–55
P3	36–40
P4	31–35

TABLE 8.6

Job Positions of Users Participating in the Study

Participant	Job Position
P1	Education
P2	Unemployed
P3	Computer science
P4	Business analysis

Middle-Eastern country. They all have a medium–high education level and quite diverse job positions (Table 8.6).

Three out of four users do not work in technical fields and/or computer science. This allows making an informed guess as to how the city population will react to the app, even if the user sample considered in the study is too small to be of any statistical significance. All users declared to regularly use smartphones in their daily activities, and one out of four declared its adoption level as *intensive*. On average, they identified their frequency of operation around 11 to 20 times a day, thus a high degree of confidence with mobile application usage and typical interaction patterns. This assumption is confirmed by the set of applications/purposes for which users declared to adopt smartphones (as reported in Table 8.7).

None of the users had previously used waste recycling support applications, thus confirming that no *a priori* bias toward some particular setup or function of the app occurred during the tests.

8.5.1.2.2 Methodology

A within-subject design was employed where eah subject performed each task in random order, to reduce ordering effects. Tasks involved different waste types and objects to avoid any possible influence due to unwanted communication between users during the session

TABLE 8.7

Surveyed Users' Smartphone Usage Patterns

Application/Activity	Number of Users Exploiting a Smartphone to Perform the Activity
E-mail/Web browsing	4
Messaging	3
Social networks	4
Maps and navigation	3
Phone calls	3

FIGURE 8.14
The usability test setup.

exchanges. Experiments were conducted in a controlled environment consisting of a simple single-room setup (Figure 8.14) with one moderator and two observers not interfering with the test activities. In general, we followed recommendations for typical user studies as reported in (Rubin, 2008).

8.5.1.2.3 Test Deployment

Every single-test session started with a short introduction of the ALMANAC project and a brief description of the citizen-centric application.

After the introductory part, the user was allowed to freely experiment with the mobile application for 2 minutes, with no suggestions and/or constraints from the moderator. After this warm-up, each user was asked to fill in a pretest questionnaire aimed at gathering first impressions and expectations for the citizen-centric application. This questionnaire, in particular, investigated the user agreement on four different statements, using a 5-level Likert scale for the allowed answer, which were as follows:

- I think the application is visually appealing
- I think the application is easy to use
- I think the application functions are clear and easy to identify
- I would like to use the application

The allowed answers were

- Completely disagree
- Disagree
- Neither agree nor disagree
- Agree
- Strongly agree

In the warm-up questionnaire, users were given space to provide feedback they might want to share after this phase. Notice that the previous statements are translated from the original Italian formulation; therefore, they just provide a means for the reader to understand the posed questions while they do not necessarily reflect the meaning nuances and the carefully formulated wording we used.

Each user was asked to perform six tasks (reported in Table 8.4), one at time. For four tasks out of six, the participant was asked to use the think-aloud protocol to describe his/her actions. User comments, suggestions, and complaints were collected by each observer, both during the task execution and at the end of the test session. For each task, observers recorded any *unusual* or *unexpected* behavior, and they kept track of time needed for task completion.

At the end of this phase, users where required to fill in a posttest questionnaire with the same questions of the warm-up questionnaire plus some additional statement to harvest the most-liked functions and any free observation and/or feedback the user wanted to provide.

8.5.1.2.4 Results

According to Nielsen's Alertbox1[*], we calculated the success rate of each participant as the percentage of tasks that users were able to complete correctly, also giving credit for partially completed tasks, that is, tasks completed with minor errors. We expected the participants to easily grasp the app's main features and to successfully accomplish all tasks. Due to the limited size of the user sample and the small number of tasks to accomplish, quantitative results are not particularly significant; however, some evidence for possible improvements did emerge. Table 8.8 reports the task success rates, where "S" indicates a successful task (score of 100%), "P" a partial success (score of 50%), and "F" a failed task (0%).

Although generally good, with an overall average success rate of 77%, results in terms of success rate on single tasks were a little lower than expected. Three tasks were identified as difficult. The participants had severe problems completing one of the tasks (T2—success rate of only 37.5%) while in another, they displayed medium-level issues (T5). By analyzing annotations taken by observers for tasks T2 and T5, we identified two main usability issues in the citizen-centric prototype.

First, regarding task T2 where users were required to identify the waste category to which a given object belonged, users were expected to exploit the barcode scanning feature

TABLE 8.8

Results of the Task Execution Test

Users/Tasks	T1	T2	T3	T4	T5	T6
P1	S	P	S	S	P	S
P2	S	P	S	F	F	S
P3	S	F	S	S	S	S
P4	S	P	S	S	P	S
Average Score	100%	37.5%	100%	75%	50%	100%

---

[*] http://www.useit.com/alertbox/20010218.html.

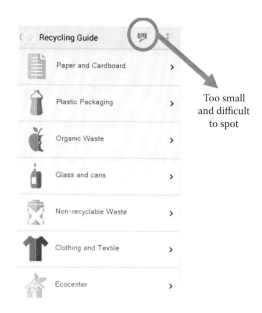

FIGURE 8.15
The usability issue hampering T2 results.

of the citizen-centric application, with the aim of stressing the *responsiveness* of the application for recycling support. Therefore, we assigned a full success (S) to waste identification through barcode scanning and a partial success to waste identification by guide browsing (P). Unfortunately, although useful and easy to understand, as confirmed by users in the cooldown part of the test, the barcode scanning feature was not easy to detect (Figure 8.15).

The low success rate of task T5 is again due to a limitation in the user interface, which did not sufficiently highlight the availability of the search by keyword feature of the recycling guide (Figure 8.16). Half of the users ended up searching each waste category for entries mentioning used tires. Moreover, one of the participants was not able to identify the right guide entry and gave up the task after a few trials.

The T5 partial success had the effect of masking to some extent the severity of the usability issue related to the *search by keyword* functionality. In fact, by inspecting the waste recycling guide, it is clear that the search function is very difficult to spot as it is hidden in a contextual menu. To mitigate this issue, common interaction patterns for Android applications suggest positioning the search icon in the place currently occupied by the barcode scanning feature. This modification is compatible with the one related to the barcode scanning issue and can therefore be easily implemented in the next application release.

The second issue relates to the absence of a *filtering* function on the waste bin map. Most users were successful at finding the bins nearby. However, identifying bins for a selected type of waste was cumbersome, as it was usually done by touching small icons on a map. The main *pattern* we imagined was through the barcode scanning feature. In fact, this feature offers disposal information and directions (map) to the nearest bin suitable for the identified waste type. However, in most cases, this was not the pattern selected by users; therefore, better interaction shall be supported, for example, by adding *filter by waste-type* functionality to the map, and by making barcode scanning more prominent in the interface.

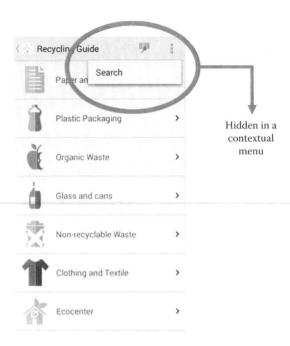

FIGURE 8.16
Usability issue on the search by keyword functionality.

Other minor issues emerged from user observations gathered both during the task execution and in the pre- and posttest questionnaires. Among these, it is worth citing two issues that can easily be fixed:

1. On the EcoCenter location activity, it would be useful to have direct access to information about accepted wastes, for example, by touching the EcoCenter icon, with some additional function for getting the EcoCenter opening hours and contacts.
2. On the Nearby Waste bin activity, the availability of a filtering function would be useful to ease the process of identifying bins for a specific waste type. Although this function is already implemented in the prototype, for example, reachable from the barcode scanning activity, it proved to be necessary in many cases thus deserving an easier access interaction pattern.

8.5.1.2.5 Discussion

Comments verbally expressed by users where in general very positive. They appreciated the notification ability and its ease of configuration (in fact, all users achieved 100% success in T6). Moreover, they judged the application easy to use and supportive of their needs, even with the issues emerging from the quantitative analysis discussed in the previous paragraphs.

8.5.2 Water Management Domain

A 10-question online survey was developed in order to get an impression of the potential end-user (private households) interest in the water management app. The survey was promoted on the ALMANAC homepage, through the project partners' contacts and on social networks such as Twitter. We may therefore assume that many respondents were from Danish households. The questions were a mix of statements where respondents had to mark the degree of their agreement with the statement and multiple choices. Two questions were contextual to find out (1) how respondents paid for water (Q9) and (2) the number of members in their household (Q10). A total of 56 surveys were returned and the overall results showed a positive interest in the water management app particularly with respect to the leakage notification functionality. The results from the survey are analyzed in more detail in the following paragraphs.

Nearly all respondents were interested in knowing more about their water consumption. When comparing this with the knowledge from Question 9 (how they pay for water), it appears that although most people (56%) get a specified water bill, they would still like to know more. Question 1 was purposely designed as a general question to assess if there is an end-user market for the water management app before moving onto questions that are more specific. Thus, Question 2 specifies that water consumption information could be made available via a mobile app. A total of 77% of the respondents were interested in an app out of which more than half (54%) agreed or strongly agreed with the statement.

A slightly different pattern compared to Question 1 is noticed, as not all respondents who want to know more would necessarily be interested in using a mobile app. In fact, one of the comments to Question 3 was: *I am not interested in the information if that means an extra app on my phone (Table 8.9). Why does everything these days have to include an app?*

A total of 13 respondents had made an entry in the *Other* field. Most respondents commented that they would like a notification or indication of a possible water leak. Other suggestions included tips on how to save water and detailed historical data: *I would like to have access to detailed historical data, so I can correlate water consumption with other data (e.g., shower after exercise and dish-washing after cooking).*

TABLE 8.9

Water Management Application Evaluation Questionnaire

Question Number	Question Text
1	In general, I would like to know more about my household's water consumption.
2	I am interested in a mobile app that gives me access to up-to-date information about my household's water consumption.
3	I would be interested in the following information consumption: current data, historical data, notification, consumption comparison, bill estimation, water supply shortage alert, other (more than one answer allowed).
4	I would be interested in receiving a notification if there is a public warning of water contamination—tap/drinking water.
5	I would be interested in receiving a notification if there is a public warning of water contamination—sea or lake water.
6	I would be interested in using the app for leakage notification at home.
7	How much would you be willing to pay to get the leakage notification system on the app, if you saved between EUR 30–50 on home insurance policy? (Nothing, 1–5, 5–10, 10–15, more [in EUR])
8	I would be interested in using the app to connect with my social media accounts in order to share my consumption data.
9	How do you pay for water consumption? (Included in my rent, flat annual rate, variable annual rate, water provider bill)
10	My household consist of: 1, 2, 3, 4, 5, 6, or more persons.

Questions 4–6 inquired into the notification functionality offered by the ALMANAC water management app. Questions 4 and 5 were related to notifications from the public authorities regarding contamination of water and respondents expressed a high interest in such a notification service.

Questions 6 and 7 were directly related to the value proposition in the Water Information App. Not surprisingly, Question 6 showed a great interest in the leakage notification functionality of the app and the answers to Q7 illustrate that respondents are willing to pay for the leakage notification functionality.

The only statement, which got an overwhelming negative response, was Question 8 on whether people were interested in using the app to share their water consumption on social media. This clearly shows that the value of Water Management app is primarily as a tool to monitor, control, and modify the individual household's water consumption; sharing water data with others is not perceived as having any value.

8.5.2.1 Discussion

Although the majority of respondents already receive a specified water bill, the overall results from the questionnaire revealed a substantial interest in being able to access detailed current and historical data on the water consumption, as well as notifications of possible water leakages and water contamination issues. Water bills are usually issued once a year and can therefore not be used as an indicator (e.g., an unusual expensive bill), for example, a water leakage in the household. Often leakages are not discovered until they cause an actual pipe breakage resulting often in extensive water damages in the home. The consequence of a water leakage is thus not simply one of an unusual high bill but also one of expensive repair work.

8.6 Conclusion

Smart City ecosystem area is characterized by their complexity and variety of actors. Even in the same city, different processes may be subject to different procedures, policies, and legal frameworks. Introducing IoT technologies in such a context can enable higher processes' efficiency and better life quality for citizens in everyday life.

The presented ALMANAC SCP has demonstrated to provide support to decision processes in both day-to-day and long-term city management, while implementing intelligent control of a large number of smart city resources.

Introducing IoT technologies in the waste domain have provided deeper understanding of waste generation dynamics, both for the municipality and for the utility service operator. IoT technologies can facilitate the optimization, planning, and control of the waste collecting and cleaning system, generating operational savings and promoting innovative and more sustainable practices in the field. Additionally, IoT technologies in the water management domain could provide further insights to citizens, water utility companies, and the city municipality. Real-time information linked to consumption habits and water contamination are of great importance to these actors, for economic, health, and security reasons. Overall, citizens are inclined to accept the use of innovative technologies to improve everyday processes in their city and their lives.

ALMANAC has paved the way for introducing new technologies and approaches in real-life settings. Results obtained during the project lifespan should be further explored within the City of Turin and in other European cities to fully achieve the potential impact of IoT technologies in smart city ecosystems, always keeping in mind that citizens should be at the center of these.

References

Bauer, M., Boussard, M., Bui, N., and Salinas, A. (2013). Final architectural reference model for the IoT v3.0. Technical report, Internet of Things Architecture IoT-A.

Bellini, P., Benigni, M., Billero, R., Nesi, P., and Rauch, N. (2014). Km4city ontology building vs data harvesting and cleaning for smart-city services. *Journal of Visual Languages & Computing,* 25(6):827–839.

Bonino, D., Delgado, M. T., Alapetite, A., Gilbert, T., Axling, M., Udsen, H., Soto, J. A. C., and Spirito, M. (2015). Almanac: Internet of things for smart cities. In *Future Internet of Things and Cloud (FiCloud), 2015 3rd International Conference on,* pp. 309–316. IEEE: Rome, Italy.

Bonino, D., Delgado, M. T., Pastrone, C., and Spirito, M. (2016). Wasteapp: Smarter waste recycling for smart citizens. In *International Multidisciplinary Conference on Computer and Energy Science–Splitech 2016.*

Fritz, P., Kehoe, M., and Kwan, J. (2012). IBM smarter city solutions on cloud. White paper, IBM.

Hogan, J., Meegan, J., Parmer, R., Narayan, V., and Schloss, R. J. (2011). Using standards to enable the transformation to smarter cities. *IBM Journal of Research and Development,* 55(12):1–10.

Rochwerger, B., Breitgand, D., Levy, E., Galis, A., Nagin, K., Llorente, I. M., Montero, R. et al. (2009). The reservoir model and architecture for open federated cloud computing. *IBM Journal of Research and Development,* 53(4):535–545.

Rubin, J. and Chisnell, D. (2008). Handbook of Usability Testing: How to Plan, Design, and Conduct Effective Tests, 2nd ed. Hoboken, NJ: Wiley.

Sma, (2012). ICT architecture supporting daily life in three smart cities. Technical report, Smart Cities EU Project.

Sma, (2015). Smart cities market by smart home, intelligent building automation, energy management, smart healthcare, smart education, smart water, smart transportation, smart security, & by services–Worldwide market forecasts and analysis (2014–2019). Technical report, marketsandmarkets.com.

Vermesan, O., Harrison, M., Vogt, H., Kalaboukas, K., Tomasella, M., Wouters, K., Gusmeroli, S., and Haller, S. (2009). The internet of things–strategic research roadmap, Cluster of European Research Projects on the Internet of Things.

Vilajosana, I., Llosa, J., Martinez, B., Domingo-Prieto, M., Angles, A., and Vilajosana, X. (2013). Bootstrapping smart cities through a self-sustainable model based on big data flows. *Communications Magazine, IEEE,* 51(6):128–134.

9

NFC in IoT-Based Payment Architecture

Alak Majumder, Shirsha Ghosh, Joyeeta Goswami, and Bidyut K. Bhattacharyya

CONTENTS

9.1 Introduction

In the era of ubiquitous computing, many wireless technologies have emerged to serve several types of communication needs. Near-field communication or NFC is a short-range wireless communication protocol, which is a successor of popular Radio-Frequency Identification or RFID technology that was patented by Charles Walton (1983). NFC was standardized as an ISC/IEC standard in 2003 and works in a very short range using 13.56 MHz frequency that makes it perfect for contactless payments. It extends the ISO 14443 RFID standard. Since its inception, NFC is on the limelight because of its versatility, inherent security, and interoperable nature. The basic mechanism of NFC is based on the Faraday's Law of Induction in which current is flown between two devices by creating a magnetic field. In the case of NFC, the reader or the active device creates a magnetic field using its antenna coil. Then, the passive devices present in that field get energy from the magnetic field and modify the properties of the incoming frequency. When the active device gets information about the passive device, a very small amount of AC current emerges from the sinusoidal waves. The AC current is converted into DC inside the tag chip using the rectifier and the passive device also generates carrier wave to pass information to the active device.

Currently NFC is involved in three major devices: Smartphones, NFC readers, and NFC tags. Based on the different working and application areas, NFC has three different operating modes known as reader/writer mode, peer-to-peer mode, and card emulation mode. NXP semiconductors have documented all these modes in a document entitled "NFC Everywhere" (2015a).

9.1.1 Reader/Writer Mode

In this mode, the communication settles between a passive NFC tag and an active NFC device such as NFC enabled smartphone or NFC terminal. The main purpose of this communication is two:

- To read prestored data from the tags. The involving steps are as follows:
  - *Read request*: The user requests data by active device such as mobile to the NFC tags that may be installed into various places.
  - *Data transfer*: The prestored data are then transferred to the active device from the passive tag to the reader.
  - *Processing by the reader device*: After getting data from the passive tag, the reader processes the data for various functions.
- To write values to the NFC tags or writer mode. In this case if some data are pre-stored in the tag, then the writer overwrites or updates those data by the latest. The steps are as follows:
  - *Write request*: First user makes a request to write data to the NFC tag.
  - *Data transfer*: Then the NFC tag endorses the success of the operation.

This operating mode has data rate up to 106 kbps. The main applications of this mode are smart posters, ticketing, accessing Internet or any media files, and so on. Figure 9.1 represents the mode.

9.1.2 Peer-to-Peer Mode

This operating mode is about the communication between two NFC active devices. This operating mode takes the advantages of two communication technology: NFCIP-1 (Near-Field Communication Interface and Protocol), LLCP (Logical Link Control Protocol), and SNEP (Simple NDEF Exchange protocol). In NFCIP-1 mode, both NFC devices must have own power source to generate magnetic fields. And in the LLCP communication, both

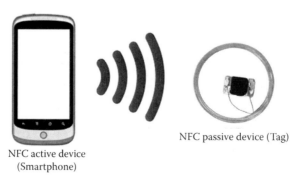

NFC active device
(Smartphone)

NFC passive device (Tag)

FIGURE 9.1
NFC reader/writer mode.

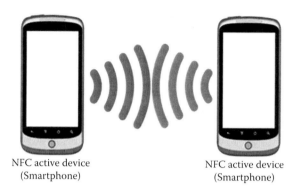

FIGURE 9.2
NFC peer-to-peer mode.

the devices stay identical. SNEP allows the exchange of NDEF messages analogous to tag operation specifications. In peer-to-peer mode, first data are exchanged between two active devices then using those data, several tasks are executed. A 424 kbps data rate is provided by this operating mode. This mode has several important tasks including exchanging data, peer-to-peer money transfer, pairing devices, and so on. Figure 9.2 shows the mode.

9.1.3 Card Emulation Mode

Card emulation mode provides a way to use an NFC enabled mobile as a substitute of contactless smart card in which the mobile devices can store different credit card information with a much secure way. The working steps of this operating mode include the following:

- *Service request*: The user establishes an NFC communication with an NFC reader and makes a request to any service provider in which the reader transfer the data collected' from the mobile phone to the service provider. Figure 9.3 shows the mode.
- *Backend service*: The service provider will run this service in backend after receiving the required data.
- *Service usage*: Finally the service provider processes the data and provides a service to the user.

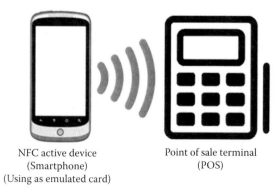

NFC active device
(Smartphone)
(Using as emulated card)

Point of sale terminal
(POS)

FIGURE 9.3
NFC card emulation mode.

The major application fields of this operating mode are Payment, Ticketing, Identity service, Smart environment, and so on with the advantage that it removes the physical objects and cash money and gives higher security.

9.1.4 NDEF Stack Description

The format by which data are exchanged between two NFC devices is known as NFC Data Exchange Format or NDEF. NFC Forum specified all the technical specification in their official documentation (2006) on NDEF. It is a binary message format in which every message contains several records, and each record consists of a payload of size 232-1 octet. Each record has header fields and a payload field, where header field contains five flags:

> *MB* (*Message Begin*): Points the first record of an NDEF message.
>
> *ME* (*Message End*): Points the last record of an NDEF message.
>
> *CF* (*Chunk Flag*): Defines the continuation of payloads from one record to its next record.
>
> *SR* (*Short Record*): Specifies the size of the payload length that is the number of octets present in the payload.
>
> *IL* (*ID Length Present*): Presence of the optional ID field and its corresponding length field.

The NFC Protocol Stack is shown in Figure 9.4.

9.1.5 Security in NFC

As NFC is widely used for payment-related applications, it is very important to take care of the security measures. Due to NFC's very small distance communication range, it is very difficult for an attacker to intersect the communication medium, but it is very important to increase the security as high as possible to eliminate the chance of data theft. NFC uses hardware secure elements that provide security at its level best.

Typically, the secure element comes included with the NFC active element such as Smartphone. Several libraries are also written in the software stack to combine NFC with several role players such as Trusted Service Manager (TSM), Mobile Network Operator (MNO), and POS. The communication is performed over a secure channel using symmetric or asymmetric cryptographic keys.

There are several types of secure elements available. They are as follows:

> *Universal Integrated Circuit Card* (*UICC*): They are mandatory for GSM applications and necessary about the network. They are coded on Java and Global Platform that allow 3rd party developer to run their applications using UICC. The main host controller uses Single Wire Protocol (SWP) to communicate with UICC.
>
> *Embedded Secure Element* (*eSE*): This type of secure element is separate hardware that is embedded in the mobile handset by the manufacturer. These will verify the legitimacy of all NFC-based Transfer. Once embedded in the hardware, they can only be removed by the manufacturers.
>
> *MicroSD*: Several 3rd parties other than the smartphone manufacturers, and Mobile Network operators, use the MIcroSD as an alternative to UICC and eSE.
>
> *Trusted Execution Environment* (*TEE*): It is a part of main processor core that can be used to store and process sensitive information.

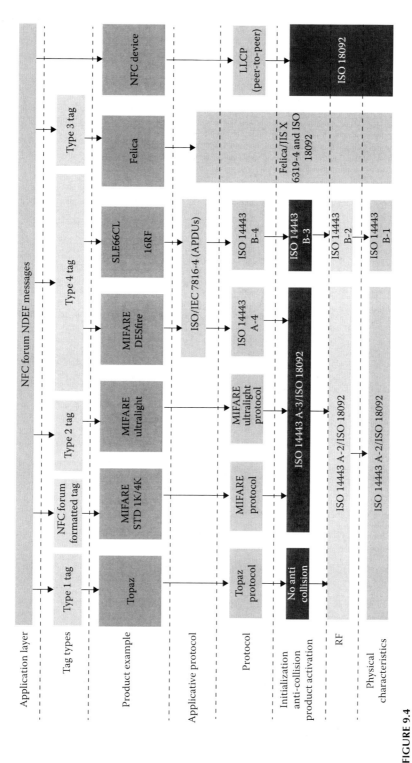

FIGURE 9.4
NDEF stack.

9.2 Basic Overview of IoT Universe

The concept of IoT-based application began from the inception of RFID technology. The primary focus of IoT development was to devise some electronics equipment that will boost the use of RFID. But soon when the need of connectivity among several peripherals was raised, researchers started to explore the use cases of other wireless technologies. Soon IoT became a much broader scope for research, and it continued to be bigger ever since.

As given by International Telecommunication Union (2005), the main motto of IoT development is very simple: "from anytime, anyplace connectivity for anyone, we will now have connectivity for anything." The vision is that there will be a world where things can communicate with Internet and provide important feedback about their environment that will rather improve the quality of human life. For proper implementation of the motto of IoT, seamless integration between several wireless technologies is very much essential. Actualization of the IoT concept into the real world is possible through the integration of several enabling technologies.

Though several technologies are integrated in the development in IoT, RFID remains the main driver technology. RFID tag is basically a microchip with an antenna used for transmitting the tag ID and receiving the reader signal. They are assigned a unique identifier. The tags are attached with different objects and using the unique identifier, different objects can be identified, and mapped into a virtual environment (Atzori et al., 2010).

Another critical technology in IoT universe is wireless sensor network or WSN. In Wireless Sensor Network, several sensor nodes are connected in a distributed or centralized and multihop fashion. The sensor data are shared between the different nodes that thereafter collected by the main server for data manipulation. WSN mainly contains the following elements:

1. *Hardware*: Comprises sensor interface, A/D Converter, transceiver, and power supply.
2. *Communication stack*: Designing the software stack for convenient topology, suitable routing and MAC Layer.
3. *Middleware*: A software infrastructure to combine the WSN hardware, communication stacks, cyber infrastructure, and application.
4. *Secure data aggregation*: It extends the lifetime of the network and ensures reliable data collection.

The recent trend is to integrating the sensor technologies with the RFID chip. This allows much more capability of sensor nodes. This system is very much useful in medical system, in which microsensors can be fitted into patients body and get essential medical data from there.

Atzori et al. (2010) compared the RFID system, WSN, and RFID Sensor Network (RSN) in their literature as shown in Table 9.1.

For an IoT system to be useful, it should be scalable. Means addition or reduction of sensor hardware should not affect the performance of the system. To make the system scalable, Uniform Resource Name (URN) system is widely used in IoT system. Several replicas of resources are made by the URN that is then accessed by URL (Gubbi et al., 2013). All the different sensor data are acquired by the central node are made addressable by URN and thereafter made accessible to the web server by URL.

TABLE 9.1

Comparison between RFID system, WSN, and RFID Sensor Network (RSN)

	Processing	Sensing	Communication	Range	Power	Lifetime	Size	Standard
RFID	No	No	Asymmetric	10 m	Harvested	Indefinite	Very Small	ISO18000
WSN	Yes	Yes	Peer-to-Peer	100 m	Battery	<3 years	Small	ISO 802.15.4
RSN	Yes	Yes	Asymmetric	3 m	Harvested	Indefinite	Small	None

The gathered sensor data are passed to a web server through internet. As the data gathered from the sensor nodes are enormous, it needs to be manipulated. Researchers are developing several algorithms using Artificial Intelligence, Neural Network, Genetic Algorithm, and so on for automatic response from the gathered data. These algorithms are executed on a centralized infrastructure to get analytics from the collected data. Nowadays, IT giants such as Google, Microsoft, Amazon, and so on are providing cloud-based server and tools for developers. These services are gaining popularity because of their reliability, uptime guaranty, and many other features for the developers.

After processing the data, the statistics of the result are sent to the users from the web server. For this process, mobile Computing plays a crucial rule. Currently developers and researchers prefer to use mobile based applications for data visualization because of huge number of smartphone users in market. Using a mobile application, important information can be visualized anywhere at any time. Smartphone also offers touch screen inputs that are very intuitive to get input from the user.

IoT has very useful practical application in our daily life. With more and more research and development on the technology, different application areas are coming up regularly. GSM Association came up with some application idea in their IoT report on July 2014. They are given below:

IoT for Smart Cities:

Dim-based smart streetlights on ambient conditions to save energy.

Smart Display board for passenger information.

Self-driving vehicles to increase safety and reduce CO_2 emission.

Smart traffic lights.

Smart camera to reduce crime and to provide quick response for emergency cases.

Smart payment system for insurance, Toll or road charges.

IoT for Health:

Connected devices to get important health parameters and updating medical reports.

Doctors can see that the treatment is properly followed or not.

Patients can take proper medicines in proper time.

Family members can ensure proper care taking.

IoT for Education:

Students carry all the materials in Smart devices.

Results are updated in real time.

Using connected device students and teachers can work collaboratively.

Self-guiding education helps adults to fill up their skill gaps.

IoT for Productivity:

Video conferencing for long distance communication.

Real time data can be shown in any meeting.

Reducing the product delivery time in e-commerce business by delivering the product from the closest inventory.

Prototyping using 3D printer.

9.3 Use Cases of NFC in IoT Universe

As discussed earlier, NFC is successor of RFID technology, which is the main role player in IoT development. NFC has the same capability as RFID but with increased security and features. In an article written by Paula Hunter (2016) of NFC Forum, she pointed out four major points for which IoT needs NFC. The points are as follows:

1. Using NFC, any unpowered device can also be connected with network. Just by taping an NFC tag, useful information can be gathered about an object.
2. Using NFC, the user can choose his own way of connection. NFC provides the simplest and intuitive way to take an action by the user.
3. Using NFC, the handshaking procedure for connection of devices can be eliminated just by using a simple tap.
4. Using NFC, the chance of eavesdropping is reduced to a great extent.

IHS Technology predicts that in 2018, 1.2 billion smartphones will be shipped with NFC embedded on them. This provides a great opportunity to developed connected devices with least amount of hardware required. Presently only using an NFC-enabled phones many IoT-based applications are developed. Some of the applications include Home Automation, Smart Meters, Healthcare, Home appliance control, Consumer Electronics, Cloud-based applications, and so on. In the whitepaper published in June 2016, NFC Forum (2016) pointed out some very useful application of NFC in Smart Home such as Water Metering, Appliance Controlling, Appliance Servicing, Ambient Setting, and so on. They also pointed out that "one of the main challenges for providers of smart home devices is providing a unified device commissioning flow that is independent of the underlying communication framework." NFC simplifies the device commissioning process by a single *tap-and-go* feature, which is completely independent of underlying communication protocol and based on standardized mechanisms. Another important factor for smart home is to connect and pair with Wi-Fi. Due to the alliance between NFC Forum and Bluetooth SIG, NFC can speed up Bluetooth pairing process by eliminating the need of time consuming processes like device discovery and device pairing. NFC forum has also alliance with Wi-Fi. Just by tapping an NFC tag that is glued with Wi-Fi router, any smartphone gets all the information such as SSID and Passcode and configure itself to connect with the Wi-Fi network. When an unpowered device with an NFC tag is tapped on the NFC module, the network parameters are written on the tag. Later, this information is used to connect the device to the network. NFC also helps for Smart home access control. The users can share their smartcard keys using smartphones with their relatives without meeting them personally. The keys then can be used with NFC to open any kind of electronic lock or smartcard based locks.

NTAG I2C Chip is another low-cost NFC module from NXP Semiconductor (2014). It provides ideal solution in the places where it is very difficult to embed any display or remote controlling features. NTAG I2C Chip works like a bridge between the appliance and users NFC-enabled mobile to make the mobile an extended display and remote controller for the appliance. The system reduces the cost due to its much simpler design and lease number of fancy hardware and also increases the security by eliminating the possibility of unauthorized access. This feature is quite useful in different applications such as Fitness Tracking Wearable, Home Appliances, and Thermostat. It also provides different features such as Self-Serve maintenance, Administrative access to Electrical Equipment, Firmware update, Bluetooth and Wi-Fi pairing, Home Automation Commissioning, and so on.

Texus Instruments also launched NFC Dynamic Transponder RF430CL330H. It provides the same features as NXP NTAG I2C Chip but has two inbuilt communication protocol: I2C and SPI.

Researchers are trying to take advantage of the features of NFC and use this technology in different fields. NFC can be a vital role player in in healthcare application. Some researches on the use of NFC in healthcare are given below. Freudenthal et al. (2007a) implanted 13.56 MHz frequency based RFID tag TI—TagitTM transponder, on animal tissues and proved the usefulness of using passive NFC to communicate with the medical devices embedded in human cadaver. Morak et al. (2007b) used NFC technology for real-time monitoring of different health parameters such as blood pressure, body weight, and so on. They can be monitored just by tapping NFC-enabled smartphone in the mobile device. Researchers came up with a system, by which using NFC-enabled phones, patients with impaired fine motor skills can get meaningful information. They have fabricated a smart poster. The patients will get information just using simple tap in the smart poster. The system also has a *physician's dashboard*, in which using a web-portal, doctors can monitor all the patient related information in real time (Prinz et al., 2011). In 2008 Deutsche Bahn, Vodafone, German rail authority, Deutsche Telekom along with some support from industry jointly launched an NFC-based ticketing system in Germany known as Touch&Travel. The system covers trains, metros busses, and a ferry. In the same year, in San Francisco, contactless payment Solution provides ViVOtech along with some other companies such as Sprint and First Data, enabled Bay Area Rapid Transit (BART) using NFC. In another literature, researchers used NFC to schedule appointment in the hospital (Sankarananrayanan and Wani, 2014). The system can also prioritize appointments based on patient's physical condition. Other than health care, NFC can be also used in *On the Go* ticketing services. In the research papers by Nasution et al. the researchers have developed a prototype of a ticketing system based on NFC enabled android mobile (Nasution et al., 2012). In a whitepaper published by NFC Forum (2011), the authors have pointed out the usefulness of NFC-based ticketing over conventional Paper- or PC-based ticketing. NFC-based mobile provides lot more facility and robustness than traditional procedures such as getting information about timetables, up-to-date weather report before travel, location map for navigation, discounted travel fares, next bus arrival time, and so on. The consumer gets many benefits such as getting the ticket electronically without the need of physically going in the office to collect the tickets, less chance of lost or theft, easy sharing, and so on. Developed countries like

London, Paris, and Madrid already have implemented automatic fare collection from the NFC-enabled mobile phone.

Educational institutes are also relying on IoT-based system to increase the relationship between the student and the management, in which NFC can play an important role. In the literature, Miraz et al. (2009) pointed out the efficiency of NFC to enhance the teaching administrative and payment-related services. The researchers also developed smart posters for different departments and faculties. Just by tapping any NFC-enabled mobile, all the information regarding the concerned department or faculty can be viewed instantly. Bueno-Delgado et al. (2012b) present a preliminary survey on smart university about the actual acceptance to the users. They have also devised some mechanism for different academic applications such as attendance monitoring and payment collection using NFC technology. The U.S. Department of Transportation also included NXP's RFID tagging technology for vehicle-to-vehicle and vehicle-to-infrastructure technology (Bodden, 2016a).

For any NFC-based project both NFC reader and Writer device are essential. If somebody is good at Android programming, then Android-based mobile phone has the capability of handling all types of NFC communication modes, that is, reader/writer mode, Host Card Emulation Mode, and peer-to-peer mode. The detailed guideline can be found on Android's official guideline on Android's official website by Google. Researchers have also published many books that provide step-by-step guideline for NFC-based Android Application development. Like, *Beginning NFC: Near-Field Communication with Arduino, Android, and PhoneGap* by Brian Jepson, Don Coleman, and Tom Igoe, *Professional NFC Application Development for Android* by Busra Ozdenizci, Kerem Ok, and Vedat Coskun; *Near-Field Communication with Android Cookbook* by Vitor Subtil and many more. The books also provide many readymade examples to start with. By using a Simple Android based mobile phone and NFC tags, anyone sitting at their home with a computer can start building NFC-based projects for different domains such as Smart poster, Access Control, and so on. NFC tags are also very cheap (Starts from 5$ at Amazon), so huge amount of cost involvement is also not required.

Hardware-based NFC projects required some basic skillset in Microcontroller programming. For any beginner, one of the best hardware platforms is Arduino. It is very popular because it is completely Open Source, huge documentation, cross platform IDE and has a large community. Anyone can visit Arduino's official website (https://www.arduino.cc/) to start with. All the documentation, hardware details, tutorials are readily available in the website. Raspberry-Pi is another very popular hardware development platform based on ARM processor. It is also an open source and has a large community base. For high performance computing, Beaglebone is mostly preferred. These are ARM Cortex-A8 based development board with 2 GB onboard Flash memory. Both Raspberry-Pi and Beaglebone has onboard Ethernet connectivity option.

For NFC communication, the most popular IC is PN532 fabricated by NXP. Many companies such as Adafruit INC, Sparkfun, and Seeedstudio have manufactured readymade NFC module based on PN532 IC. This module is very easy to use with any Microcontroller board. They either use Serial Communication or Serial Peripheral Interface (SPI) protocol communicating with the host microcontroller. There are also other NFC transceiver ICs such as TRF7970A by Texas instrument and ST95HF by STMicroelectronics, and so on.

Though NFC is widely used for different application, Payment remains the main use of NFC. As the need of market is increasing rapidly, the Payment-related technologies are growing exponentially. In the next part, the usefulness of NFC in payment architecture in discussed in detail.

9.4 NFC-Based Payment Systems

Due to its very less communication range, NFC is preferred for payment-related needs. In 1995, Seoul Bus Transport first implemented contactless payment technology. In 2007, Barclay card first embedded NFC in its payment card to support contactless payment. Since then, NFC development took a huge leap, and contactless transaction is increasing every day. Many researchers are exploring the use cases of NFC for different payment-related issues. Figure 9.5 shows a typical NFC-based payment system architecture.

NFC provides several advantages over other payment options like

1. *Ease of Use*: NFC is very easy and intuitive to use. Users can pay just by using a single tap. NFC can be also integrated with mobile wallet based app that provides more convenience to customers in digital payment domain.

2. *Versatility*: As NFC is embedded on Smartphone, it can be used I a wide range of application such as POS and peer-to-peer Payments, Ticketing, Reward system access control, and so on. So NFC is one of the most versatile technologies of present day.

3. *Security*: NFC provides hardware level security using inbuilt/external secure element. NFC-based payment cards are most secure. And as NFC is contactless, retailers do not get physical card information. And as NFC can be integrated with digital payment easily, NFC-based transactions can be made more secure using digital cryptography.

Google integrated NFC in its Nexus-S device in 2011, and launched their first Digital Wallet known as Google Wallet that uses Host Card Emulation. After that due to a tremendous development in Mobile Computing platform, different Mobile wallets are coming up in the market, most of which are based on NFC or Contactless technology.

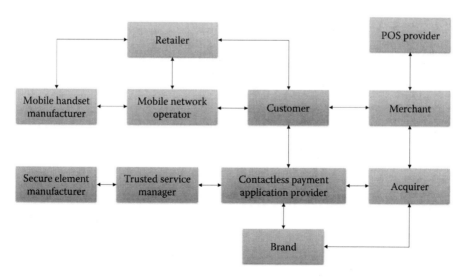

FIGURE 9.5
NFC card emulation mode.

Monteiro et al. (2012) have made a mobile application, which can transfer mobile phone credit between two phones using NFC and Bluetooth. In the research paper (Mainetti et al., 2012), the authors have devised a mechanism, by which all the credit card information are securely stored in the mobile device using the secure element. Apple and Samsung both released their payment systems in 2014 and 2015, respectively. Both of them are highly secure using Biometric Authentication and Tokenization. Apple Pay can only work on NFC-enabled Point of Sale terminals, in which Samsung pay works in traditional Magnetic Stripe based Point of Sale terminals using a technology called Magnetic Secure Transmission (Wallner, 2014). Google also launched Android pay in 2015 using Biometric authentication and Tokenization.

Researchers are predicting that the pervasive computing is field of next generation computing in which IoT plays a vital role. So soon people started to develop payment solutions for IoT-based products. The areas such as Public Transport ticketing, Smart University, Mobile Commerce, Logistics, peer-to-peer payments, and so on need a flawless, robust, and scalable payment architecture. NFC can be a crucial element in this context. The main advantage of NFC is its very less power consumption and very small antenna size, which makes it perfect to be mounted on several things that need to be connected within the IoT network.

In the article "How to Integrate Payments in IoT Devices?" (2015), it is found that several fintech startups are focusing on IoT-based payments to provide innovative payment solution to the consumer using NFC, Mobile Apps, Sensors, and so on. One of the most payment companies, PayPal has made developed payment app solely for IoT development, which is featured on Samsung's wearable devices. Pebble is another very popular wearable. Different companies are also launching payment app based on Pebble. Semiconductor companies are also embedding NFC functionalities in the core of their processor. For example, Qualcomm is embedding NXP's NFC and Embedded Secure Element in some of their Snapdragon-based processors. Intel Corporation & Ingenico Group are jointly launching a tablet that will serve the purpose of EMV and NFC payments. Card Payment Network Processing companies such as VISA, American Express are also investing on IoT-based product for payment.

When Apple launched Apple Pay in 2014, they also introduced Apple Pay for their wearable Apple Watch. Using Apple Watch users can have a seamless payment functionality across contactless point of sale terminals.

NFC is also introduced in Automated Vending machine. USA technologies introduced NFC-based contactless payment on their vending machine in about 100,000 locations across the United States.

In the article "Gemalto and Visa roll out contactless payment wristbands at Eurovision," Security giant Gemalto also launched contactless payment wristband that can be used at any VISA contactless terminal around the globe (Bodden, 2016b). In another article by the same author, it is found that Chinese Ecommerce giant Alibaba, in partnership with SAIC Motor Corp, has introduced *internet car*. In this car, the driver will be able to make a purchase using company's mobile payment platform Alipay (Bodden, 2016c). In 2016, prototype of an NFC-based digital card is invented by Alak et al. that serves all kinds of Payment needs as well as Multiple ID Card Virtualization. The card also uses Tokenization and Biometric Security (Majumder et al., 2016). This card is based on NFC peer-to-peer communication mode. Researchers have shown a very interesting way to make a transaction without using any smartphone. One digital card sends the transaction details to other

TABLE 9.2

Different Stack Holders in NFC Ecosystem

Stake Holder	Role
NFC Chip Manufacturer	They design the silicon chip for NFC communication. They may also include secure element in the NFC chip, for example, NXP, TI, and so on
Secure Element Manufacturer	They are behind the designing of secure elements by which every transaction is validated. The secure element may be hardware or Cloud based service. For example, NXP and Apple manufactures secure element chip whereas Google provides Cloud-based Secure element solution for their Google Wallet service
Mobile Phone Manufacturer	They manufactures the mobile handsets for consumer and integrate NFC payment solution into their hardware and software design, for example, Samsung, Apple, and so on
Trusted Service Manager (TSM)	One of the most important role players. As Payment services are of heterogeneous structure, it is very important to bring all different users, mobile networks, handset manufacturers under a common platform so that the system works seamlessly. TSM plays a very crucial role by combining all under one place
Mobile OS Provider	They provide Operating system to run over the mobile hardware, such as Google's Android, Apple's iOS, Microsoft's Windows, and so on
Application Developers	Application developers are behind the top notch application for a mobile device. The applications should be intuitive, easy to use, and should provide great security

Card using NFC peer-to-peer communication protocol. These details are then forwarded to cloud server using GSM and a message server. Server processes the data and makes the transaction happen. The card can also be used in any POS terminal using Host Card Emulation feature, though it is not required as because one card can alone work as a POS terminal. The digital card is also biometric protected which increases the security of the system.

As discussed in the previous section, NFC may be a very vital role player for *On the Go* ticketing services. It provides several useful features than traditional ticketing services. Using automated machines, tickets can be purchased and also validated. User now has the freedom to get the tickets as per their convenience and can also share tickets over Internet to their friend or relatives. All the tickets are purchased from a central database. The administrators can easily get the statistics from the central database. Based on user demands, personalized advertising can be directly sent to the user's mobile.

In any NFC-based payment system, there are several role players. For proper functioning of the payment network, it is very essential to perform their role efficiently. Ghosh et al. (2015) discussed the role of the stake holders in their literature. The stake holders for an NFC-based payment network are listed in Table 9.2.

Figure 9.6 describes the whole NFC ecosystem. For Development of IoT-based product, first consideration is that the module must have connectivity peripheral inbuilt. The device should be small such that it can be mounted anything to get information from it. Other than Arduino, Raspberry-Pi and Beagle bone, any companies are manufacturing development board customized for IoT-based application only. Some of them are listed below:

- *Panstamp*: These are very small development boards based on Atmega328p MCU and provides RF connectivity via Texas Instruments CC1101 RF interface.
- *RFduino*: These are fingertip sized mini development boards based on ARM Cortex-M0 processor and provide on-board Bluetooth connectivity.

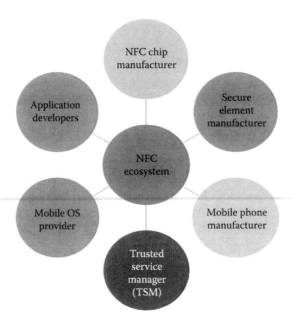

FIGURE 9.6
Stack holder in NFC ecosystem.

- *Pinoccio*: These are microcontroller development boards based on Atmel ATmega256RFR2. They provide the provision to use them for mesh networking as well as web connectivity.
- *mbed*: These are based on ARM Cortex M3 processor and provide Ethernet connectivity.
- *Wi-Go Module*: These modules are based on Freescale KL25Z (ARM Cortex M0+) processor. They also have many sensors inbuilt on the board such as Accelerometer, Magnetometer, Altimeter, and Ambient Light Sensor. These modules provide Wi-Fi connectivity and many more.

All these modules are perfect to kick off any IoT-based project. For any NFC-based project, it has to be verified that whether the board has the provision for either serial connection or SPI connection. For NFC peer-to-peer communication mode, it is necessary to use SPI protocol due to constraints in the software stack. To use the peer-to-peer communication, two Protocols have to be followed. One is SNEP, and the other is LLCP. The description of both can be found from the NFC Forum's website. Researchers and NFC Module manufacturing companies have written LLCP and NDEF software stack for popular development environment such as Arduino and Raspberry-Pi. With minor modifications, they can be run in any environment.

For processing and visualizing the data from different nodes, one can use the standard web development languages such as HTML, PHP, and JavaScript. But nowadays, there are many readymade solutions available, which are efficient and reliable, such as Thing speak, IoT Toolkit, Berg Cloud, and Electric Imp. They are also very easy to learn and implement in the real world.

9.5 Common Threats and Vulnerabilities of NFC-Based Payment System and the Process to Eliminate Them

The fact that criminals are trying to hack mobile phones for fraudulent transaction in unquestionable. As the technology is becoming harder to crack, the attacks are also becoming more refined. To make any payment system more sustainable, it is very important to understand the current and possible future threats and vulnerabilities and devise a full proof mechanism to reduce down the chance of them.

In a Whitepaper by Mobeyforum (2013), they pointed out the present vulnerabilities of NFC-based payment system. They are as follows:

- *Untrusted Communication Channel*: As it is impossible to authenticate both the POS and NFC devices in an unprotected channel, there is a possibility of Identity Hijacking.

- *Denial of Service in the Physical Layer of Communication*: If a huge amount of data can be sent from any device using same 13.56 MHz frequency, then it may create a Denial of Service at the communication channel.

- *Denial of Service in Payment Cards*: An attacker can also hack the POS terminal causing it to exceed the limit of failed transaction.

- *No Encryption in Communications*: As there is no encryption in the channel between the NFC device and the POS, with proper hardware any attacker can intercept in the communication.

The common threats in NFC-based payment systems are listed below:

- *Eavesdropping*: With proper antenna if any attacker intercepts the message in the channel between the NFC device and the POS terminal, he may get sensitive financial information. But the attacker's antenna should be very near to the NFC device for successful eavesdropping.

- *Data Corruption*: Using same frequency antenna, at the time of data transmission, attacker may insert or modify the data sent over NFC.

- *Phishing*: If the NFC tags are replaced by any attacker, then instead of getting useful information, users may be forced to visit malicious websites. The user may also download malicious software or virus in their device.

- *Card Cloning and Skimming*: Another common mode of attack in contactless systems. Using a hidden electronic device called skimmer, attackers try to get the card information.

Researches are going on to secure the NFC-based transactions. Here are some common practices recommended by NFC Forum.

- *Eavesdropping*: As NFC has very short communication range (<3 cm), it is very difficult to plot an eavesdropping attack. To make it even more secure, it is better to use any standard encryption protocol to secure the channel between the POS and NFC device.

- *Data Modification*: NFC devices have the ability to check for RF signals in the channel. So if a device is programmed to check for an unknown RF signal, it can sense data modification attack. This attack can be also reduced by securing the communication channel.
- *Man in the Middle*: Same as eavesdropping, the chance of this type of attack is also very less. Man in the middle attack can be reduced by using Half Duplex communication mode. It means, only one device will transmit data one time. The devices may also detect any disturbance in the channel.

It can be seen that from the above discussion, the main target or objective of designing security measure for NFC devices is securing the channel. Researchers are developing new encryption algorithms based on hash Function. In the literature (Chang et al., 2011c), the researchers have developed a method by using XOR operation and one-way hash functions using which many drawbacks of RFID communication such as Replay attack, DoS attack, and so on are addressed. The authors also have theoretically proved their claims. Sun and Zhao (2012) found some loop holes in the previous literature and proposed a new scheme, by which the loop holes of the previous paper are properly overcome. Zhuang et al. (2014) improvised previous algorithm and applied for NFC Ecosystem by optimizing the power consumption, reducing the errors, and optimizing the storage usage. Urien (2014) proposes a new model to use NFC P2P model for Internet of Things. This model is basically a TLS security layer written over NFC Forum's LLCP which provides data integrity.

9.6 Conclusion

One of the oldest relationships that exist from the evolution of the earth is between the Payer and Payee. The concept of exchanging goods came into picture with the need basic commodities in daily life. Soon it turned to the concept of payment that provides high priced goods such as precious metals, gems, and so on. With the steady development of technologies and human needs, the payment methods changed enormously. With the progress of the society and technology, people changed the means of payment. In the recent years, with the adaption of POS payments, Contactless payments, Cloud-based payments, the growth of digital money transfer is increasing exponentially. Continuous progress in technologies and quality of life, has made a strong impression on the development on payment techniques. After the evolution of NFC technology, contactless payment has recently got a huge attention because of its short-range conducive nature. Again, as mobile computing took a great leap due to enormous development of smartphone platform, companies such as Google, Samsung, and Apple have embedded NFC in smartphone to provide *On the Go* payment eliminating the need of payment cards. The digital payments are becoming more secure with the invention of technologies such as biometric security, 3D secure algorithm, Tokenization, and so on. In all cases, the customers opted for a solution, which is secure and more than that, which is convenient for the user. Big IT giants such as Apple, Google, and Samsung are continuously trying to be a market leader in the digital payment domain, by embedding more features in their flagship smartphones. People are also slowly adapting them in countries such as the United States, United Kingdom, and so on. For example, 20% of users that uses iPhone have tried

Apple Pay in their daily life. Though Developing countries are still far behind from the first world country, they are gradually adapting digital payment means as services such as, Mobile Wallets, Electronic Fund transfer, and so on.

In this chapter, we have mainly focused on the pros and cons of NFC technology and why we must choose it for IoT universe, mainly in e-payment infrastructure.

References

A Security Analysis of NFC Implementation in the Mobile Proximity Payments Environment (2013). Available: http://www.mobeyforum.org/w/wp-content/uploads/NFCSecurityAnalysis_Part1_FINAL1.pdf

Amit, How to Integrate Payments in IoT Devices? (2015). Avaliable: https://letstalkpayments.com/how-to-integrate-payments-in-iot-devices/

Atzori, L., Iera, A., and Morabito, G. (2010). The internet of things: A survey. *Computer networks*, 54(15), 2787–2805.

Bodden, R. (2016a). US Smart City Challenge adds NXP's RFID tagging solutions NFC World. Available: https://www.nfcworld.com/2016/05/18/344832/us-smart-city-challenge-adds-nxps-rfid-tagging-solutions/

Bodden, R. (2016b). Gemalto and Visa roll out contactless payment wristbands at Eurovision. Available: https://www.nfcworld.com/2016/05/10/344604/gemalto-visa-roll-contactless-payment-wristbands-eurovision/

Bodden, R. (2016c). Alibaba Unveils Connected Car for Alipay Mobile Payments in China. Available: https://www.nfcworld.com/2016/07/19/346300/alibaba-unveils-connected-car-alipay-mobile-payments/

Bueno-Delgado, M. V., Pavón-Marino, P., De-Gea-García, A., and Dolón-García, A. (2012b). The Smart University experience: A NFC-based ubiquitous environment, *Sixth International Conference on Innovative Mobile and Internet Services in Ubiquitous Computing* (IMIS), 799–804.

Chang, Y. F., Lin, S. C., and Chang, P. Y. (2011c). A location-privacy-protected RFID authentication scheme. *IEEE International Conference on Communications* (ICC), pp. 1–4.

Freudenthal, E., Herrera, D., Kautz, F., Natividad, C., Ogrey, A., Sipla, J., Sosa, A., Betancourt, C., and Estevez, L. (2007a). November. Suitability of NFC for medical device communication and power delivery. *Engineering in Medicine and Biology Workshop, 2007 IEEE Dallas*, pp. 51–54.

Ghosh, S., Goswami, J., Kumar, A., and Majumder, A. (2015). Issues in NFC as a form of contactless communication: A comprehensive survey. In *Smart Technologies and Management for Computing, Communication, Controls, Energy and Materials* (ICSTM), pp. 245–252.

Gubbi, J., Buyya, R., Marusic, S., and Palaniswami, M. (2013). Internet of Things (IoT): A vision, architectural elements, and future directions. *Future Generation Computer Systems*, 29(7), 1645–1660.

Hunter, P. (2016). IoT + NFC: Four reasons why IoT needs NFC. Available: http://internetofthingsagenda.techtarget.com/blog/IoT-Agenda/IoT-NFC-Four-reasons-why-IoT-needs-NFC

Mainetti, L., PatronoI, L., and Vergallo, R. (2012). IDA-Pay: An innovative micro-payment system based on NFC technology for Android mobile devices, *20th International Conference on Software, Telecommunications and Computer Networks* (SoftCOM), pp. 1–6.

Majumder, A., Bhattacharyya, B. K., Ghosh, S., and Goswami, J. (2016). A digital card serving identity and payment purpose, filed as *Indian Patent at IPO Kolkata, Application No. 201631004666*.

Miraz, G. M., Ruiz, I. L., and Nieto, M. Á. G. (2009). How NFC can be used for the compliance of European higher education area guidelines in European universities. *First International Workshop on Near Field Communication*, pp. 3–8.

Monteiro, D. M., Rodrigues, J. J. P. C., and Lloret, J. (2012). A secure NFC application for credit transfer among mobile phones. *International Conference on Computer, Information and Telecommunication Systems* (*CITS*), pp. 1–5.

Morak, J., Kollmann, A., Hayn, D., Kastner, P., Humer, G., and Schreier, G. (2007b). Improving tele-monitoring of heart failure patients with NFC technology. In *Proceedings of the Fifth IASTED International Conference: Biomedical Engineering*, pp. 258–261.

Nasution, S. M., Husni, E. M., and Wuryandari, A. I. (2012a). Prototype of train ticketing application using near field communication (NFC) technology on android device. *International Conference on System Engineering and Technology* (*ICSET*), pp. 1–6.

NFC Everywhere (2016). Available: www.nxp.com/documents/brochure/939775017634.pdf

NFC Forum (2016). Simplifying IoT: Connecting, commissioning, and controlling with near field communication (NFC) NFC Makes the Smart Home a Reality. Available: http://nfc-forum. org/wp-content/uploads/2016/06/NFC_Forum_IoT_White_Paper_-v05.pdf

NFC Forum (2011). NFC in public transport. At http://nfc-forum.org/wp-content/uploads/2013/12/ NFC-in-Public-Transport.pdf

NXP Semiconductors (2014). NFC for embedded applications: Your critical link for the internet of things, Available: https://cache.nxp.com/documents/brochure/75017587.pdf

Prinz, A., Menschner, P., Altmann, M., and Leimeister, J. M. (2011). inSERT—An NFC-based self reporting questionnaire for patients with impaired fine motor skills. *3rd International Workshop on Near Field Communication* (*NFC*), pp. 26–31.

Sankarananrayanan, S. and Wani, S. M. A. (2014). NFC enabled intelligent hospital appointment and medication scheduling. *2nd International Conference on Information and Communication Technology* (*ICoICT*), pp. 24–29.

Sun, X. and Zhao, Z. (2012). An improved location-privacy-protected RFID authentication scheme. *Journal of Hangzhou Dianzi University*, 32(5), 57–60.

Texus Instrument NFC in the Internet of Things. https://e2e.ti.com/blogs_/b/connecting_ wirelessly/archive/2013/09/17/nfc-in-the-internet-of-things-iot

The Internet of Things (2005). *ITU Internet Reports*. Avaliable: https://www.itu.int/net/wsis/tunis/ newsroom/stats/The-Internet-of-Things-2005.pdf

Urien, P. (2014). LLCPS: A new secure model for internet of things services based on the NFC P2P model. *IEEE Ninth International Conference on Intelligent Sensors, Sensor Networks and Information Processing* (*ISSNIP*), pp. 1–6.

Wallner, G. (2014). System and method for a baseband nearfield magnetic stripe data transmitter. *U.S. Patent 8,628,012.*

Walton, C. A. (1983). U.S. Patent No. 4,384,288. Washington, DC: *U.S. Patent and Trademark Office.*

Zhuang, Z. J., Zhang, J., and Geng, W. D. (2014). Analysis and optimization to an NFC security authentication algorithm based on hash functions. *International Conference on Wireless Communication and Sensor Network* (*WCSN*), pp. 240–245.

10

Trust Management in IoT

Avani Sharma, Pilli Emmanuel Shubhakar, and Arka Prokash Mazumdar

CONTENTS

10.1 Introduction

The advent of new technologies and innovation in networking enabled communication to be wireless, pervasive, and ubiquitous across the globe. A new generation of World Wide Web, that is, Web 3.0 (Hendler, 2009), is bringing the *intelligent web* to reality by providing Internet-based services using the semantic web, microformats, machine learning, and so on. With an increasing number of Internet users, research is being focused on developing new technologies that can fulfill user's demands with their satisfaction level. This has led to the evolution of Internet of Things (IoT) (Atzori et al., 2010) which is an emerging paradigm that facilitates communication between uniquely identifiable heterogeneous physical objects seamlessly through the Internet. These physical objects, called IoT devices, have embedded intelligence and decision-making power which makes them smart enough to communicate with each other to get data, information, or services. Deployment of IoT for real-world application suffers various challenges such as addressing the huge number of devices, analyzing heterogeneous information, open communication channel, and so on (Borgia, 2014; Stankovic, 2014). Security is another critical issue that adversely affects the performance of IoT applications (Sicari et al., 2015). Characteristics of IoT environment such as wireless channel, resource constraint devices, limited bandwidth, Internet connectivity, and so on make it vulnerable toward various kinds of attacks. Managing trust is an important aspect of security in uncertain IoT environment, in which devices may behave maliciously with time, to interrupt the services and performance. A trusted environment, in which devices can share, process, and access information without bothering about security and privacy, is needed to effectively utilize IoT services. This chapter is intended to introduce the readers about the importance of trust while building a secure IoT system. Describing the major characteristics and challenges for IoT environment, this section gives a glance on the problem of *Privacy, Security,* and *Trust* in IoT environment. Section 10.2 presents detail description about *trust* covering various properties of *trust* and its management. In the next Section 10.3, various phases associated with the trust-management system (TMS) are explained and commonly used computation models for the trust evaluation are discussed. Finally, Section 10.4 introduces various attacks which are possible to be launched against TMS. These attacks continue to thwart performance of IoT even after embedding trust-management solutions.

10.1.1 Internet of Things: Introduction, Characteristics, and Challenges

IoT was first introduced by Kevin Ashton of Massachusetts Institute of Technology at Auto-ID Center in the year of 1999 with the integration of radio-frequency identification with sensor technology. Today, IoT has become the foremost technology to achieve a smart vision of the world. According to Gartner's Hype Cycle 2015, IoT took place on the top of emerging IT technologies. Cisco Internet Business Solutions Group forecasts connectivity of 50 billion devices to the Internet by the end of 2020 (Evans, 2011).

Various organizations define IoT but no standard definition of IoT exists in literature that can be used globally. Cisco (Evans, 2011) states that "IoT is simply the point in time when more things or objects were connected to the Internet than people." According to ITU vision of IoT (Pena and Ismael, 2005), "From anytime, anyplace connectivity for anyone, we will now have connectivity for anything." In support of connecting various heterogeneous devices to the Internet protocol (IP) based network, IPSO (IP for smart Objects) Alliance forum (Dunkels and Vasseur, 2008) was formed in September 2008. The forum brings the importance of wise adaptation in IP together with IEEE 802.15.4 for full deployment of IoT across the globe. Collaborative efforts have been made by Internet Engineering Task Force and IEEE 802.15.4 working group to standardized IoT. They provide a paradigm shift from traditional IP network-based protocol stack to the smart protocol stack with power efficiency that can run on low-power IoT devices. Figure 10.1 presents layered architecture of IoT with various enhanced communication protocols and standards.

The major driving force behind the success of IoT is the integration of various technologies such as wireless sensor networks, radio-frequency identification, IPv6 & enhanced communication protocol, cloud computing, nanotechnology, identification and tracking (Tan and Koo, 2014). Some of the important characteristics of IoT network inherited from the above technologies are heterogeneous and light-weight devices, low power consumption, limited computation capacity, limited memory, battery operated, embedded intelligence, decision making, and distributed computing. Deploying IoT for real world applications cope up various challenges that need to be addressed for effective realization of IoT (Borgia, 2014; Stankovic, 2014). Some of the major challenges are as follows:

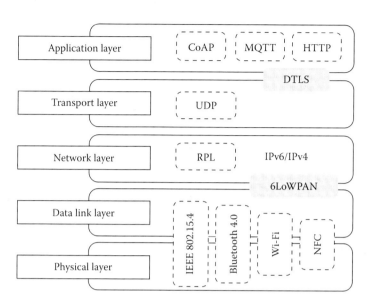

FIGURE 10.1
Layered architecture for Internet of Things.

- *Addressing and identification*: IoT is a collection of large number of devices. To cope up with the existing IP-based Internet functionality, a larger address space is required for IoT devices which uniquely identify these devices. IPv6 overcomes this issue but present challenge is how to get these addresses and how to integrate addresses with the existing solution is a challenge in IoT.

- *Architecture*: A generic architecture to permit easy connectivity, control, and communication between heterogeneous devices are required. Designing such a standard architecture with heterogeneous devices is a nontrivial task.

- *Creating knowledge*: Huge amount of data is generated by heterogeneous IoT devices as they are continuously sensing information. Building knowledge from this uncertain data to create wisdom is a key issue.

- *Robustness*: Dynamic behavior of IoT environment makes it susceptible toward asynchronous activities like clock synchronization, which is undesirable. A robust design which overcomes issue of reconfiguration with the dynamically changing situation is the need of ours.

- *Security*: Collaboration of different technologies in IoT environment makes it vulnerable toward various security issues insured with underlying technologies. An unauthenticated or malicious device in the system can compromise system performance by launching various types of attacks.

- *Openness*: Most of the communication between IoT devices is performed wirelessly. Openness in IoT refers to the freedom of device to configure itself with network infrastructure. IoT devices do not need standard protocols stacks to join the network. This openness introduces uncertainty in the system.

- *Interoperability*: Supporting operability between heterogeneous devices is a major task that should be handled by IoT infrastructure carefully. Cross platform design of protocol is required to support heterogeneous communication.

10.1.2 Security, Privacy, and Trust in IoT

Security is a measure to define system's reliability, robustness, and effectiveness. It is defined as the "quality of remaining secure from the risk" (Merriam-Webster, 2007). For a system to be secure, five security measures are defined: Confidentiality, Integrity, Availability, Authentication, and Authorization (ISO, 2005). One important aspect of security is privacy (Sicari et al., 2015). *Privacy* deals with data protection and confidentiality of user's information. In open environment like IoT, where information access can be done using the Internet, the probability of the disclosure of a user's personal information among the others is higher than traditional information systems. *Privacy* solutions are generally concerned with whom and when information must be released or disclosed. Mechanisms like cryptography and access control are used to deal with the problem of security and privacy. These mechanisms are termed *Hard Security* by Rasmusson and Janssen (1996) in which a security policy is always assumed to exist for authorization of entities in the system and to give them right for accessing the information. However, existing security mechanisms do not ensure the reliability of a system in the presence of malicious activities performed by an insider entity due to its behavioral change. An authenticated entity can be compromised to disrupt the system performance. Even with an embedding security solution, a system can behave in an untrustworthy manner with respect to certain context.

Trust (Eder et al., 2013; Josang, 2007) solves this problem by maintaining legitimacy between the entities by analyzing their behavior with respect to particular context. It defines a measure of confidence about the expected behavior of entities. The concept of Trust comes under the category of *Soft Security* (Rasmusson and Janssen, 1996). Mechanisms to implement soft security follow ethical norms of a system for assessing the behavior. They basically deal with maintenance of reputation that defines general perception about behavior of an entity seen globally in a system.

In IoT, traditional solutions for security and privacy such as cryptography, digital signature, key management, cannot be directly applied because of resource constraint nature of IoT devices. Involvement of different protocol stacks, technologies, and standards need a unified approach that can be embedded with a heterogeneous environment. Moreover, scalability is essentially refined by a huge number of interconnected heterogeneous devices. There is a need to embed advanced security mechanism like *trust* and *reputation* (Yan et al., 2014) that can capture dynamic behavior of a system.

10.1.3 Motivation

IoT facilitates various applications where information or services can be accessed ubiquitously through the Internet. A lot of work has been proposed to optimize and use these applications in the best possible way. However, vulnerability assessment, which is the foremost aspect that should have been considered for these latest technologies, is still far from the required progress. Malicious devices across the globe are able to compromise the security primitives of IoT. Behavior-based analysis of devices is required that can predict the device performance over the time. A trustworthy system is needed to prevent from unwanted activities conducted by malicious devices. *Trust* is an important aspect of security which encompasses security and privacy with goodness, strength, ability, and availability. Trust management provides analysis on the dynamic behavior of devices considering their past behavior, reputation, or recommendation received from the other's opinion. After analyzing the importance of trust management in IoT, following conclusions can be drawn:

- Trust management facilitates reliable data fusion and mining, qualified service in heterogeneous IoT environment.
- Trust management is required in various decision-making processes like intrusion detection, management of keys, authentication, and so on.
- In uncertain conditions, where a device can leave or join the network at any moment of time, trust management is useful to make the decision with the limited amount of knowledge available.
- Trust management helps users to overcome the risk engaged in acceptance and consumption of IoT services.
- Trust management accelerates access control by isolating misbehaving devices in the system.

10.2 Trust: Properties and Management

The concept of trust is not confined to a single domain. Trust is a multidomain, multidimensional concept having importance in every aspect of life like social science, economics, philosophy, psychology, and so on. (McKnight and Chervany, 1996). Trust defines reliance about the behavior of an entity. Cook (2003) defines trust as the degree of subjective belief about the behaviors of a particular entity. Various definitions for trust have been proposed in the literature (Castelfranchi and Falcone, 2000; Cook, 2003; Gambetta, 2000). These are:

> Trust is an attitude that we have toward people whom we hope will be trustworthy, where trustworthiness is a property, not an attitude (Castelfranchi and Falcone, 2000).
> Trust basically is a mental state, a complex mental attitude of an agent 'x' towards another agent 'y' about the behaviour/action 'a' relevant for the result (goal) 'g' (Cook, 2003).
> Trust (or, symmetrically, distrust) is a particular level of the subjective probability with which an agent assesses that another agent or group of agents will perform a particular action, both before he can monitor such action (or independently of his capacity ever to be able to monitor it) and in a context in which it affects his own action (Gambetta, 2000).

Wanita et al. (2013) present importance of trust in a social scenario which derives various facts about the theory of trust to be used in domain of communication and networking. IoT is all about making ubiquitous network of physical devices where different devices can communicate with others to share information or to request services from other devices anywhere and at any time. Trust between the devices is required to create a reliable environment where various devices can share their thoughts, opinions, and experiences without being bothered about the privacy of their information.

10.2.1 Properties of Trust

Adaptation of trust in IoT environment should fulfill certain properties. These properties, given in Figure 10.2, define the necessary condition on trust management (Cho et al., 2011; Lopez et al., 2010; Yu et al., 2007).

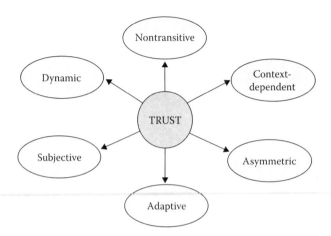

FIGURE 10.2
Properties of trust.

- *Asymmetric*: In heterogeneous IoT environment, in which each device with different capacity (computation power and energy) plays different roles, trust remains asymmetric (Rasmusson and Janssen, 1996; Yan et al., 2014). A device A, trusting another device B, may not receive the same level of trust from B. For example, a provider of a service can trust on requester of service but the same may not be true because opinion, perception, and expectation of requester may vary.

- *Adaptive*: For a dynamic environment like IoT where devices can quit or join the network at any moment of time, trust should follow adaptability with the environment. To capture dynamicity of IoT environment, trust must be expressed in continuous form with regular update in its value so that uncertainty in the network can be mapped to current state (Wanita et al., 2013).

- *Dynamic*: Value of trust may be changed according to a new observation. It may increase or decrease with time as experienced by the devices. With each interaction, opinions of devices may change, which results in change in trust value for the device.

- *Context dependent*: Scope of trust is context specific. For example, a device A, trusting on other device B which is providing weather information, will not have the same degree of trust on the other device C providing temperature or pressure information (Lopez et al., 2010; Yu et al., 2007).

- *Nontransitive*: Trust follows nontransitivity. The principle of nontransitivity says that if a device A trusts on a device B and the device B trusts on another device C, then it is not necessary that the device A will trust the device C (Cho et al., 2011; Lopez et al., 2010).

- *Subjective*: Trust is subjective with respect to the context in which it is applied (Lopez et al., 2010; Yan et al., 2014). Opinions of different trustors (evaluator of trust) may have the different level of trust on the subject's behavior. Trustworthiness of a subject not only depends on its behavior but also on the perception of trustors about subject's behavior.

10.2.2 Management of Trust

Trust management integrates individual's views about behavior of an entity with the global opinions seen by the other members of a system. These global opinions are called *reputation*. Blaze et al. (1996) were first toward bringing the concept of trust for security computation. He introduced the term *Trust Management* and highlighted its importance as a separate component for building secure systems. According to Blaze, management of trust in a system opens the way to interpret and specify security policies, relationships, and credentials. For IoT, the importance of trust management is not confined to a single layer of communication stack (Hoffman et al., 2009; Lacuesta et al., 2012; Yan et al., 2014). All the three layers: physical, network, and application are subjected to achieve security goals while embedding trust management. These goals are as follows:

- *Physical layer/device layer*: At the physical layer, efficient and scalable identity management (IM) for the devices is required in IoT because of handling numerous heterogeneous devices. These identities can be any real-world identifier or may be assigned by the users. The objective of trust management is to prevent the system from outcomes of acquiring faked or multiple identities.
- *Network layer*: At the network layer, solutions for reliable data fusion, transmission, mining, and communication are required. The objective of trust management is to analyze and process huge amount of sensor data generated from heterogeneous devices for the purpose of receiving accuracy and reliability.
- *Application layer*: At the application level, ensuring quality of services of the requested information is a major concern with the expectation *Only me, only now and only here*. Trust management should ensure that requested services should precisely deliver to the right person at a right time without losing the quality. Moreover, preserving the privacy of a user's information according to the expectation and policies of user should be there.

10.3 Trust-Management System

TMS can be represented as a framework of multiple processes which iteratively evaluate, disseminate, and update trust score of an entity on the basis of information gathered to maintain trustworthiness between the entities. Definition of the entity may vary according to application scenario in which they are used. In the case of IoT, an entity is an end device who is involved in the process of communication with the embedded decision-making capacity. Various components are associated with management of trust like experience, knowledge, reputation, recommendation, and so on. Bringing trust to the domain of computation is closely related to the term *reputation* (Eder et al., 2013). Although *Trust* represents subjective view about the trustworthiness of an entity, *reputation* defines general perception about the behavior of an entity computed globally (Dobelt et al., 2012). The generic structure of the TMS involves four phases as shown in Figure 10.3.

10.3.1 Information Gathering

The first phase of TMS deals with defining preliminary requirement to compute trust in a system and acts as input to the process of trust computation. For example, what parameters

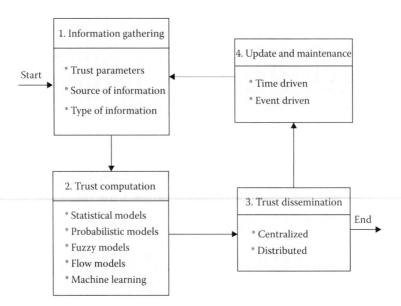

FIGURE 10.3
Phases involved in the designing of trust-management system.

should be chosen to define trust in a system, how the information is collected, and how to represent the information in computational form?

Strategies adopted for TMS may vary according to the application scenario. In a social network, trust can be built by measuring the degree of *friendship* or *community of interest* between the members of a group. Similarly, in some applications like banking or finance, *honesty* and *fairness* play an important role in building trust (Wanita et al., 2013). These qualitative attributes are called trust parameters which can be represented in quantitative forms like binary, discrete, or continuous (Ruan and Durresi, 2016). Before evaluating trust score for an entity, one should be aware of about what trust parameters should be chosen and what mathematical representation should be used to maintain this information in the computational form. To understand the process of trust management, we imply *trustor*, as an entity who is an evaluator of trust, and *trustee*, as an entity for which trust is calculated.

1. *Trust parameters*: Figure 10.4 depicts various trust parameters that can be used to compute trust. Three essential attributes that are used to derive these parameters are knowledge, experience, and reputation (Ruan and Durresi, 2016). Reputation about an entity can be maintained based on the knowledge and experience about the behavior of that entity. General parameters which are involved in the process of defining trust are as follows:

   - *Honesty* refers to belief or loyalty of trustee toward trustor.
   - *Behaviors* defines different patterns of individual with each interaction.
   - *Attitude* represents positive and negative views of an entity.
   - *Friendship* represents state of intimacy with someone.
   - *Fairness* represents quality of treating each entity equally in a reasonable way.
   - *Benevolence* defines desire of an entity for being good or kind toward others.

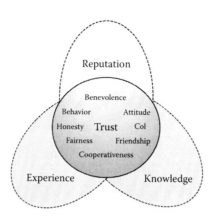

FIGURE 10.4
Trust parameters.

- *Cocooperativeness* represents compliance of trustee with trustor in a social application.
- *Community of interest* represents similar characteristics or capacity of entities in same community.

2. *Source of information*: Collecting raw information for trust computation is a major task in designing of any TMS. There are two measures on the basis of which information can be obtained (Hoffman et al., 2009; Sabater and Sierra, 2005). These are as follows:

- *Direct measure*: A trustor can rely on own observations or findings about the behavior of an entity, like analyzing the number of successful or failed transactions, resource utilization for each transaction, and so on. Such type of information is interpreted as first-hand information.

- *Indirect measure*: Information received from third party or obtained through transitive relation is called second-hand information or recommendation. Recommendation is the most common way to utilize second-hand information where reputation or trust evaluation of an entity depends on global view observed by the other entities in the system.

3. *Scaling of information*: Information collected for trust computation must be represented in a form in which computation can be carried out. Various numeric formats, such as continuous, discrete, and intervals, exist that can be used to represent gathered information in mathematical computation form (Ruan and Durresi, 2016; Wanita et al., 2013). Presenting the information as a discrete value can take any of the following two forms: binary discrete and multinomial discrete. In binary discrete form, information is presented in the binary form of 0's and 1's. Binary values are easy to understand and can easily represent behavior of an entity by using 0 for distrust and 1 for trust. Multinomial discrete values are used when trustor does not have binary opinion about the behavior of an entity. To scale more trustworthiness of an entity, trustor can use different scales of measurement like *High Trust, Trust, Low Trust, Distrust,* and so on. This type of representation provides more accurate information toward the degree of trust of an entity having more opinions. Another way to represent trust information is to use continuous values

which are generally used in the situations where probability density functions (PDFs) are used for input distribution. One more possible representation is to use interval value, in which information is maintained in the form of intervals which are used by most fuzzy models.

10.3.2 Trust Computation

The process of trust computation deals with the selection of an appropriate computational model for trustworthiness evaluation of entities. Various trust computation models exist in literature that can be used to build TMS for IoT. These models take input from the previous phase according to the requirement of the application. On the basis of the approach adopted for output calculation, computation models can be classified as *deterministic* and *probabilistic* (Hoffman et al., 2009). In *deterministic* approach, an output of any computation process can be determined having knowledge of input parameters. Such kinds of approaches are suitable where size of the system is small and computation depends on the accuracy of individual views instead of considering global perspective. Performing simple weighted sum or average of the feedback ratings falls under this category. In *probabilistic* approach, input variables used to derive output possess random values. Computation is performed over the random set of values instead of having discrete input values. These are also called stochastic algorithms where predictive outputs have certain error bounds. Error bounds define approximation in predicted values with an acceptable level of error. Bayesian model (Josang, 2001; Josang and Ismail, 2002) and Belief model (Josang, 2001) are the most commonly used probabilistic models in the domain of trust computation. Some models are designed based on the machine learning algorithms to capture dynamic behavior of entities. Hidden Markov Model (HMM) (Anjum et al., 2015) and Artificial Neural Network (ANN) (Zong et al., 2009) are the examples that use machine learning as a tool to compute trust. The other models discussed in the literature to compute trust are fuzzy models (Chen et al., 2011; Manchala, 1998; Mahalle et al., 2013; Sabater and Sierra, 2002), graph-theory-based models (Theodorakopoulos and Baras, 2006) and flow models (Levien, 2009; Page et al., 1999).

- *Simple statistical models*: The easiest and straight-forward form of computing trust of an entity is to perform simple mathematics over the feedback provided for each transaction like sum of ratings (positive and negative rating) and averaging the rating to measure the degree of trustworthiness. eBay's reputation management (Resnick and Zeckhauser, 2002) deploys such approach to maintaining trust over its service providers and customers. Other commercial sites that use such type of computation are Epinion (Massa and Avesani, 2005) and Amazon (Massa and Avesani, 2007). The model efficiently computes trust score; however, the systems deploying such model lacks in maintaining the global reputation of the entities.

- *Probabilistic models*: Instead of taking the direct discrete value of input, these models take the probability of occurrence of an event. Though various algorithms consider probability as a measure to derive trust score, most commonly used probabilistic models which use probability distribution functions are as follows:

  - *Bayesian models*: In this, the concept of probability and evidence is considered to compute trust score (Josang and Ismail, 2002). Binary values with beta PDF (Falls, 1974) are taken as input for the model. These binary ratings, depicting positive and negative behavior of entities, are used to upgrade trust score. To calculate the posterior probability of an event, Baye's rule is used in which prior probability of occurrence of an event derives predicted posterior probability. Representation of trust score with beta PDF considers tuple of two parameters, α and β, in which α denotes positive behavior and β denotes negative behavior of an entity. The beta PDF, represented by beta$(p|\alpha,\beta)$, can be expressed in the form of gamma function, Γ, as

$$\text{beta}(p|\alpha,\beta) = \frac{\Gamma(\alpha+\beta)}{\Gamma(\alpha)\Gamma(\beta)} p^{\alpha-1}(1-p)^{\beta-1} \quad \text{Where } 0 \le p \le 1, \alpha, \beta > 0$$

The expectation value of beta distribution, which is used as trust score, is given as

$$E(p) = \frac{\alpha}{\alpha+\beta}$$

Using Bayesian model in a system provides effective trust or reputation computation as compared with the simple statistical models because of the inference of prior information in posterior behavior prediction. However, Bayesian model does not give any information about the selection of prior distribution. Moreover, these models consider constant probability distribution over each transaction which restricts their efficiency for deploying dynamic behavior of the system. To overcome this problem, some algorithms use *decay factor* or *forgetting factor* while building trust score (Bao and Chen, 2012; Bao et al., 2013; Chen et al., 2015). Decay factor is a kind of scaling variable which is used to give more weight to recent information as compared with the previous behavior seen.

- *Belief model*: These types of models too are related to the theory of probability, but summation of probabilities seen with the all possible outcomes of an event does not necessarily give 1. Residual probability is treated as *uncertainty* that can be anything depending on the application scenario. For example, one basic idea given by Josang (2001) is to model trust with the integration of three components: *belief*, *disbelief*, and *uncertainty*. A tuple of four parameters (b, d, u, and a) is used to derive opinion about an entity where b represents *belief*, d represents *disbelief*, and u represents *uncertainty* with the summation $b + d + u = 1$. a is called relative atomicity which is used to derive probability expectation value for the opinions as $b + au$. A well-known example of reasoning with uncertainty is Dempster–Shafer Theory (Shafer, 1992).

- *Fuzzy models*: In Fuzzy models, computation of trustworthiness or reputation is done using the fuzzy set of values instead of having deterministic values. Input variables in such type of model have randomly assigned value (Chen et al., 2011; Mahalle et al., 2013). A membership function is used with the observable variables to describe degree of trust in the interval ⟨0, 1⟩. Instead of computing trust using simple statistical calculations, fuzzy logic uses set of rules incorporated with IF-Else conditions. These rules are defined to deal with uncertainty in the network and to perform reasoning with fuzzy measurements. The schemes which fall under this category are reputation model for gregarious societies (REGRET) (Sabater and Sierra, 2002) and the reputation system given by Manchala (1998).

- *Machine learning-based models*: Machine-learning algorithms can be used to analyze future behavior of entities based on analyzing their previous transactions.

Most common machine-learning algorithms to capture the dynamic behavior of an entity are HMM (Anjum et al., 2015) and ANN (Zong et al., 2009). HMM predicts dynamic behavior of an entity using the concept of states. Each interaction of trustor with trustee is associated with some contextual information such as number of successful transactions, number of failure transactions, and delay in receiving services. The output of interaction defines levels of trust rating where each level represents the state of an entity for that transaction. For a HMM (λ), trustor can infer next state of trustee, L, using a sequence of observation $F_m + 1 = F_0 F_1 F_2 ... F_m$ with the probability function, $P(S = L | F_m + 1)$. A forward algorithm is applied to recursively derive this probability. Finally, the trustor is able to decide whether to interact with the target entity having the highest probability based on trust rating. In ANN based algorithms, learning is applied to map input vector into output values. These algorithms can work with any nonlinear relationships between input and output by applying iterative training to data sets.

Linear discriminate analysis (LDA) and decision tree (Liu et al., 2014) are other two most common machine learning-based data analytics techniques which can be used to derive trustee's future behavior based on historical data of past interaction. These two techniques are application dependent, in which interaction between trustor and trustee is described by a set of features and two classes, *successful* and *unsuccessful*, define trustworthiness of entities. In LDA, a trustor after classifying the historical data into *Successful* and *unsuccessful* performs LDA on two groups to obtain a linear classifier. This classifier is then used to estimate classification of potential interaction in successful group. In decision tree algorithms, first a tree is constructed using training data. After that, classification algorithms are applied to the potential transaction by starting from the root of the tree and moving down until a leaf node is found.

- *Flow models*: In these models, trust or reputation is computed iteratively considering the long chain of transitive relationship between the entities. Computation of trust score considers opinions of other entities in the system. Incoming flow and outgoing flow concentrated on an entity is used to update true score of that entity. For example, ranking about an entity can be determined by considering how many of others referring that entity. Application of these models can be seen in Google's page rank (Page et al., 1999) and reputation scheme given by Levien (2009). In Google's page rank, hyperlinks toward the page rise its ranking and hyperlink from the page leads to decrease its rank on the web. Another example is EigenTrust model (Kamvar et al., 2003) used with peer-to-peer networks. In EigenTrust, repeated and iterative aggregation and multiplication of trust score derived along with transitive chain is used to make final trust score.

- *Graph theory models*: In Graph Theory, *centrality* and *distance* are generally used measures to calculate degree of trust for an entity. *Centrality* defines importance of an entity by considering in-degree (transaction received) and out-degree (transaction initiated) on that entity. *Distance* between two entities defines connectivity between them. Most popular methods for graph-based trust measure are Google page rank (Page et al., 1999) and EigenTrust (Kamvar et al., 2003).

- *Discrete models*: Sometime it is better to use human's general perception instead of using any existing computational model to measure degree of trustworthiness of an entity. The situation arises in the systems unable to handle complex information lacking with computation resources. Some heuristic approach can be used to compute trust score similar to look-up table (Carbone et al., 2003; Liau et al., 2003). Disadvantage of using such approaches is that they do not present generalized structure for trust management.

10.3.3 Trust Dissemination

Communicating trust information between the entities of a system can opt following two approaches to design the architecture of a system (Hoffman et al., 2009; Josang, 2001):

- Centralized
- Distributed

1. *Centralized architecture*: In this architecture, a central entity is chosen, who takes in charge of all the computation, storage, and decision making. Information about the behavior of entities is collected by those entities who have some previous interaction with the evaluating entity. A central authority is chosen which takes responsibility of maintaining reputation of all the entities and to make them available for public use. To know the trust score of a device, trustor requests centralized authority which, in turn, will do all the needful and return trust score about the device to the requesting entity. In this approach, trust computation has global opinions instead of having bias value. However, the approach suffers from potential single point of failure. Most common application scenario which implements this architecture is wireless sensor networks in which *sink node* act as base station to perform all the centralized computation and decisions on the behalf of other nodes in the network.

2. *Distributed architecture*: In this type of architecture, there is no central entity for computing and managing trust scores. Instead, each entity contributes toward the process of information exchange and stores opinions about past behavior. A well-suited environment for distributed architecture is peer-to-peer networks in which every node plays the role of both client and server. In the case of IoT, where devices are able to perform computation and decision making on their own without the need of any centralized server, use of distributed architecture is more appropriate compared with the centralized architecture. The advantage of using distributed approach is that no single point of failure affects trust management. At the same time, memory requirement and computation cost per entity increases. Moreover, such approaches have higher convergence time compared with the centralized architecture.

10.3.4 Update and Maintenance

Final stage of TMS is the maintenance of trustworthiness of entities in the network. To maintain recent behavior of the entities, their trust score must be updated and maintained with time. Specially, in distributed and dynamic scenario similar to IoT, in which a device's behavior is not constant throughout the time, computed trust score needs to be updated. Two basic approaches toward updating trust scores are as follows (Guo and Chen, 2015):

1. *Event driven*: Trust of a device can be updated upon encountering an event. In such a scenario, each activity is considered as an event and trust value is updated on each activity. For example, wherever a trustor receives a service from trustee, the trustor may update previously computed trust score according to current behavior seen.

2. *Time driven*: Trust of a device can be updated periodically. Counter-based strategies are adopted to implement such kind of approach. Here, trust score is updated with each time unit instead of waiting for occurrence of an event.

10.4 Attacks against Trust Management

Trust management comprises of managing and building the reputation of entities locally (direct measure) and globally (indirect measure) in the system. In open systems similar to IoT in which devices can join and leave the network at any moment of time, maintaining trust and reputation between heterogeneous devices is a challenging task. Even after embedding trust management strategies, the system can be compromised by insider adversaries. These adversaries can launch the attack either for self-benefit or to perform malicious activity (Koutrouli and Tsalgatidou, 2012). A selfish attacker subverts the system performance with the intention of increasing own benefit, whereas a malicious attacker targets other devices to degrade the system performance. These attackers can work individually or may collude with other devices in the system (Marmol and Perez, 2009). In later category, attacking device forms a group with the other devices to perform a malicious activity and boost the effect of attack on the system performance. One major aspect that affects the process of trust management is IM. Identity belonging to an entity (device) plays an important role while building the reputation of that entity. These identities can be any real-world identity or can be defined by the users. Pseudonym in assigning identities allows attackers to bear multiple identities. If a system is compromised with faked identities, any trust management strategy can be vulnerable toward attacks. Therefore, a system is assumed to embed efficient IM scheme for managing authentication between heterogeneous identities. We classify the attacks (Figure 10.5) into two categories where an attacker can disrupt either reputation management or IM system.

10.4.1 Attacks Based on Reputation Management

Effectiveness of reputation management depends on the information gathered about the behavior of devices using their past history and recommendations received from the other devices in the system. An attacker can behave in inconsistent manner or may provide the dishonest recommendation about a device to maintain false reputation. By showing conflicting behavior, an attacker can spoil the process of information gathering which in turn results in maintaining spurious trust score about the devices. Such type of behavior makes system prone toward Denial of Service attack where system resources can be depleted

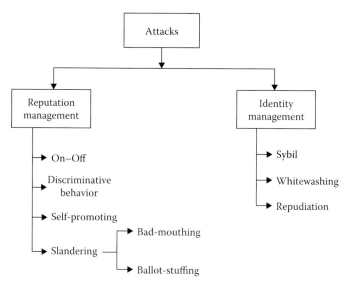

FIGURE 10.5
Attacks classification against trust management.

because of processing and storing false information about the devices. Attacks that come under this category are as follows:

1. *On–off attack*: In this attack, the attacker switches its behavior between good and bad by alternatively providing good or bad services (Koutrouli and Tsalgatidou, 2012). Here, the attacker shows inconsistency in the time domain. As the reputation goes down during the course of bad behavior, the attacker regains its reputation by providing good services. This alternative behavior makes the presence of attacker undetected and difficult to update and maintain effective reputation timely. The attack is also called *traitors*, because of the sudden change in the behavior of an attacker. Most of he TMS are vulnerable toward on–off attack, because of considering the only recent behavior of devices instead of combining past behavior with the current one comprehensively.

   Solutions to deal with on–off attack consider decay factor or forgetting factor while computing trust score (Bao et al., 2013; Chen et al., 2015). This decay factor is used to give more weight for recent information over the past information. Another approach is to consider bad behavior in the form of *distrust* along with the *trust* while building the reputation of a device because only a few bad transactions can drastically affect reputation over the consistent long term good transactions (Sun et al., 2008).

2. *Discriminative behavior attack*: In this attack, the attacker shows inconsistency in user domain instead of targeting time domain (Josang et al., 2007). An attacker may show good behavior toward one group of devices and may misbehave with another small set of devices. Such discrimination leads to increase reputation of attacker significantly by the group who experienced good behavior compared

with decrease in reputation by the group who experienced bad behavior. For example, a service provider can provide good services to most of the requesters and can target one or few requestor with bad services. While building reputation, recommendation received with the targeted requesters (who experienced bad services) has less or no significance as compared with the recommendation received from the large pool of users.

One possible way to deal with these situations is to consider direct measure only for trust computation instead of using global recommendation values. Using global recommendation increases chances of introducing false reputation values in building trust.

3. *Self-promoting attack*: Here, an attacker tries to increase one's own reputation by falsely reporting positive feedbacks or rating (Josang, 2001). This attack infects the systems in which trust computation uses simple mathematics over the positive feedback received. An attacker can be an individual device or may be the group of malicious devices who collude to fabricate reputation value. Main targets for such kind of attacks are the systems which do not implement any authentication scheme to distinguish between legitimate and fabricated feedbacks. Systems, lacking with proper authentication schemes, are susceptible toward acquiring faked identities (Sybil attack). Having multiple identities, an attacker can significantly promote own importance.

Mitigating solutions for these attacks include efficient authentication scheme with a proof of accountability and rate limiter in acquiring multiple identities (Hoffman et al., 2009).

4. *Slandering attack*: In slandering attacks, the attacker or group of attackers provide a false recommendation about other devices in the system to maintain false reputation about them (Sun et al., 2008; Josang et al., 2007). Here, attackers are the recommenders who are involved in process of deriving global reputation of devices. Targets of these attacks are the systems where trust computation is carried out using indirect measure. The attack exploits the transitive property of TMS where trust is computed on the behalf of recommendations received from the third party. Recommenders can show slandering behavior in the following manner:

   • *Bad-mouthing attack*: Attacker has bad mouth about the victim device and provides dishonest recommendations for the legitimated victim device to reduce its reputation. By reducing other's repudiation, attacker increases own chances to be selected as an honest entity in the system.

   • *Ballot-stuffing attack*: Meaning of ballot stuffing is to provide rating more than the required legitimated number. In this, malicious recommender colludes with other dishonest devices in the system and increases their reputation by providing good recommendation about them.

Methods to alleviate the effect of slandering attacks include statistical analysis on the recommendations received from multiple recommenders such as similarity-based filtering. Moreover, instead of limiting number of recommenders, aggregate the recommendations from a large pool so that the effect of false recommendation can be diminished (Hoffman et al., 2009; Sun et al., 2008).

10.4.2 Attacks Based on Identity Management

Identity scheme adopted for any system deals with registration and authentication of entities in the system. Attackers can exploit identities either during registration or during the authentication process to spoil system performance. Attacks under these categories are:

1. *Sybil attack*: Attacker can associate with multiple identities. These identities take the blame of being malicious in the system (Hoffman et al., 2009; Marmol and Perez, 2009). With these fake identities, an attacker can perform differently each time reputation is generated or provided. This different behavior creates anonymity in computing reputation value. Sybil attack is a form of the initiator to launch subsequent attacks on reputation management. For example, attacker having fabricated identity can degrade reputation of another device to increase own reputation that results in self-promotion attack. Moreover, providing negative information about an entity results in bad-mouthing attack.

2. *New comer/whitewashing attack*: In such a kind of attack, attacker subverts the system performance for some duration and then after knowing about degradation in owns reputation value, whitewashes its bad history by disappearing from the system (Koutrouli and Tsalgatidou, 2012). The attacker can rejoin the system having a new identity to take a fresh start. Attacker behaves as a new comer in the system with a different identity which allows regaining the same reputation even after abusing the system with the previous identity.

3. *Repudiation attack*: An attacker, providing or receiving recommendations, may deny involvement in such kind of information (Koutrouli and Tsalgatidou, 2012). The attack is against authentication scheme in which the sender of bogus information remains unidentifiable because there is no prove of blaming a device in the network.

To defend against identity-related attacks, a strong authentication scheme is required where recommenders are imposed with proof of their legitimacy while providing recommendations about the devices. Use of digital signature, unique identity, and restriction in identity generation can thwart the process of acquiring multiple identities (PKI Page; Jurca and Faltings 2003).

10.5 Conclusion

Embedding security solutions in IoT environment is of prime concern to get reliable services. Despite using traditional security mechanisms, there is a need to propose some advanced solutions that can be integrated with resource constraint IoT devices while supporting heterogeneity. Trust management is a major aspect toward mitigation of secure issues. Importance of trust and its management for IoT applications to achieve the goals of security and privacy has been presented in this chapter. This chapter intends to describe the design of a TMS while introducing the various computational models that can be used to evaluate trust in IoT. Vulnerability assessment of the process of trust management is to be considered while building robust TMS. This chapter familiarizes the readers with the various kinds of attacks that can be launched to compromise the management of trust.

References

Anjum, B., Rajangam, M., Perros, H., and Fan, W. (2015). Filtering unfair users: A hidden Markov model approach. *Information Systems Security and Privacy*, pp. 147–154.

Atzori, L., Iera, A., and Morabito, G. (2010). The Internet of things: A survey. *Computer Networks*, 54(15), 2787–2805.

Bao, F. and Chen, I. R. (2012). Dynamic trust management for Internet of things applications. *Proceedings of International Workshop on Self-Aware Internet of Things*, pp. 1–6.

Bao, F., Chen, I. R., and Guo, J. (2013). Scalable, adaptive and survivable trust management for community of interest based Internet of Things systems. *International Symposium on Autonomous Decentralized Systems (ISADS)*, pp. 1–7.

Blaze, M., Feigenbaum, J., and Lacy, J. (1996). Decentralized trust management. *IEEE Symposium on Security and Privacy*, pp. 164–173.

Borgia, E. (2014). The Internet of things vision: Key features, applications and open issues. *Computer Communications*, 54, 1–31.

Carbone, M., Nielsen, M., and Sassone, V. (2003). A formal model for trust in dynamic networks. *BRICS Report Series*, 10(4), 54–61.

Castelfranchi, C. and Falcone, R. (2000). Trust is much more than subjective probability: Mental components and sources of trust. *IEEE Hawaii International Conference on System Sciences*, pp. 1–10.

Chen, D., Chang, G., Sun, D., Li, J., Jia, J., and Wang, X. (2011). TRM-IoT: A trust management model based on fuzzy reputation for Internet of things. *Computer Science and Information Systems*, 8(4), 1207–1228.

Chen, I. R., Bao, F., and Guo, J. (2015). Trust-based service management for social Internet of things systems. *IEEE Transactions on Dependable and Secure Computing*, 13(6), 684–696.

Cho, J. H., Swami, A., and Chen, R. (2011). A survey on trust management for mobile ad hoc networks. *IEEE Communications Surveys & Tutorials*, 13(4), 562–583.

Cook, S. (2003). Trust in Society. *Russell Sage Foundation Series on Trust*. Retrieved from https://www.russellsage.org/.

Dobelt, S., Busch, M., and Hochleitner, C. (2012). Defining, understanding, explaining TRUST within the uTRUSTit project. *White Paper*.

Dunkels, A. and Vasseur, J. P. (2008). IP for smart objects. *Internet Protocol for Smart Objects (IPSO) Alliance, White Paper*.

Eder, T., Nachtmann, D., and Schreckling, D. (2013). Trust and reputation in the Internet of things. Retrieved from https://web.sec.uni-passau.de/projects/compose/papers.

Evans, D. (2011). The Internet of things, How the next evolution of the Internet is changing everything. *Cisco Internet Business Solutions Group*, pp. 4–11.

Falls, L. W. (1974). The beta distribution: A statistical model for world cloud cover. *Journal of Geophysical Research*, 79(9), 1261–1264.

Gambetta, D. (2000). Can we trust trust? *Trust: Making and Breaking Cooperative Relations*, Oxford, UK: University of Oxford, pp. 213–237.

Guo, J. and Chen, R. (2015). A classification of trust computation models for service-oriented Internet of things systems. *IEEE International Conference on Services Computing (SCC)*, pp. 324–331.

Hendler, J. (2009). Web 3.0 emerging. *Computer*, 42(1), 111–113.

Hoffman, K., Zage, D., and Nita-Rotaru, C. (2009). A survey of attack and defence techniques for reputation systems. *ACM Computing Surveys (CSUR)*, 42(1), 1.

ISO (2005). Code of practice for information security management. *ISO/IEC IS17799-Information Technology*.

Josang, A. (2001). A logic for uncertain probabilities. *International Journal of Uncertainty, Fuzziness and Knowledge-Based Systems*, 9(3), 279–311.

Josang, A. (2007). Trust and reputation systems. *Foundations of Security Analysis and Design IV*, Berlin Heidelberg: Springer, pp. 209–245.

Josang, A. and Ismail R. (2002). The beta reputation system. *Proceedings of the 15th Bled Electronic Commerce Conference*, Slovenia, Balkans, pp. 2502–2511.

Josang, A., Ismail, R., and Boyd, C. (2007). A survey of trust and reputation systems for online service provision. *Decision Support Systems*, 43(2), 618–644.

Jurca, R. and Faltings, B. (2003). An incentive compatible reputation mechanism. *Proceedings of the IEEE Conference on ECommerce*, pp. 285–292.

Kamvar, S. D., Schlosser, M. T., and Molina, H. G. (2003). The EigenTrust algorithm for reputation management in P2P networks. *International World Wide Web Conference*, pp. 640–651.

Koutrouli, E. and Tsalgatidou, A. (2012). Taxonomy of attacks and defence mechanisms in P2P reputation systems-Lessons for reputation system designers. *Computer Science Review*, 6(2), 47–70.

Lacuesta, R., Navarro, G., Cetina, C., Penalver, L., and Lloret, J. (2012). Internet of things: Where to be is to trust. *EURASIP Journal on Wireless Communications and Networking*, 1, 1–16.

Levien, R. (2009). Attack resistant trust metrics. *Computing with Social Trust*, London: Springer, pp. 121–132.

Liau, C. Y., Zhou, X., Bressan, S., and Tan, K. L. (2003). Efficient distributed reputation scheme for peer-to-peer systems. *International Conference Human Society at Internet*, Berlin, Germany: Springer, pp. 54–63.

Liu, X., Datta, A., and Lim, E. P. (2014). *Computational Trust Models and Machine Learning*. Boca Raton, FL: CRC Press.

Lopez, J., Roman, R., Agudo, I., and Fernandez-Gago, C. (2010). Trust management systems for wireless sensor networks: Best practices. *Computer Communications*, 33(9), 1086–1093.

Mahalle, P. N., Thakre, P. A., Prasad, N. R., and Prasad, R. (2013). A fuzzy approach to trust based access control in Internet of things. *International Conference on Wireless Communications, Vehicular Technology, Information Theory and Aerospace and Electronic Systems (VITAE)*, IEEE, pp. 1–5.

Manchala, D. W. (1998). Trust metrics, models and protocols for electronic commerce transactions. *Proceedings of the 18th International Conference on Distributed Computing Systems*, IEEE, pp. 312–321.

Marmol, F. G. and Perez, G. M. (2009). Security threats scenarios in trust and reputation models for distributed systems. *Computers & Security*, 28(7), 545–556.

Massa, P. and Avesani, P. (2005). Controversial users demand local trust metrics: An experimental study on epinions.com community. *Proceedings of the 20th National Conference on Artificial Intelligence*, pp. 121–126.

Massa, P. and Avesani, P. (2007). Trust-aware recommender systems. *Proceedings of Conference on Recommender Systems*, pp. 17–24.

McKnight, D. H. and Chervany, N. L. (1996). *The Meanings of Trust*. Minneapolis, MN: University of Minnesota.

Merriam-Webster (2007). Retrieved from http://www.m-w.com/.

Page, L., Brin, S., Motwani, R., and Winograd, T. (1999). The PageRank citation ranking: Bringing order to the web. *Technical Report, Stanford Digital Library Technologies Project*.

Pena, L. and Ismael. (2005). Internet report 2005: The Internet of things. *ITU*.

PKI Page, *An Overview of Public Key Infrastructures*, Retrieved from http://www.techotopia.com/index.php/An_Overview_of_Public_Key_Infrastructures_%28PKI%29.

Rasmusson, L. and Janssen, S. (1996). Simulated social control for secure Internet commerce. *Catherine Meadows, Proceedings of the New Security Paradigms Workshop*.

Resnick, P. and Zeckhauser, R. (2002). Trust among strangers in Internet transactions: empirical analysis of eBay's reputation system. *The Economics of the Internet and ECommerce, Advances in Applied Microeconomics*, 11(2), 23–25.

Ruan, Y. and Durresi, A. (2016). A survey of trust management systems for online social communities–Trust modeling, trust inference and attacks. *Knowledge-Based Systems*, 106, 150–163.

Sabater, J. and Sierra, C. (2002). Social ReGreT, a reputation model based on social relations. *SIGecom Exchanges*, 3(1), 44–56.

Sabater, J. and Sierra, C. (2005). Review on computational trust and reputation models. *Artificial Intelligence Review*, 24(1), 33–60.

Shafer, G. (1992). Dempster–Shafer theory. *Encyclopedia of Artificial Intelligence*, 330–331.

Sicari, S., Rizzardi, A., Grieco, L. A., and Porisini, A. (2015). Security, privacy and trust in Internet of things: The road ahead. *Computer Networks*, 76, 146–164.

Stankovic, J. A. (2014). Research directions for the Internet of things. *IEEE Internet of Things Journal*, 1(1), 3–9.

Sun, Y., Han, Z., and Liu, K. R. (2008). Defence of trust management vulnerabilities in distributed networks. *IEEE Communications Magazine*, 46(2), 112–119.

Tan, J. and Koo, S. G. (2014). A survey of technologies in Internet of things. *IEEE International Conference on Distributed Computing in Sensor Systems* (DCOSS), pp. 269–274.

Theodorakopoulos, G. and Baras, J. S. (2006). On trust models and trust evaluation metrics for ad hoc networks. *IEEE Journal on Selected Areas in Communications*, 24(2), 318–328.

Wanita, S., Surya, N., and Cecile, P. (2013). A survey of trust in social networks. *ACM Computing Survey*, 45(4), 1–33.

Yan, Z., Zhang, P., and Vasilakos, A. V. (2014). A survey on trust management for Internet of things. *Journal of Network and Computer Applications*, 42, 120–134.

Yu, H., Shen, Z., Miao, C., Leung, C., and Niyato, D. (2007). A survey of trust and reputation management systems in wireless communications. *Proceedings of the IEEE*, 98(10), 1755–1772.

Zong, B., Xu, F., Jiao, J., and Lv, J. (2009). A broker-assisting trust and reputation system based on artificial neural network. *IEEE International Conference on Systems, Man and Cybernetics* (SMC), pp. 4710–4715.

Section III

Interdisciplinary Aspects of the Internet of Things

11

Highway Safety of Smart Cities

Sándor Szénási

CONTENTS

11.1 Introduction

The main area of road safety engineering is the development of methods used to prevent road users from being killed or seriously injured. The goal of the eSafe initiative (declared by the European Union in November 2002) is to halve the number of public road accidents. The target of a similar project (called the ITS Ten-Year-Plan and announced in January 2002) in the United States is attempting to decrease the number of public road accidents by 15%.

Beyond several other factors (changes in legislation, construction or reconstruction of road objects, etc.), an important part of these projects is to utilize the benefits that come from new tools that have appeared in the last few decades (GPS, the Internet, wireless communication, cheap sensors, etc.). The public road network has fairly large physical dimensions, so the direct connection between the installed devices is impracticable. That is the point at which the Internet of Things (IoT) paradigm can help by creating a loosely coupled, decentralized network of nodes.

These nodes can be the following:

- Sensors embedded into the surface of the road, traffic signs, and so on
- Sensors built into vehicles
- Devices in vehicles to communicate with the driver
- Applications in the cloud to collect analyze and distribute data

Although most of the technical requirements are fulfilled, there are not any publicly available, well-functioning, massive interconnections between these devices that take advantages of every opportunity. Even the required standards are missing to start building these networks tomorrow. But there are several smaller projects and pilot applications already in use. Further integration of these or a new system based on the experience from their use may bring about the expected results.

This chapter contains three main sections regarding the fields where the IoT will become essential in the near future:

- *Accident prevention*: IoT devices in vehicles and traffic control objects will help decrease the number of accidents.

- *Accident data recording*: When an accident has already occurred, road accident scene investigators have to gather and record all information about the crash, vehicles, persons, environment, and so on; the IoT paradigm is widely used in this process.
- *Accident analysis*: A substantial portion of road safety engineers' work is the deep analysis of accidents; the information given by IoT devices makes this work more comfortable and more accurate.

11.2 Accident Prevention

According to the IoT design, various nodes (wireless sensors, RFID sensors, satellite networks, and other intelligent technologies) can be integrated into a new heterogeneous network to bring many benefits to highway users. Predicting dangerous situations and decreasing the number of accidents is one of the major goals of these developments; however, there are several additional benefits, including reducing road congestions, managing traffic, decreasing environmental pollution, and so on.

11.2.1 Available Services

In the near future, there should be a large network containing all road-safety-oriented objects (sensors, road signs, etc.) and vehicles controlled by intelligent services in the cloud (He et al. 2014). Although the technological conditions are given, only some independent applications currently exist (Bennakhi and Safar 2016). But these show that it is worth investing in these ideas; most of the existing systems are easy to use and—directly or indirectly—help one to reduce the number of road accidents.

11.2.1.1 Intelligent Navigation Systems

It is obvious that using printed maps is obsolete; GPS-based navigation systems mostly do a better job in route planning and turn-by-turn navigation. But we are already one step ahead. It is possible to use online traffic data to avoid traffic jams and roadworks. Navigation systems based on this philosophy are one of the best examples of the IoT design. Millions of sensors (traffic counters, cameras, etc.) and the vehicles themselves send a lot of data to the cloud.

These are just small information crumbs (the speed of a given vehicle, etc.) and useless alone, but collecting and deep analyzing these together gives a very clear picture of the traffic situation. This knowledge makes it possible to create more appropriate route and travel time estimations. As the communication is of two-way, in return for their information, clients are informed about any unexpected events submitted by other users (accident, roadworks, bad weather, etc.) that can help avoiding accidents.

11.2.1.2 Warning Messages

Unexpected situations are the most dangerous situations. Driving in bad weather conditions (ice, snowy road surface, heavy rain, fog, etc.) is not an issue if the driver can prepare for it. There are various sensors to detect any of those dangerous situations. Using the IoT paradigm, these sensors can send a signal directly to the approaching vehicles, or more likely to the cloud. The onboard system of the car can access all data concerning it and warn the driver to be careful.

11.2.1.3 Voice Recognition

Driving any vehicle in traffic requires full attention from the driver. However, even the most careful driver occasionally has to watch elsewhere (the gauges of the car, the entertainment system, etc.). From a road safety perspective, it would be better to minimize these moments; voice recognition can partially solve this problem.

Nowadays, this service is available only in the premium segment because of its high resource demand. But these services are also available without high onboard computation power using cloud services. Voice recognition is just an example; there are several comfort or safety issues easily solved by appropriate cloud applications. IoT design helps using this heterogeneous architecture.

11.2.1.4 Monitoring Vehicle Status

Modern cars are fully equipped with electronic sensors. The control unit of the onboard system needs a lot of information from different components to efficiently coordinate their operation. Using the IoT paradigm, it is possible to send some essential information back to the car manufacturer. They have the necessary experience and computation power to analyze these and send back a warning message if it is necessary to avoid further dangerous technical problems.

11.2.1.5 eCall

Although the main topic of this section is the prevention of accidents, the eCall system is worth mentioning. It is activated after a crash, but it can help one to decrease the severity of an unavoidable accident. The eCall system is intended to bring rapid assistance to any participants in an accident anywhere in the EU. The initiative aims to deploy a small device in all vehicles that will automatically dial 112 in the event of a serious road accident and send some information (GPS coordinates, etc.) to local emergency services. The eCall system is not Internet-based, but in the case of an already-available continuous online connection, the development of an IoT-based similar service is expected (Jankó and Szénási 2013).

11.2.1.6 Self-Driving Cars

Robotics is one of the fastest growing businesses today (Takács, és mtsai 2016). At the time of writing this chapter, the development of self-driving cars is one of the most popular research areas (Atzori et al. 2010). Today, these projects are in an early stage; however, some results are very promising. Google's driverless cars have already driven millions of kilometers.

At the same time, some manufacturers, such as Tesla Motors, have already enabled the self-driving feature in their cars. The combination of IoT and self-driven cars may be the pathway to a world without any road accidents (Fazekas and Gáspár 2015). But nowadays, it is just a possibility in the distant future.

11.2.1.7   Additional Services

There are various additional services based on IoT that are loosely connected to the road safety area (Kanoh 2013; Ashokkumar et al. 2015). Every application that decreases the traveling time (and consequently the number of vehicles on the road) or the frustration of the driver can indirectly help one to decrease collisions. A lot of research papers present results about automatic parking systems, and so on.

11.2.2   Internet Access in Vehicle

As the name suggests, IoT cannot exist without continuous Internet access. It is obvious that we can provide an online connection for static objects in urban or interurban areas, but it is more difficult to do the same for a moving vehicle. There are several protocols to solve this, usually based on the cellular phone network or on local Wi-Fi networks (Chen 2012).

11.2.2.1   Using Cellular Network

The cellular network infrastructure allows Internet access for users from the 1990s. The main advance of this method is that the coverage is especially good in industrial countries, and the access is relatively cheap. In the last decade, these networks have prepared to serve broadband requests, making mobile Internet access usable for IoT purposes. The speed of a common 3G network is enough for the necessary communication. Movement of clients does not cause any problems; the protocols used are able to switch between relay stations if necessary. Although there is a speed limitation, it does not affect the everyday usage.

To connect a vehicle to the cellular network, one of the following options is used:

- *Built-in connectivity*: In this case, a cellular communication module is integrated into the onboard system of the car. The Internet connection relies on this built-in module. It has several advantages (there is no limit for the size of the radio antenna, and this module can be an integral part of the onboard entertainment system), but it is expensive. Therefore, this is used by premium brands only.

- *Brought-in connectivity*: In this case, the driver or any of the passengers share the Internet connection of his or her smartphone with the car. MirrorLink technology standardizes the protocol for this two-way connection. Persons in the car have access to the entertainment system of the car by their smart devices, whereas the car gains immediate access to the Internet.

11.2.2.2   Using Wi-Fi Networks

Using Wi-Fi networks may be a cheaper solution, and it is also available in some smart cities. Although the efficient range is significantly smaller in this case, but, in built-up areas, the density of access points is quite high. Recent research shows that it is possible to access the Internet inside a moving vehicle within the coverage of Wi-Fi hotspots.

This method is called *drive-thru Internet*. It has several limitations (the coverage of Wi-Fi hotspots, the time required to make a connection to a Wi-Fi hotspot for high-speed vehicles), but there are several ideas to solve these. For example, a vehicle connected to a Wi-Fi hotspot can act as an access point to share that connection with the nearby clients.

11.2.2.3 Communication between Vehicles

Intelligent Transport Systems (ITS) try to solve the issues caused by the high traffic rate of our century. Several large projects begun in Europe, Japan, and the United States since the mid-1980s focus on the following areas (Tsugawa 2005):

- *Traffic safety*: To decrease the number of accidents (or at least decrease the number of killed persons).
- *Efficient transportation*: To decrease the overall traveling time, and consequently the environmental footprint, fuel cost, and others.

Although the goals to be achieved are very different, the requested tools are quite identical. Both kinds of projects need a lot of sensors collecting data about vehicles, traffic, and the environment; and one of the key points is the effective communication between the nodes. It is usual to distinguish the following communication subtypes (Lu et al. 2014):

- *Intervehicle communication*: The main communication form between two participants in traffic (also often referred to as *vehicle-to-vehicle communication*); this can be an interaction between drivers or at a lower level between onboard computers.
- *Road-to-vehicle communication*: The communication between some passive nodes (like temperature sensors) and participants in traffic.

11.2.2.4 Intervehicle Communication

The requirement for intervehicle communication is not new. There were some early projects in the 1980s in Japan (such as the Association of Electronic Technology for Automobile Traffic and Driving) to share information between cars. It became a heavily researched area in Europe and the United States to make systems improving car-to-car communication to relay incidents or emergency information from a proceeding car or just to send location data to nearby participants to avoid accidents caused by low visibility.

Several wireless communication protocols are known, but the intervehicle communication has some very specific requirements:

- The nodes of the network are not stationary. Individual cars are following different routes; therefore, the network of nearby vehicles is constantly changing.
- The speed of the vehicles is an important factor. Fast-moving nodes have not got enough time to execute slow interconnection procedures.
- Sensors are connected to the main battery of the car charged by the electric system; therefore, there is no energy constraint (Figure 11.1).

There were several attempts to solve this issue using various wireless techniques, such as Bluetooth (two of the constraints of this technology are the limited number of connecting devices and the limited communication distance, etc.), radio-frequency identification (the two major challenges are the large power loss at some locations and the unreliable data transfer among simultaneous transmissions), ultra-wide band (UWB) radio technology, and so on.

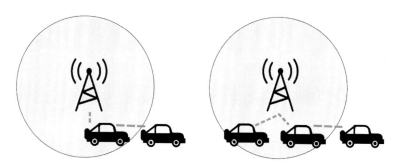

FIGURE 11.1
Example for intervehicle communication.

11.2.2.5 IoT Design in Intervehicle Communication

The rapid spread of wireless Internet opens a whole new path in this area. Instead of the direct connection of vehicles, each car connects to some application in the cloud individually. Vehicles can send and receive all corresponding data to and from the cloud via the appropriate services.

The only constraint is the necessary bandwidth on both sides. On the vehicle side, the limiting factor is the speed of the wireless network; it may be not enough fast to enable this two-way data stream. On the server side, the biggest challenge is the high number of clients. In the case of thousands or millions of cars, it can be very difficult to store and process all of the input data.

Using this IoT design has several advantages.

- The direct connection between devices becomes unnecessary, avoiding the previously mentioned issues. Each vehicle has to connect only to the cloud and keep this connection alive.

- In the case of multiple neighboring vehicles, it is not necessary to send the same message multiple times. It is enough to send it to the service, and the service will distribute it to the affected clients.

- Clients have the ability to get messages from faraway locations. For example, messages from all nodes affecting the planned route can help avoid further accidents.

Some applications have been already developed and widely used on the basis of the previously presented architecture. For example, Waze is a very popular geographical navigation application developed by Waze Mobile (purchased by Google in 2013) that won the Best Overall Mobile App award in 2013. It differs from other navigation application software in that it is community driven. Drivers can report incidents (such as accidents, traffic jams, and police traps), and other drivers receive these events.

Thanks to the IoT design, this is the biggest step forward since the appearance of online maps on smart devices. This step turns the navigation application into an interactive tool that is no longer based on a static map but instead is able to instantly adapt the planned route to current real-world events and to warn the driver to be careful in the case of dangerous places.

11.2.3 Risks

Every new technology has some disadvantages beyond the advantages. This is true for IoT devices; it is worth considering the potential risks of using online devices. The two major risks are personal data theft and the potential for the onboard systems to be hacked.

11.2.3.1 Personal Data Theft

Data from sensors providing weather-related information or general traffic information is usually publically available for all clients. The accessibility of data from devices owned by road operators, such as camera systems, is more confusing. Nowadays, there are several publically accessible camera systems (webcams); however, it is possible to track vehicles using these online video streams.

Data gathered and submitted by vehicles of private persons are more sensitive. No one wants to let everyone know the details of his or her trips (source, destination, GPS location, speed, etc.). The information itself is valuable for the community, but the default attitude is to hide the identity of the source. Most available applications promise that they use anonymous data and do not store any unnecessary information about the drivers.

11.2.3.2 Hacking

The more complex a computer system is, the more vulnerable it is to attack. In the world of the Internet, computer hacking is a permanent real threat. In the case of wireless networks, this becomes extremely dangerous, because the client does not have any effective physical method of prevention. Moving vehicles connected to the Internet with a two-way connection can be very risky in the near future. Hacking the system of a self-driving car can lead to a tragic end.

11.3 Recording Road Accident Data

Although there are several new techniques to prevent highway accidents, it is a fact that road accidents happened, happen, and will happen. It is how we should accept the world, and we have the methodology to handle these accidents. It is important to determine the reasons for each accident, because they can lead to some kind of potential solution (Yang et al. 2013; Elmaghraby and Losavio 2014). Theoretically, it is not possible to eliminate all further accidents, simply because of the exorbitant cost and general inconvenience. But it is necessary to do our best.

11.3.1 Main Steps of Scene Investigation Work

It is hard to determine the cause of an accident, but it is usually one—or some combination—of the following:

- Bad driving habits
- Bad weather conditions
- Vehicle error
- Engineering of roads

In accident analysis, we have to keep in mind all these factors. The most important is the last one, because traffic engineers can easily and quickly change the environment (maintain/build/rebuild objects, etc.). On the contrary, it is very hard and time-consuming to change general driving habits.

Nevertheless, it is very important to record as much data as possible about the crash. It is important to understand that most of these data are available on the accident scene only. There is no way to reproduce the circumstances of the accident later. Therefore, it is very important for accident scene investigators to have the appropriate tools to efficiently record all accident data (Wan et al. 2014). This requires instruments for the following steps:

- Accident location identification.
- Taking some photos of the environment.
- Recording all additional data about the accident.
- Submitting all information to the cloud.

Traditionally, accident scene investigators had to use a special form to record all accident data. This led to several errors: it was very hard to determine the exact location of the accident (they had to measure the distance from the nearest milestone), and it was usual to skip some fields of the form or to write in contradictory data (e.g., the accident type was a head-on collision, but only one participant was recorded).

The developments of the last decade significantly changed the methodology to be followed. New hand-held devices allow the investigators to solve all the mentioned problems using a single device with higher accuracy than has ever been available. The IOTs' point of view helps one to merge and utilize all the available tools.

11.3.2 Accident Location Identification

The location of the accident is the most crucial data for accident analysis. First, to analyze the environment, we need a precise location to check the road signs, road surface quality, and so on. In the case of an error of dozens of meters, the result of the analysis can be very different and misleading. Second, the location is very important for the hotspot search algorithms, and these are the first steps of most higher level accident analysis procedures.

11.3.2.1 *Using Global Positioning System Technology*

Global Positioning System (GPS) technology allows us to determine accident location in an appropriate and quick way as never experienced before. Using traditional handheld GPS devices, the coordinates can be read from the screen after only a few minutes. Although it seems very unambiguous, this leads to serious issues in practice.

- There are several coordinate systems for coding a location on the surface of the Earth. Police officers sometimes accidentally change the settings of the device and use another format. For example, the following values are all coding the same location: N43°38'19.39", 43°38'19.39"N, 43 38 19.39, 43.63871944444445, and so on.
- The other problem is that numerical coordinates are meaningless for users (except some well-trained experts). It is easy to find the difference between a right and a mistyped street name (*Neuannsreet* versus *Neumann street*), but it is easy to erroneously copy the coordinates from the GPS device to the paper form (and later, from the paper to the personal computer of the officer).

11.3.2.2 Using IoT to Gain More Accurate Data

A smart device with an Internet connection is the perfect solution for this problem. It is possible to develop software that is able to show the actual position of the device based on the built-in GPS receiver and other sensors (available Wi-Fi hotspots).

Using the Internet, the smart device is able to upload this location data to the server without any manual work. This makes the entire process error-free, meaning that there is no chance to make any typos in the coordinate pair.

However, it is worth keeping in mind that the precision of GPS technology is not always satisfactory. Especially inside built-up areas, the deviation between the real location and the supplied coordinates can be 10–30 m. This can lead to unacceptable results such as coding the accident to the neighboring street.

That is the other point in which the online connection can help the investigators work. It is possible to use online map services to show the accident location. It is possible to display the following data (Figure 11.2):

- The GPS coordinate given by the smart device.
- The corresponding topographical identification (road number and road section or street name and house number).
- A small map centered to the position given by the coordinates.

For a human, it is hard to directly check the validity of GPS coordinates; but thanks to online maps, that is no longer necessary. Based on the given street name + house

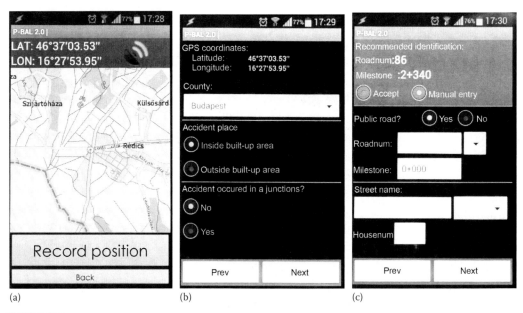

(a)   (b)   (c)

FIGURE 11.2
Screen captures from the presented Android-based data recording application: (a) show current position; (b) enter general information about the accident (outside or inside built-up area, etc.); and (c) give the traditional location definition (road name and section).

FIGURE 11.3
Screen capture from the presented location checker application.

number pair and the displayed map, it is easy to decide whether the GPS coordinate is correct or not.

After uploading the coordinates to the server, there are more advanced tools to analyze data location. For example, Figure 11.3 shows the position given by the GPS and the position according to the given road number and section. This program can check the uploaded data and instantly warn the investigator via the online connection to recheck the given data if the distance between the two positions is too high.

11.3.3 Taking Photos about the Accident Scene

It is ordinary to take photos for documentation purposes in every field. In the accident scenes, there are some special requirements:

- The photos must contain every detail of the scene environment (road signs, vehicles, other objects, road surface, etc.).
- Depending on the legislation, these photos cannot contain any personal data. There must not be the face of any person or any special identification objects (such as registration plate of the participants' cars).
- The environmental conditions are often very unsuitable for taking photos. In other fields of photography, it is possible to choose the correct time for shooting; but in the case of accident investigations, there is no time to wait. The investigator must be as fast as possible to minimize the traffic jam caused by the crash and roadworks.

Using analog cameras is outdated, and nowadays it is common to use digital ones. These are very easy to use, flexible, and give the opportunity to check the photos taken on the scene (and shoot more if necessary) (Rövid et al. 2016).

Advanced smartphones are able to take pictures in high resolution. Maybe, these are not applicable for advanced artwork, but the result is totally satisfactory for these purposes.

The software given to scene investigators has a function to take a photo and upload it to the cloud instantly.

This is very convenient compared with the traditional method (taking a photo, sending it to the laboratory, cataloging the final pictures, etc.). The uploaded photos are instantly attached to the appropriate accident, have location geocodes, and if the smartphone makes it available (with a built-in magnetic sensor), also contains the shooting direction data.

11.3.3.1 Using IoT Services to Retouch Photos

The role of the cloud is much more than just storing uploaded photos. It opens a connection between the scene investigator and his or her colleagues in the headquarters. Someone can check and evaluate all uploaded photos and ask for corrections if they are necessary.

It is also important to note that speed is a very important factor (Kertész and Vámossy 2015). The accident scene must be kept untouched until the investigation ends; therefore, the investigator has to take all the photos as fast as possible. On the other side, it is obvious that there is not any other chance to take these pictures.

All necessary facts must be recorded and checked before the investigator leaves the scene. The mentioned automatic two-way communication enables this (Figure 11.4).

Some other necessary operations must be taken with the photos. It is also true for normal photos and aerial images (Kiss 2014; Molnár et al. 2016). It is necessary to blur all parts of the image that make it possible to identify persons or vehicles. This is very tedious and tiresome work for humans, but it is easily automated (e.g., it is well known that Google Street View has similar algorithms to blur some parts of the taken images). This process is very resource intensive (Windisch and Kozlovszky 2015); therefore, it is hard to execute on smart devices. The IoT point of view can help in this problem. It is possible to upload the images to the cloud and use some service to perform the necessary modifications.

This methodology is also usable in the case of taking photos under unsuitable circumstances. Contrary to general photography, it is common to shoot pictures at night or in

FIGURE 11.4
Retouched accident photo.

wrong environmental conditions (because these are usual accident reasons). It is also possible to use some cloud services to improve the image. It is a big advance that this is available on the scene because in the case of failure, the investigator can take some additional photos.

11.3.4 Recording Additional Accident Data

Beyond the location and photos, there are several other pieces of information to collect at the accident scene. One of them in particular is especially important: the estimated time of the crash. But there is a lot of additional data to find from the following data sources:

- The accident itself
- The vehicles
- The persons
- The road and environment

Table 11.1 contains the main attributes of a typical road accident (Table 11.1).

Using the IoT, all of these data can be instantly uploaded from the accident scene. On the server side, there are several systems and intelligent algorithms to check and evaluate the uploaded data. These can send back a warning message if there is any missing or erroneous information. This instant response is very useful, because the investigator can check all warnings at the scene and modify the uploaded data (Figures 11.5 through 11.8).

11.3.5 Sending Traffic Message Channel Messages

Traffic Message Channel (TMC) is a technology for distributing information about unexpected events in the public road network. The main topology is many to many: messages from multiple sources received by the Traffic Information Center (TIC) and forwarded to all receivers. Obviously, the permissions of the users are strictly regulated. Only some privileged sources can send messages (road accident scene investigators are typically in this group), but the broadcasted messages are usually available for everyone.

TABLE 11.1

Main Attributes of Road Accidents

Level	Attribute	Example
Accident	Time	2016.06.12. 22:20
	Location	LAT 47°48′22″, LON 18°45′38″
	Nature of accident	Collision between cars and bicycle
Vehicles	Category	Motor-car
	Brand	Opel
Persons	Age	23
	Injury	Light
	Vehicle ID	1
Road and environment	Road surface	Wet
	Weather	Rainy
	Visibility	Below normal

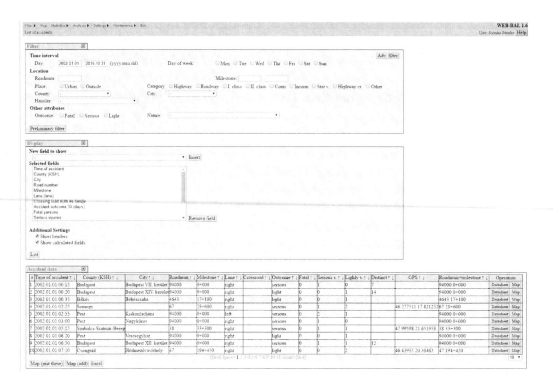

FIGURE 11.5
Select accidents using given filtering rules.

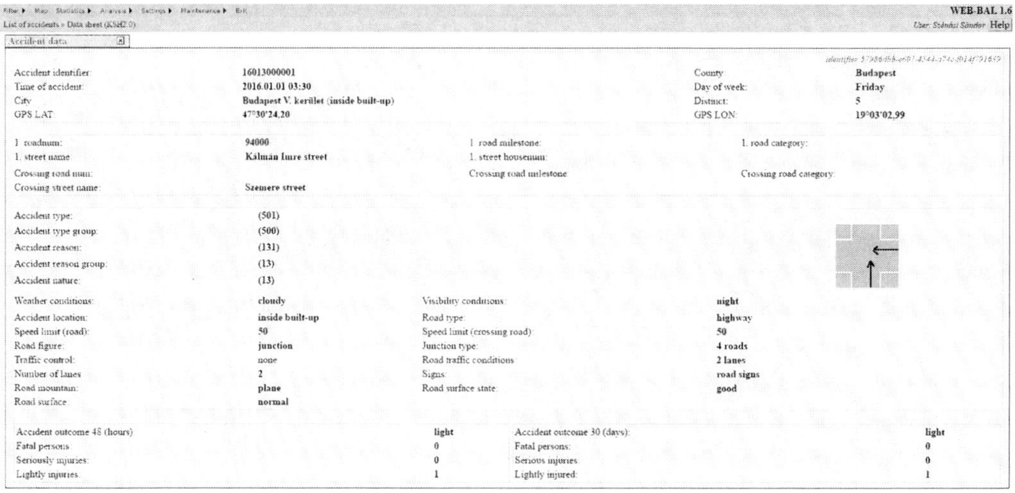

(a)

FIGURE 11.6
Check the details of a given road accident: (a) accident data.   (*Continued*)

Vehicle data

				identifier: 39caa91a-ac9b-49d5-a77d-cd5cb4c71a87
Participant vehicle identifier:			Vehicle type:	car
Vehicle type:		2	Production data:	
Location (country or county):		Opel	Danger:	
Movement of vehicle:		moving	Moving direction:	same
First object		none	Left the scene:	no
				identifier: 2ab653a0-c31c-4a12-ad09-c8948c1b12fd
Participant vehicle identifier:		1	Vehicle type	car
Vehicle type:		Volkswagen	Production data	
Location (country, vagy county)			Danger:	
Movement of vehicle:		moving	Moving direction:	same
First object		none	Left the scene:	no

Personal data

				identifier: 631ceb7c-d3b1-48e9-91a8-a14292a29a5e
Participant identifier:	2		Role:	participant
Vehicle identifier:	2			
Age:	46		Gender:	male
Nationality:	Hungarian		Role:	driver
Safety eq.:	safety belt		Airbag:	
Driving licence:	yes		Driving licence data	
Driving licence 2:	yes			
Alcohol:	no		Drug test:	negative
Pedestrian location:	no		Bicycle:	no
Injury outcome:	not injured		Injury outcome 30 days later:	not injured
Hospital:				
				identifier: 631ceb7c-d3b1-48e9-91a8-a14292a29a5e
Participant identifier:	2		Role:	participant
Vehicle identifier:	2			
Age:	46		Gender:	male
Nationality:	Hungarian		Role	driver
Safety eq.:	safety belt		Airbag:	
Driving licence:	yes		Driving licence data	
Driving licence 2:	yes			
Alcohol:	no		Drug test:	negative
Pedestrian location:	no		Bicycle:	no
Injury outcome:	not injured		Injury outcome 30 days later:	not injured
Hospital:				

(b)

Road data

						identifier: 0aaaaa5-7a59-45ff-93dd-ac4a1aa55c8e
Road number:	67		Lane:	unknown side	Category	IL cat
Interval begin identifier:	C140370		Interval begin km milestone:	22	Interval begin m milestone:	650
Interval end identifier	C140373		Interval end km milestone:	33	Interval end m milestone:	137
County KIG:	Somogy Megyei KIG.					
Validity begin:	2000.01.01 00:00		Validity end:	2016.06.03 00:00		

Traffic data

						identifier: 00000000-0000-0000-0000-000000001000
SuSum vehicle:	3626 vehicle/day	Sum motorcycle:	3620 vehicle/day	Bicycle:	6 vehicle/hour	
Yearly average traffic:	4070 E/day	Hourly traffic:	366 vehicle/hour	Yearly average traffic (szgkoj):	252 szgkoj/day	
Sum car:	3287 vehicle/day	Personal car traffic:	2803 vehicle/hour	Light truck:	482 vehicle/hour	
Sum bus:	61 vehicle/day	Long bus:	0 vehicle/day	Single bus:	61 vehicle/day	
Sum truck:	235 vehicle/hour	Middle weight truck:	66 vehicle/hour	Heavy truck:	100 vehicle/hour	
Truck:	48 vehicle/hour	Long truck:	20 vehicle/hour	Special vehicle:	1 vehicle/hour	
Sum heavy vehicle:	230 vehicle/day	Slow vehicle:	10 vehicle/hour	Motorcycle:	27 vehicle/hour	
Confidence: d>	14 %	Source: Viewer1SmbRSource:	F	Days	4	
Traffic type category 1:	b	Traffic type category 2:	2	Station ID		
Validity begin:						

(c)

FIGURE 11.6 (Continued)
Check the details of a given road accident: (b) vehicle and personal data, (c) road and traffic data.

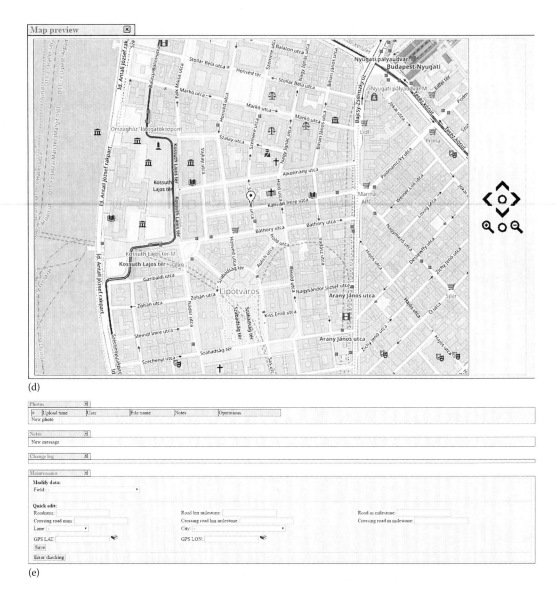

(d)

(e)

FIGURE 11.6 (Continued)
Check the details of a given road accident: (d) map preview, and (e) notes/photos and maintenance tab.

Parameters [x]

Calculation parameters
Interval length: 1000 m ▼
Display data: Accident data ▼
Additional data: none ▼

Display Settings
Columns num : 8
Summary columns: yes ▼

Schematic pointmap

Schematic pointmap [x]

Fatal	Serious	Light	Right lane	from	to	Left lane	Fatal	Serious	Light
0	3	1		18+012	19+000		0	0	0
0	0	2		19+000	20+000		0	0	0
0	1	1		20+000	21+000		0	1	0
0	0	0		21+000	22+000		0	0	0
0	0	4		22+000	23+000		0	0	0
0	0	2		23+000	24+000		1	0	0
0	0	1		24+000	25+000		0	0	0
0	0	1		25+000	26+000		0	0	0
0	0	1		26+000	27+000		0	1	0
0	0	2		27+000	28+000		0	0	0
0	0	0		28+000	29+000		0	1	0
0	1	2		29+000	30+000		0	0	0
0	0	0		30+000	31+000		0	0	0
0	0	0		31+000	32+000		0	0	1
0	0	0		32+000	33+000		0	0	1
1	0	0		33+000	34+000		0	0	0
0	0	0		34+000	35+000		0	0	1
0	0	1		35+000	36+000		0	0	0
0	0	0		36+000	37+000		0	0	0
0	0	0		37+000	38+000		0	1	0
0	0	0		38+000	39+000		0	0	0
0	0	0		39+000	40+000		0	0	0

FIGURE 11.7
Visualize the accidents of a given road interval.

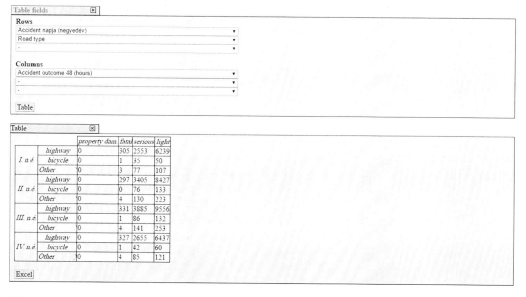

Table fields [x]

Rows
Accident napja (negyedév) ▼
Road type ▼
- ▼

Columns
Accident outcome 48 (hours) ▼
- ▼
- ▼

Table

Table [x]

		property dam.	fatal	serious	light
I. n.é	highway	0	305	2553	6239
	bicycle	0	1	35	50
	Other	0	3	77	107
II. n.é	highway	0	297	3405	8427
	bicycle	0	0	76	133
	Other	0	4	130	223
III. n.é	highway	0	331	3885	9556
	bicycle	0	1	86	132
	Other	0	4	141	253
IV. n.é	highway	0	327	2655	6437
	bicycle	0	1	42	60
	Other	0	4	85	121

Excel

FIGURE 11.8
General statistics of a given road interval.

Traffic incidents are binary-coded in TMC messages. The main data fields of a message are the following:

- Event code—one of the codes from the predetermined code table
- Location code—in this case, the location of the accident crash site
- Expected duration—how much time is needed to clean up the area
- Other additional data—not used in this case

It was a good idea to assign codes for events and not to send plain text. Using this methodology, it is easy to translate the message code to the language of any user. Moreover, the database of events is directly available for statistical queries.

The event itself takes 11 bits from the message. This limits the number of available events to 2^{11}, but that is absolutely enough to send the most frequent occurrences. These message codes cover a variety of events. Some of the codes describe an unexpected and short duration event (like car accidents), whereas some of them describe the opposite (e.g., road construction work; Table 11.2).

TABLE 11.2

Road Accident Related TMC Codes

TMC Code	Description
857	Unprotected accident area(s)
858	Danger of unprotected accident area(s)
12	Accident(s), traffic being directed around accident area
141	All accidents cleared, no problems to report
201	Accident(s)
202	Serious accident(s)
203	Multivehicle accident (involving Q vehicles)
204	Accident involving (a/Q) heavy lorr(y/ies)
205	Accident(s) involving hazardous materials
206	Fuel spillage accident(s)
207	Chemical spillage accident(s)
208	Vehicles slowing to look at accident(s)
209	Accident(s) in the opposing lanes
215	Accident(s). Stationary traffic
222	Accident(s). Queuing traffic
228	Accident(s). Danger of queuing traffic
229	Accident(s). Slow traffic
240	Road closed due to accident(s)
241	Accident(s). Right lane blocked
242	Accident(s). Center lane blocked
243	Accident(s). Left lane blocked
244	Accident(s). Hard shoulder blocked
245	Accident(s). Two lanes blocked
246	Accident(s). Three lanes blocked
247	Accident. Delays (Q)
248	Accident. Delays expected
249	Accident. Long delays (Q)
250	Vehicles slowing to look at accident(s). Stationary traffic

11.3.5.1 *Sending Traffic Message Channel Using a Smart Device*

At the accident scene (after some more important safety procedures), the investigator should send a TMC message to the TIC. Using the Internet, this can be done using his smart device. It depends on the officer when to send this message. If it is important, it is possible to send the message out before any other operation. Another way is to input some of the previously mentioned accident data (location of the accident, nature of the accident, etc.), and based on this information, the software on the smart device can determine the appropriate message code.

It is very auspicious that the scene investigator can directly handle these TMC messages from the scene. First, he or she is an expert; therefore, he or she can send a reliable and trustworthy opinion about the actual situation. The TMC message contains a field about the expected duration of the event. It is just an approximation; the scene investigator has to guess how much time he or she needs to collect all necessary information and perform the requested cleaning operations.

Of course, it is possible to manually stop the event. The investigator has to remain on the crash site until all the previous operations have been completed. At the end, he or she is able to send a closing TMC message, indicating to all vehicles that the road is opened for traffic.

11.4 Analyzing Road Accident Data

11.4.1 Advantages of the New IoT Devices

As previously mentioned, the reason or reasons for an accident usually belong to one or more of the following categories (Paridel et al. 2011):

- Bad driving habits
- Vehicle errors
- Engineering of roads

The main task of road safety experts is to find the critical areas of the public road networks, analyze these, propose some possible improvements, and estimate the expected positive effects.

11.4.1.1 *Using the Data from IoT Sensors*

Traditionally, experts have to use one database containing accident data (Gellerman et al. 2016). This database is controlled by the scene investigators; they were the source of all available accident information.

In addition to the resource-intensive human work, this leads to some inaccuracy. A lot of attributes are estimated by the scene investigators several hours after the accident.

Nowadays, it is possible to install several kinds of sensors recording various data (Perera et al. 2014). These are usually cheap devices, and it is possible to use a lot of them in vehicles, traffic objects, and so on. However, this new possibility raises some new questions. First, how can we store this enormous amount of data? And second, how can we process this data?

Fortunately, data storage devices have become more and more cheap, and it is very easy to record all available data in a convenient way using cloud technology. Processing this amount of data is a more complex problem, but there are several data-mining algorithms for this purpose.

11.4.1.2 Sensor and Data Types

Here are some examples of how embedded devices can assist in data gathering.

- *Traffic measurement sensors*: There are several ways to measure the traffic at a given section of the highway. There are camera systems with this additional feature, and there are special target hardware developments for this purpose (sensors built into the road surface using induction to count the number of vehicles). These embedded devices can send all recorded data to the cloud.
- *Speed measuring sensors*: These sensors are able to measure the speed of a given vehicle. It is possible to upload all recorded data to a central warehouse.
- *Analyzing weather conditions*: The weather has a very big impact on road accidents. However, it is very rare that the weather is the direct cause of an accident; but inclement weather conditions can change a semiaccident situation into a real crash. It is possible to measure the following using various sensors:
  - Temperature
  - Light conditions
  - Atmospheric pressure and so on
- *Camera systems*: The biggest help for analyzing the reasons is a short video about the accident. Nowadays, it is possible to install camera systems in dangerous places, and thanks to advanced compression algorithms and fast wireless Internet connections, it is possible to upload all recorded data to the cloud.
- *Sensors inside vehicles*: It is also possible to gather several kinds of information from the built-in vehicle sensors. For example, the speed of the vehicle can be inferred by the built-in GPS receiver or the smartphone of the driver (Zarronbashar and Mahmud 2010; Isen et al. 2013; White et al. 2011).

It is possible using multiple data sources to determine the appropriate value of a given parameter. In the case of accidents, the most important information is the location. Using multiple sources (a smart device used by the accident scene investigator, a GPS device built into the vehicle, camera images, etc.) makes it possible to have a more accurate result.

11.4.2 Traditional Method to Find Accident Hotspots

Accident hotspots are the area of the public road network in which the density of accidents is higher than expected. As previously mentioned, the new devices make it available to assign GPS coordinates to accident locations (Reshma and Sharif 2012). This has several advancements (see Section 11.4.1) but also raises some interesting questions (Figure 11.9).

11.4.2.1 Sliding Window Method

Traditional hotspot searching algorithms (Sorensen and Pedersen 2007; Montella 2010; Costa et al. 2015) use road number + road section based geocoding. It allows the usage

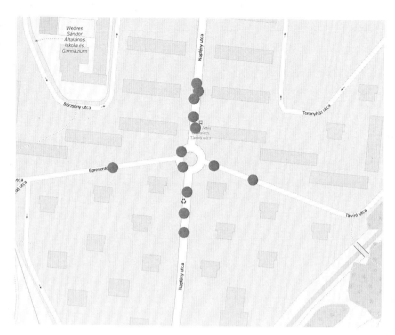

FIGURE 11.9
Hotspot in built-up areas.

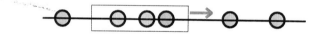

FIGURE 11.10
Traditional one-dimensional sliding-window method.

of some basic methods such as the well-known sliding window technique (Lee and Lee 2013; Szénási and Csiba 2014). This method contains the following consecutive steps:

- Select all accidents on the examined road (this method is not applicable for analyzing accidents occurring on multiple roads).
- Sort these accidents by the road section number.
- Slide a window with a given size (e.g., 300 m) over the road.
- Count the number of accidents inside the window according to its actual position.
- If the number of accidents is higher than a given threshold value, the position can be considered as a hotspot candidate (Figure 11.10).

This method has several advantages:

- It is easy to use.
- There are several already-existing applications.
- The results are easily readable for humans.

But it has some disadvantages in the case of built-up areas:

- Using the given technique, it is important to know the distance between two accidents. Outside urban areas, when we use road number + road section pairs, it is easy to calculate. The distance between two accidents equals the difference between the two road section values.
- But in the case of built-up areas, the street name and house number pair is the common identification method. There is not any simple way to calculate the distance between two accidents based on these.
- Public roads constitute a network of road sections and junctions. These differ in their geometric characteristics and traffic technological design. Within built-up areas, a significant amount of accidents can be linked to junctions. In such cases, based on the decision of the scene investigator, it differs to which street the accident will be allocated.
- In the case of larger junctions, it is possible to have up to four or five different street names appearing at the same junction (and sometimes the junction also has a square name). Thereby, these accidents in the database appear under different street names, and this makes it harder to find the hotspots.

11.4.2.2 Adaptation for Two-Dimensional Coordinates

There are several papers about the sliding window technique and its adaptations (fixed window size, flexible window size, using several special weighting factors, etc.). It was widely used in the past, but the appearance of GPS coordinates mean that it will soon be out-dated. All geocoded information from the embedded sensors uses GPS coordinates. The appropriate format (WGS'84, EOV, etc.) does not really matter because it is easy to convert from one to the other.

But all of these use a significantly different approach compared with the road number + road section identification. These are all some kind of two-dimensional spatial coordinates, and it is not possible to use the sliding window method in this case (it is possible to use a two-dimensional window instead of the one-dimensional one, but it is not widely popular; Figure 11.11).

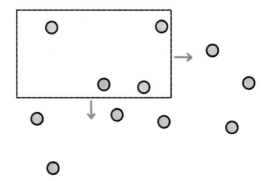

FIGURE 11.11
Two-dimensional sliding-window method.

11.4.3 Data-Mining Approach-Finding Accident Hotspots

It is necessary to develop new hotspot searching algorithms based on the newly available data sources. GPS coordinates give unequivocal and accurate locations, but it is necessary to use novel spatial methods instead of the old-fashioned sliding window.

11.4.3.1 Clustering Algorithms

One of the most promising methods is from the field of data-mining. Data-mining is a special database application used to discover hidden correlations in data. As a multidisciplinary field, it combines fields such as database management, statistics, and artificial intelligence. The main advantage of this methodology is that it tries to automate certain processes and expert activities helping in processing a large amount of data, which requires disproportionately large efforts by experts. The main difference between traditional statistical operations and data-mining is that in the latter case, it is possible to find correlations between the entities inside an existing database, which are not already known.

A special field of data-mining concerns clustering algorithms. Clustering means that we have to find groups (also known as clusters) of items as follows: All items inside a cluster are similar to each other (according to a prespecified criterion), whereas objects outside the cluster (or in different clusters) are quite different from each other (according to the same criterion).

In the case of accident analysis, the requested output is the groups of accidents, which are similar to each other (according to location or another accident attribute). As in the case of similar accidents, we can assume that the reason was the same, it is possible to eliminate this factor in order to significantly decrease the number of future accidents.

11.4.3.2 Density-Based Clustering of Applications with Noise Algorithm

There are several well-known clustering algorithms. One of these, the Density-Based Spatial Clustering of Applications with Noise (DBSCAN), seems ideal for accident hotspot searching (Szénási and Jankó 2016). The purpose of this algorithm is to find areas in the multidimensional search space where the density of elements is higher than a given threshold. One of the advantages of this method is that it is able to work effectively in a noisy environment.

In the case of accidents, the search space is the examined country/city/etc. The elements inside this search space are the occurred accidents. The position of a given element in the search space accords to the location of the corresponding accident. Hotspots are areas of the search space where the density of accidents is higher than a given threshold. Every accident not belonging to any hotspot is considered noise.

Another advancement of the DBSCAN algorithm is the capability to find clusters of any shape, in contrast to the two-dimensional sliding window method, which always uses a rectangular-shaped window. In the case of junctions, irregularly shaped areas can lead to significantly higher accident density.

11.4.3.3 Details of the Density-Based Clustering of Applications with Noise Method

The main input parameters of the DBSCAN algorithm are the following:

- ε: A radius type variable (meters)
- *MinPTS*: Minimal number of accidents within a cluster (number of accidents)

To understand the mechanism of the DBSCAN algorithm, it is necessary to introduce some definitions.

- *ε-environment*: Space radius of ε of an element (it is necessary to define a distance concept).
- *Internal element*: There are at least MinPTS number of elements in the ε-environment of this element.
- *Directly dense accessibility*: One element is directly dense accessible from another if it is in the ε-environment of it.
- *Dense reachability*: One element is accessible from another through a chain of directly densely accessible ones.
- *Dense connection*: Two elements are densely connected if there is any element from which both of them are densely reachable.
- *Density-based cluster*: A domain of densely connected elements that shows maximum accessibility of density.

Based on these definitions, the goal of the DBSCAN algorithm is to find domains of accidents in which all elements are densely connected, and no further expansion is possible (the number and the size of clusters are maximal).

It is also necessary to define the concept of distance. In the case of accidents identified by GPS coordinates, this can be considered to be simply the Euclidean distance between the two locations. However, it is possible to use more sophisticated methods.

The main steps of the DBSCAN algorithm are the followings (Figure 11.12):

1. Select a starting point (any point of the search space).
2. If there are not at least MinPTS accidents in the ε-environment of this point, it is possible to skip the remaining steps and select another starting point.
3. If there are at least MinPTS accidents, it is worth starting the further cluster analysis. This is an internal point so that we can create a new cluster-candidate containing this and the other elements in its ε-environment.
4. The ε-radius of these new elements is also investigated, and if any of the existing elements meet the necessary growing conditions, they are added to the cluster candidate.

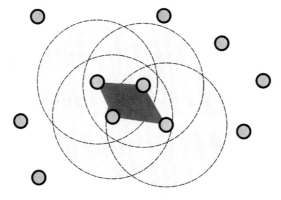

FIGURE 11.12
DBSCAN method to find accident hotspots.

5. The search continues using these new points with Step 4.

6. If it is unable to select a new point to continue the search, the recursion finishes.

Based on the goal of the algorithm, there are several variations of the previous algorithm:

- In the simplest variation, the presented algorithm will find clusters, in which the number of accidents is higher than a given threshold.

- It is also possible to use some weighting factors. The outcome of the accident is known, and it usually belongs to one of the following classes: fatal accident, accident with serious injury, accident with light injury, accident without personal injury. For example, here is a possible weighting factor configuration:
  - Fatal accident: 10
  - Accident with serious injury: 3
  - Accident with light injury: 1
  - Only property damage: 0.5

11.4.3.4 Calculating Accident Density

The presented variations are accounting for only the number of accidents. But in the case of hotspot identification, we are looking for areas with high accident density. Density means the number of accidents divided by the bounding area. If the accidents are located on multiple roads, it is necessary to use a spatial area definition based on the shape spanned by the accidents.

To find the convex shape with the minimal area containing all accidents (all of them have to be inside the shape or in the contour/edge of the shape). If this polygon is known, it is possible to calculate its area using the Shoelace formula (also known as the Gauss area formula) that requires only the appropriate sequence of the two-dimensional edges. The outer points of the polygon must be in a given order (e.g., in a clockwise direction).

The production of this sequence is very resource-intensive; therefore, the DBSCAN algorithm maintains this automatically when a new accident is inserted into the cluster.

- The first two accidents are automatically inserted into the sequence.

- The third accident is also inserted into the sequence, but if necessary the algorithm reorders the points to ensure their clockwise order.

- If the fourth or any further element is outside the already-existing polygon, then it is necessary to expand it and to insert the accident into the contour sequence.

- If the fourth or any further element is inside the already-existing polygon, there is nothing to do. This will change the accident density, but not the area of the minimal convex bounding polygon (Figure 11.13).

At some point, the DBSCAN algorithm will not be able to find any new accidents to expand the actual cluster. It calculates the area of the minimal convex bounding polygon and the accident density of the cluster. If this density is higher than a previously given threshold, then this is an interesting area for further processing. If not, then it restarts the DBSCAN algorithm from another starting point.

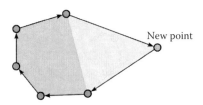

FIGURE 11.13
Area calculation method.

11.4.4 Analysis of Hotspot Candidates

The result of the data-mining algorithm is a set of clusters. We usually call these hotspot candidates. The word *candidate* makes it clear that these are only accidents grouped together using some kind of mathematical method. It is possible that these are totally independent accidents without any common attributes; in this case, this is not a real hotspot, but just an unfortunate distribution of accidents. Because of this, all hotspot candidates require deeper analysis.

11.4.4.1 Using Information Gathered by IoT Devices

That is the point in which all data recorded by the IoT devices can help in understanding the situation:

- *Traffic measuring sensors*: The above-described method gives only a raw analysis of accidental data. The result is the areas of the road network in which the density of accidents is high. The expected number of accidents in road sections with high traffic is higher than in road sections with low traffic.

    Therefore, it is worth taking into consideration the traffic information of the scene (in the most simplified case, as a divisor).

- *Temperature sensors*: It is possible to check the recorded temperature data. The average temperature on a daily/hourly basis is already known. Based on this, the expected distribution of temperatures at the time of accidents is also available. If it significantly differs from the actual distribution, it is worth finding out the reasons. For example,

    - If most of the accidents occurred in low-temperature environments, it is worth checking the surface of the road in cold situations.

    - If most of the accidents occurred in high-temperature environments, it is worth checking the surface of the road in hot situations.

- *Weather sensors*: These sensors can give us additional information about the weather conditions at the time of the crash. For example, the atmospheric pressure, rain, snow, sunset, and so on. It is also worth checking if there is any uncommon pattern in these data series. For example,

    - If most of the accidents occurred in rainy situation, it is worth checking the road surface. Maybe, it becomes unusually slippery when it is wet.

    - If most of the accidents occurred in sunny weather, it is worth checking the routing of the road. Maybe, the sun shines directly into drivers' eyes.

- If most of the accidents occurred on snowy days, it is worth considering limiting the speed or performing other preventive actions.

- *Speedometers*: The speed of the participant vehicles can be very important in the analysis phase. Speeding is one of the most frequent reasons for an accident. On the one hand, it is one of the things which is really hard to influence (although there are some methods, such as speed limit road signs, fixed installation of speedometers, etc.). But on the other hand, it is worth knowing that the origin of the problem is bad driving habits and not road engineering issues.

- *Camera images*: If there are videos of the crash events, these can help one to understand the reasons for the accident. It is worth watching these videos one by one. If there are any common patterns among the occurred accidents, it is a real hotspot, and it is necessary to take preventive actions.

11.5 Conclusion

As can be seen, IoT can help in several points of the data analysis process. It makes it available using accurate location information about accidents, which leads to usable clustering methods. It also can help in the next phase of the analysis project, the deep investigation, in which the main goal is to uncover the common patterns among accidents.

The IoTs paradigm provides two additional advancements for experts:

- The data collection process becomes comfortable and automated. For example, in the past, it was possible to count the number of cars by using inductive devices built into the road surface; but these devices have only a very limited storage, and someone has to manually read out and clear the recorded data weekly.

- Using IoT, it is not necessary to use local storage; devices can upload all data instantly to the cloud. It is able to read all recorded information only seconds later. It is more important for traffic control applications, but it also helps the work of accident analysts.

- Using the new devices also makes the gathered information even more precise. Embedded systems allow for the determination of vehicle speed and direction in a very accurate way. The distribution of camera systems gives essential information about traffic situations.

References

Ashokkumar, K., B. Sam, R. Arshadprabhu, and Britto. 2015. Cloud based intelligent transport system. *2nd International Symposium on Big Data and Cloud Computing*, pp. 58–63.

Atzori, L., A. Iera, and G. Morabito. 2010. The Internet of things: A survey. *Computer Networks* 54: 2787–2805.

Bennakhi, A., and M. Safar. 2016. Ambient technology in vehicles: The benefits and risks. *The 6th International Symposium on Frontiers in Ambient and Mobile Systems*, pp. 1056–1063.

Chen, Y.-K. 2012. Challenges and opportunities of Internet of things. *17th Asia and South Pacific Design Automation Conference.* IEEE, Sydney, pp. 383–388.

Costa, S. D., X. Qu, and P. M. Parajuli. 2015. A crash severity-based black spot identification model. *Journal of Transportation Safety and Security* 7 (3): 268–277.

Elmaghraby, A. S., and M. M. Losavio. 2014. Cyber security challenges in smart cities: Safety, security and privacy. *Journal of Advanced Research* 5: 491–497.

Fazekas, Z., and P. Gáspár. 2015. Computerized recognition of traffic signs. *Acta Polytechnica Hungarica* 12 (5): 35–50.

Gellerman, H., E. Svanberg, and Y. Barnard. 2016. Data sharing of transport research data. *Transportation Research Procedia* 14: 2227–2236.

He, W., G. Yan, and L. D. Xu. 2014. Developing Vehicular Data Cloud Services in the IoT Environment. *IEEE Transactions on Industrial Informatics* 10 (2): 1587–1595.

Isen, L., A. Shibu, and M. S. Saran. 2013. Evaluation and treatment of accident black spots using geographic information. *International Journal of Innovative Research in Science, Engineering and Technology* 2 (8): 3865–3873.

Jankó, D., and S. Szénási. 2013. Expected impacts of the implementation of ITS on congested periods regarding the Hungarian national road network. *Moving Toward an Integrated Europe Pan-European eCALL Service in the HeERO Countries.* Budapest, pp. 72–77.

Kanoh, H. 2013. Development of MMRS (Mind Map and Relief System), an information sharing system for children's safety. *Procedia Computer Science* (22): 762–771.

Kertész, G., and Z. Vámossy. 2015. Current challenges in multi-view computer vision. *10th Jubilee IEEE International Symposium on Applied Computational Intelligence and Informatics.* Timisora, Romania.

Kiss, D. 2014. Eigenvector based segmentation methods of high resolution aerial images for precision agriculture. *5th ICEEE-2014 International Conference: Global Environmental Change and Population Health: Progress and Challenges.* Budapest, Hungary, pp. 155–162.

Lee, S., and Y. Lee. 2013. Calculation method for sliding-window length: A traffic accident frequency case study. *Easter Asia Society for Transportation Studies* 9: 1–13.

Lu, N., N. Cheng, N. Zhang, X. Shen, and J. W. Mark. 2014. Connected vehicles: Solutions and challenges. *IEEE Internet of Things Journal* 1 (4): 289–299.

Molnár, A., D. Stojcsics, and I. Lovas. 2016. Precision agricultural and game damage analysis application for unmanned aerial vehicles. *International Journal of Applied Mathematics and Informatics* 10: 38–43.

Montella, A. 2010. A comparative analysis of hotspot identification methods. *Accident Analysis & Prevention* 42 (2): 571–581.

Paridel, K. et al. 2011 Teamwork on the Road: Efficient Collaboration in VANETs with Context-based Grouping, *Procedia Computer Science* (5): 48–57.

Perera, C., A. Zaslavsky, P. Christen, and D. Georgakopoulos. 2014. Sensing as a service model for smart cities supported by Internet of things. *Transactions on Emerging Telecommunications Technologies* 25 (2): 81–93.

Reshma, E. K., and S. U. Sharif. 2012. Prioritization of accident black spots using GIS. *International Journal of Emerging Technology and Advanced Engineering* 2 (2): 117–122.

Rövid, A., Z. Vámossy, and S. Z. Sergyán. 2016. Thermal image processing approaches for security monitoring applications. In S. László Nádai, S. József Padányi (Eds.), *Critical Infrastructure Protection Research: Results of the First Critical Infrastructure Protection Research Project in Hungary,* pp. 163–175. Switzerland: Springer International.

Sorensen, M., and S. K. Pedersen. 2007. Injury severity based black spot identification. *Annual Transport Conference.* Aalbor, pp. 1–10.

Szénási, S., and P. Csiba. 2014. Clustering algorithm in order to find accident black spots identified By GPS coordinates. *14th GeoConference on Informatics, Geoinformatics and Remote Sensing.* Albena, pp. 497–503.

Szénási, S., and D. Jankó. 2006. Black spot treatment system using a "hunting for irregular pattern" process and a safety knowledge-base. *On safe roads in the XXI Century*. Budapest, Hungary, pp. 100–105.

Szénási, S., and D. Jankó. 2016. A method to identify black spot candidates in built-up areas. *Journal of Transportation Safety & Security* 9 (1): 1–25.

Takács, Á., D. Á. Nagy, I. Rudas, and T. Haidegger. 2016. Origins of surgical robotics: From space to the operating room. *Acta Polytechnica Hungarica* 13 (1): 13–30.

Tsugawa, S. 2005. Issues and recent trends in vehicle safety communication systems. *IATSS Research* 29 (1): 7–15.

Wan, J., D. Zhang, S. Zhao, L. T. Yang, and J. Lloret. 2014. Context-aware vehicular cyber-physical systems with cloud support: Architecture, challenges, and solutions. *IEEE Communications Magazine* 52 (8): 106–113.

White, J., C. Thompson, H. Turner, B. Dougherty, D. C. Schmidt, and Wreckwatch. 2011. Automatic traffic accident detection and notification with smartphones. *Mobile Networks and Applications* 16 (3): 285–303.

Windisch, G., and M. Kozlovszky. 2015. Parallel image sharpness measure for supercomputing environment. *13th International Symposium on Applied Machine Intelligence and Informatics*. Slovakia, Herlany: IEEE, pp. 283–288.

Yang, L., S. H. Yang, and L. Plotnick. 2013. How the Internet of things technology enhances emergency response operations. *Technological Forecasting and Social Change* 80 (9): 1854–1867.

Zarronbashar, E., and A. R. Mahmud. 2010. Intelligent GIS-based road accident analysis and real-time monitoring system. *International Journal of Engineering* 2 (1): 1–7.

12

Connected Bicycles: Potential Research Opportunities in Wireless Sensor Network

Sadik Kamel Gharghan, Rosdiadee Nordin, Nor Fadzilah Abdullah, and Kelechi Anabi

CONTENTS

12.1 Introduction

Wireless sensor networks (WSNs) can be used to monitor sports activities as suggested in several recent research works. In the area of high performance cycling, cyclist performance can be considered as one of the most important application. BWSNs (bicycle

WSNs) have been identified as a technology candidate that is suitable for the mobility model, energy model, and real-time monitoring of a cyclist. A few key WSN technologies that have been utilized are Bluetooth, ZigBee, Wi-Fi, and Advanced and Adaptive Network Technology (ANT). For example, in the work of Balbinot et al. (2014) and Casas et al. (2016), the Bluetooth wireless protocol was used to monitor the three-dimensional forces applied on the bicycle crankset. The monitoring system was achieved based on a data logger. Similar research work adopted ZigBee wireless protocol to monitor the location of a cyclist on a cycle track (Shin et al., 2013), bicycle biomechanical parameters (Olieman et al., 2012), bicycle location estimation and radio reachability (Hayashi et al., 2012), and knee and ankle angles monitoring (Marin-Perianu et al., 2013). The ANT proprietary wireless protocol was used by Baca et al. (2010) to monitor the biomechanical and physiological parameters of the bicycle and cyclist. By utilizing the infrastructure of the mobile and Internet networks, the cyclist parameters can be transmitted to a remote location via a framework system that consists of wireless protocol and mobile phone device. Although Wi-Fi technology consumed more power and larger in size compared to Bluetooth, ZigBee, and ANT, it was also used to monitor the cyclist performance in some research work (Yu et al., 2009).

In terms of available commercial products, the Schoberer Rad Messtechnik device can be considered as the most widely used to monitor the professional cyclist parameters such as power, cadence, speed, heart rate, and temperature (Bini et al., 2011).

Previous research works and commercial products mostly focus on methods of measuring cycling performance and transferring these parameters from the bicycle sensor nodes to the monitoring device that is fixed on the bicycle handlebar as shown in Figure 12.1. With the advancement of sensors technology, wireless communication technologies and cloud computing, the BWSN is expected to join the Internet of Things (IoT) hype.

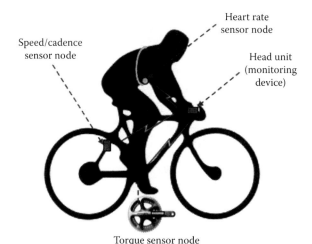

FIGURE 12.1
The general configuration of the BWSN with related sensor nodes.

12.2 Problem Background

High performance cycling requires a high fitness level with sufficient training program and teamwork. Cycling is a highly competitive sport, where the margin between victory and defeat is in the order of milliseconds. The biomechanical and physiological parameters of the bicycle and cyclist can be used to develop and refine the cyclists and team tactic and training programs to become more competitive in international competitions. These parameters can be measured and wirelessly transmitted to the coach or remote location that can be used to analyze the performance and progress of the cyclist. In addition, the measured parameters can be transmitted to the monitoring device that mounted on the handlebar of the bicycle to display on the liquid crystal display to be seen in real time by the cyclist.

BWSNs are unlike other types of WSNs. Several limitations can be highlighted as below:

1. *Limited energy resources*: All sensor nodes are portable on the bicycle and use the battery as the main power source. Recharging or replacing batteries when the bicycle in competition case is impossible, especially in the long distance event. Accordingly, the ANT and ZigBee wireless protocols that consume less energy during operation can be used. Besides that, energy-efficient techniques are an essential demand.

2. *Limitation on size, weight, and memory size of sensor nodes*: The sensor node needs to be small as possible to reduce the weight and size. In addition, the mathematical relationship between the measured parameters of the bicycle needs to be merged into one sensor node. A task-division technique through a coordinator node can also be used to avoid exhaustive calculation in sensor nodes.

3. *Inaccurate distance estimation*: Radio communication is extremely sensitive to the location of the bicycle on the cycle track or road due to the changing distance and varying channel conditions between the cyclist and the coach. For efficient energy utilization, the RF-transmitted power needs to be modified based on distance or signal-to-noise-ratio (SNR) estimation.

4. *Difficulty to use energy harvesting as an alternative to the battery source*: The harvesting apparatus adds extra weight and the biker is faced with rolling resistance or aerodynamic resistance.

Of the limitations stated above, it can be seen that energy is the most critical source in BWSNs. Therefore, reducing power consumption and extending battery life are mandatory because bicycle sensor nodes use battery power as a primary energy source. Several schemes of power-reduction techniques toward prolonging the battery life of BWSNs will be presented in Section 12.3.

Wireless connectivity supports wearable equipment that allows for unobtrusive and noninvasive monitoring. The development of wireless technologies has helped many practical applications. In sports application, WSNs allow the gathering and use of data, possibly in real time, and potentially performing the comparison or combination of the athletes' performance. As stated in the previous section, Bluetooth, ZigBee, ANT, and

TABLE 12.1

Comparison of Wireless Technologies Used in Bicycle Monitoring

Parameters	Wireless Technology		
	Bluetooth (IEEE 802.15.1)	ZigBee/IEEE 802.15.4	ANT
Frequency band(s)	2.4 GHz	868/915 MHz, 2.4 GHz	2.4 GHz
Modulation type	GFSK	BPSK and OQPSK	GFSK (Zhang et al., 2013)
Data rate	1 Mbps[a] (Wong et al., 2013); 1.2 Mbps[b] (Georgakakis et al., 2011); 3 Mbps[c] (Decuir, 2014); 24 Mbps[d] (Georgakakis et al., 2011)	20[h], 40[i], and 250 kbps[j] (Cavallari et al., 2014)	250 kbps, 1 and 2 Mbps
Output power	0[e], 4[f], and 20 dBm[g] (Liu et al., 2012)	(−25 to 0) dBm	−18, −12, −6, and 0 dBm (Kohvakka et al., 2010)
Transmission range	10–100 m (Abbasi et al., 2014)	(10–100) m	30 m (Caballero et al., 2011)
Power consumption	(40–100) mW	1 mW	0.183 mW
Battery life (coin cell)	(1–7) days	(4–6) months	3+ years (Adibi, 2012)
Latency	<10 sec	(20–30) ms	≈ Zero (Maharjan et al., 2014)
Spreading type	FHSS	DSSS	Adaptive isochronous (Gasparrini et al., 2013)
Number of devices in network	7 (active)/255 (total) (Padgette et al., 2012)	2^{16}	2^{32}
Number of RF channels	79	One[h], 10[i], 16[j] (Mraz, Cervenka, Komosny, & Simek, 2013)	125 (Zhan & Yu, 2013)
Network topology	Ad-hoc, point-to-point, and star	Ad-hoc, peer-to-peer, mesh, and star	Broadcast ad-hoc peer-to-peer, mesh, and star (Rawat et al., 2014)

[a] Bluetooth LE, [b] Bluetooth 1.2, [c] Bluetooth 2.0 + EDR, [d] Bluetooth 3.0 and Bluetooth 4.0, [e] Class 3, [f] Class 2, [g] Class 1, [h] for 868 MHz, [i] for 915 MHz, [j] for 2.4 GHz.

Wi-Fi have been considered for BWSNs. However, Wi-Fi technology is unsuitable for BWSN due to the high power consumption and large node size; thus, it is not considered in the next sections.

The main dissimilarities between the Bluetooth, ZigBee, and ANT wireless protocols used in bicycle monitoring are summarized in Table 12.1 (Hassan, 2012; Khssibi et al., 2013; Li & Zhuang, 2012; Rault et al., 2014). Bluetooth and ZigBee protocols are based on an IEEE 802 family of standards while ANT is a proprietary standard. Particularly, Classic Bluetooth, ZigBee, and ANT use a low data rate. Bluetooth and ZigBee are designed for a wireless personal area network communication distance of nearly 100 m, whereas ANT has shorter transmission distance of approximately 30 m (Casamassima et al., 2013). The RF-transmitted power level can be modified for all three wireless technologies. As seen in the table, the power consumption is the lowest power for ANT, followed by ZigBee and Bluetooth.

ZigBee protocol more typically is used in WSNs or control applications that have a lower requirement on data rate and complexity (Amzucu et al., 2014) as compared to Bluetooth protocol. In addition, ZigBee can support 65,535 nodes in the network, with simple procedures for removal or addition of nodes to the network, supports mesh configuration such that the failure of one node does not bring down the rest of the network, and lower power consumption. ZigBee/IEEE 802.15.4 has a data rate of 250 kbps (Abbasi et al., 2014; Wang, 2013) compared to ANT that caters higher data rate options of 250 kbps, 1 and 2 Mbps. Therefore, ANT is capable of transmitting over the air for a shorter duration compared to ZigBee for the same amount of data. ANT has been designed to stay in an ultra-low power mode (sleep mode) for long periods, quick wake up, speedily transmit for 150 µs, and return to sleep mode. In terms of latency, the ANT offers near zero latency (Smith, 2011; Tabish et al., 2013), ZigBee latency is below 30 ms, and Bluetooth has the highest latency of up to 10 s. Latency is a very crucial problem for race-cycling applications because the cyclist parameters must be transmitted and monitored in real time. ZigBee employs direct sequence spread spectrum (Nagarajan & Dhanasekaran, 2013), whereas Bluetooth uses frequency-hopping spread spectrum (Langhammer & Kays, 2012) schemes, to ensure the security of the wireless channel. However, the ANT employs an adaptive isochronous (Gasparrini et al., 2013) network technology to ensure coexistence with other wireless protocols.

From the table, it is clear that the classic Bluetooth wireless protocol is inappropriate for wireless monitoring of sports cycling performance due to the limitation of communication range, power consumption, and number of nodes in the network. ANT is more appropriate when the power consumption is a critical demand, whereas ZigBee is more suitable in terms of communication distance. In addition, latency is worse for Bluetooth and ZigBee compared to ANT.

12.3 Toward Battery-Free Internet of Bicycle

This section provides a taxonomy view of several energy savings and harvesting approaches specific to the bicycle wireless sensor nodes as seen in Figure 12.2. Advantages and limitations of each technique will be highlighted. At the end of the section, several techniques related to battery-free Internet of Bicycle (IoB) will be presented.

12.3.1 Sleep/Wake Strategy

The sensor nodes radio can be put in sleep mode with the *sleep/wake* strategy to reduce the power consumption of WSNs. In sleep mode, no data communication occurs. The sensor nodes wake to collect and transmit their data to a portion of the time and then return to sleep mode to conserve energy. The *sleep/wake* strategy can be performed with duty cycling (Anchora et al., 2014; Carrano et al., 2014), medium-access control protocols (Chen et al., 2014; Jing Liu et al., 2015; Khanafer et al., 2014), topology protocol (Karasabun et al., 2013; Zebbane et al., 2015), and passive wake-up radios (Ba et al., 2013; Farris et al., 2016). In the duty cycling method, sensor nodes were alternating between active and sleep modes, according to network activity. A low duty cycle saves a huge amount of energy relative to the traditional operation in the order of dozens of milliamperes but significantly increases communications latency (Rault et al., 2014). Therefore, a trade-off between energy consumption and latency should be taken into consideration when designing a WSN.

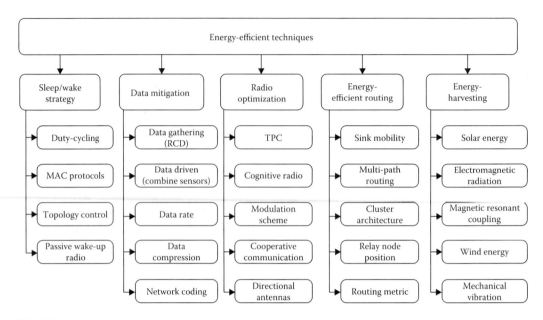

FIGURE 12.2
Energy-efficient techniques in WSN.

A number of research works have proposed effective duty cycle settings to enhance the power consumption of WSNs (Anchora et al., 2014; de Paz Alberola & Pesch, 2012; Rasouli et al., 2014). Topology control cannot be taken into account in BWSN because the existence of redundant sensor nodes added extra weight to a bicycle, increases hardware size and complexity, and increases aerodynamic resistance. In addition, this method is expensive, specifically when the sensor nodes are deployed in a vast area. Also, the passive wake-up radios method is inappropriate for BWSN because it has short communication distance of 3 to 6.7 m.

12.3.2 Data Mitigation

Data mitigation is another type of power consumption solutions for WSNs. This method aims to reduce the amount of transmitted data from source nodes to destination node. Data mitigation can be achieved with data gathering (Dhasian & Balasubramanian, 2013; Talele et al., 2015), data rate, data-driven techniques (Mesin et al., 2014), network coding (Qin et al., 2014; Rout & Ghosh, 2013), and data compression (Incebacak et al., 2015; Srisooksai et al., 2012). Among the data mitigation types, the data gathering and data rate were used to minimize the power consumption of the bicycle sensor and mobile nodes for track cycling application has been proposed in our research works (Gharghan et al., 2014, 2015, 2016). For data gathering, the reduction is accomplished by using the proposed redundancy and converged data algorithm for speed/cadence sensor node (Gharghan et al., 2014) and the average torque data for the bicycle torque sensor node (Gharghan et al., 2015). Moreover, by taking advantage of the data link between bicycle speed and cadence measurement method, the bicycle speed and cadence sensor nodes are merged into one node. Consequently, the transmitted data volume is minimized. Thus, the power consumption, cost, size, and complexity of the bicycle speed/cadence sensor node will also be reduced. The data rate of BWSN wireless protocol can be adjusted by bicycle mobile nodes to reduce their power consumption based on accurate distance estimation between the coach and the bicycle on the track. High data rate translates to lower receiver sensitivity and lower communication distance. Therefore, a trade-off between data rate and communication range is necessary.

12.3.3 Radio Optimization

Previous research works have shown that most of the power are dissipated in the RF components of WSNs compared to data processing units, such as microcontrollers and microprocessors (Mesin et al., 2014). Radio optimization schemes were used to reduce the power consumption of the RF components of sensor nodes using schemes such as transmission power control (TPC) (Cotuk et al., 2014; Ramakrishnan & Krishna, 2014; Yan et al., 2013), adaptive modulation (Rosas & Oberli, 2012; Sendra et al., 2011), cognitive radio (Mansoor et al., 2015; Naeem et al., 2013), directional antennas (Karapistoli et al., 2009; Khatiwada & Moh, 2015), and cooperative communication (Dong et al., 2015; Garcia et al., 2013; Jung et al., 2011). TPC scheme is considered for cycling application in Gharghan et al. (2016), where the RF-transmitted power of the ANT mobile node was modified based on accurate distance estimation between the cyclist and the coach. A combination of two power-reduction schemes (i.e., *sleep/wake* and TPC) was presented in Castagnetti et al. (2014) and Dai et al. (2012).

Note that adaptive modulation schemes cannot be used in the cycling sensor nodes based on 2.4 GHz ZigBee and ANT wireless protocols because only single modulation scheme is supported, specifically offset quadrature phase-shift keying (Setiawan et al., 2015) and Gaussian frequency-shift keying (Zhang et al., 2013) for ZigBee and ANT, respectively. It is also difficult to implement cooperative communication in BWSN because the installation of adjacent nodes in a bicycle equates to extra weight. In addition, the directional antennas approach cannot be employed in cycling application because of the adopted wireless protocols (i.e., ZigBee and ANT) have been designed with the use of omnidirectional antennas.

12.3.4 Routing Protocols

Routing protocols in WSNs are unlike other wireless networks due to the limited energy resources of sensor nodes, data are gathered from several nodes to a sink node and a single global address cannot be adopted in the random positioning of the sensor nodes (Arora et al., 2016). Due to the fact that sensor nodes are commonly battery powered (Hayes & Ali, 2016), the routing protocols in BWSNs are mainly designed to improve power consumption, prolong the network lifetime, use of cluster architecture, and sink mobility (Rault et al., 2014) that is suitable when the cyclists are moving in group.

12.3.5 Energy Harvesting

A major limitation of sensor nodes is limited battery capacity. An alternative method that has been used to address the problem of limited nodes lifespan is to employ energy harvesting. Numerous technologies have been developed to allow sensor nodes to harvest energy from different sources such as solar (Yongtai et al., 2012), electromagnetic radiation (Cortés-Sánchez et al., 2014), magnetic resonant coupling

(Li et al., 2015), piezoelectric microgenerators (Nechibvute et al., 2012), mechanical vibration (Yang et al., 2012), and kinetic and wind energy (Sudevalayam & Kulkarni, 2011). Compared with conventional sensor nodes, rechargeable sensor nodes can work continuously, ideally for an infinite lifetime. The ambient energy can be converted to electrical energy and can be employed directly to supply the sensor nodes or can be stored and used later on. The energy of the sensor nodes needs to be estimated to evaluate the battery-charging cycle. Hence, the sensor nodes vital parameters can be optimized such as duty cycling, sampling rate, and TPC to adjust their power consumption consistent with the periodicity and amount of the harvesting source. The energy harvesting mechanisms can be operated alongside batteries of the sensor nodes. For example, sensor node using solar energy to charge the batteries can work extensively during the daytime, whereas the sensor node can use their batteries based on power-reduction techniques such as sleep mode (i.e., duty cycle) in the night to save energy. Furthermore, a sensor node can enter higher sleep periods (i.e., low duty cycle) and low transmission power when the batteries of the sensor node have low residual energy (Nintanavongsa et al., 2013). Recent developments in wireless power transfer (WPT) are expected to significantly increase the lifetime of WSNs and make them continuously operational, as the WPT techniques can be employed to transmit power between transmitter and receiver devices without any contact between them. Many research works used WPT to charge the sensor nodes for different applications such as medical sensor nodes for implantable devices, unmanned aerial vehicles or drones, and WSNs (Griffin & Detweiler, 2012; Kong & Ku, 2016; Nintanavongsa et al., 2013; Xie et al., 2013; Yoshida et al., 2013).

Energy harvesting is not suitable for use in race cycling because the harvesting equipment adds extra weight. However, it is possible for use in day-to-day cycling.

12.4 Why Connected Bicycles?

The market value of data in the age of the IoT was estimated at US$19 trillion per year by the CEO of Cisco in 2014 (Ravindranath, 2014), of which $4.6 trillion was public sector savings because cities are working better through interconnectivity. This section explores several aspects of IoT end-user application in the bicycle sector herein referred as IoBs.

12.4.1 Smart City Driven

Cycling contributes significantly to the concept of smart city initiatives by preventing congestion because cyclist takes up less space than parked cars. Cycling also contributes to the health of the society and promotes the concept of green energy. Therefore, it is no surprise that there has been an increase in the number of people using a bicycle as their preferred mode of transportation (Wegman et al., 2012). Studies conducted in Monon Trail in Indianapolis concluded with the report that homes closer to urban bicycle greenways are 11% more marketable compared to homes without such facility (Fischer et al., 2010). As part of the Connected Streets project, London-based Future Cities Catapult has come up with innovative technologies that could improve cyclists' route choices, wayfinding and safety on the urban road, by using the IoT to complement existing cycling infrastructure.

These technologies that are currently in the prototype stage, include (1) a handlebar-mounted device to measure air quality and suggest cleaner roads, (2) a *route-rectifying* attachment for bike-sharing schemes, and (3) a helmet visor that offers alternative directions utilizing the idea of Head-Up Displays to overlay real-time data that provide a *visual nudge* to the user, offering safer directions based on the immediate streetscape around them, an adaptation to the Google Glass.

12.4.2 Intelligent Transport System Concept

The intelligent transport system involves the incorporation of artificial intelligence unit resting on the upper layer of wireless sensor application protocol stack. In reality, many projects have been initiated under the intelligent transport system concept driven by wireless sensors. SafeSpot, a project of INFRASENS, deploys cameras, laser scanners, and RFID-systems to detect dangerous situations. The advanced driver assistance systems use the in-vehicle sensor to increase the safety of cyclist. The Dutch InnoCross project aiming to decrease the number of accidents involving the cyclist by decreasing the number of red light violence of the cyclists with the aid of green light assistance (Thielen et al., 2012). With the rise of driverless or autonomous cars, streets will be made safer and air quality experienced by cyclists will also be improved. In addition, with advanced sensor technology and a networked traffic system, it is possible that in the future all mobility devices will be connected, eliminating traffic accidents almost completely (Mayne, 2016a).

In recent years, there have a lot of hype on electronic bicycle, or commonly known as e-bike. The time has come to integrate the e-bike industry into the smart mobility platforms. Using the capability of location-tracking capability, bike share is a particularly powerful tool as an information source about people's mobility habits. The cycling data collection could be used by the city for transport planning and infrastructure design. Furthermore, it could also be used as a marketing avenue for cities and businesses. For example, the mobility data could be used to plan targeted advertising to a rider individualized journeys, transmitted to the bike via a smartphone or to a docking station where the ride ends (Mayne, 2016b).

12.4.3 Economic Value Chain

The connected bike will open up a host of revenue opportunities for entrepreneurs wanting to create floating e-bike-rental schemes in cities. A similar service, but for scooters, has been introduced in Paris to decongest the city and boost the tourism industry, called Cityscoot (Hayashi et al., 2012). With Cityscoot, you can locate an available electric scooter via the Cityscoot smartphone app, and book the scooter for an emission-free drive around the city whenever you want, without any commitment and the downsides of owning them.

In May 2013, Citi Bike (Citi Bike NYC), a privately owned public bicycle-sharing system in New York City has been launched. To date, there are 8000 bikes and 500 stations across Manhattan, Brooklyn, Queens, and Jersey City, which marks as the largest bike share program in the United States. As of March 2016, Citi Bike has a total of 163,865 annual subscribers, with an average of 27,287 rides per day in 2015. In March 2014, the VeloCitta project (VeloCittà) was launched, where five participating European cities (Krakow, London, Szeged, Padua, and Burgos) are looking to develop a new approach to the organizational and financial aspect of the bike-sharing systems. On further analysis, the direct and indirect spill-over effects accruing from bicycling industry are \$46.9 billion on food, transportation, lodging, gift, and entertainment (Flusche, 2012). Another possible business opportunity is hosting organized rides and races events, which will draw thousands of people to the host communities and help boost the local economy.

12.4.4 Cyclist Safety Driven

Worldwide, more than 1.2 million people lose their lives in road accidents annually (WHO, 2015). According to U.S. Department of Transport study (USDOT, 2010), bicycle and pedestrian crashes accounted for more than 15% of traffic fatalities. As modern urban planning concepts emphasize and encourage nonmotorized modes of transportation, there is an important need for innovative and proactive methods of analyzing and improving cyclists' safety and flow efficiency. A cyclist is generally considered as vulnerable road users and exposed to both fatal and reduced injury (Thielen et al., 2012). To mitigate or at least reduce the occurrence of these accidents, several candidate solutions have emerged. In Goldhammer et al. (2012), cooperative sensor network has been proposed to reduce cyclist accident at road intersections. The idea behind their proposals is to reduce the cyclist accident probability by increasing their field of view that creates a bird's eye view of their surroundings. Besides that, there is considerable research conducted on cyclist safety focused on analyzing collision data and factors that contribute to the collision. These factors include attributes such as bicycle helmet usage, cyclist–vehicle-driving characteristics (e.g., speed limit and gap acceptance), conflicts at different road topologies, and separated cycle paths infrastructure design (Dill et al., 2012).

An inventive invisible cyclist safety helmet, called the Hövding, has been invented by students at the University of Lund in Sweden (Hövding). It is an airbag helmet that is actually a collar, worn around the neck, which inflates instantly during a crash using an algorithm that can distinguish normal cycling from accidents.

12.4.5 Technology Driven

Many cyclists use heart rate monitors strapped to their chests or wrists, and even the lactate threshold work on the calf to measure the level of lactic acid builds up in the bloodstream due to intensive exercise, causing discomfort and forcing the athlete to slow down. Such devices require the rider to look down to view their data on different displays, or take their hands off the bike. These smart wearables can present a safety issue, particularly when riding in a close pack of cyclists, or navigating crowds. The Solos smart cycling glasses (Solos) have been used for U.S. track cyclists training for the 2016 Olympic Games, which displays the real-time performance metrics culled from the athletes' bike sensors, heart rate monitors, and other self-tracking devices and provides micro-display of these at the right-hand corner of the cyclist's glass. This enables riders to view their critical stats without taking their eyes off the road.

The smart pedal by Connected Cycle (Smart Pedal) is an add-on to any normal bike, which comes with GPS, GPRS, Bluetooth, Accelerometer, and a backup battery that recharges itself through the bike electric circuit. The pedal automatically records the speed, route, incline, and calories burned on each bike trip. These statistics are sent to the cloud and made available to users through the Connected Cycle application available on smartphones. It provides a simple solution to bike theft, where the bike owner is instantly notified via smartphone when the bike is moved. The location capability not only prohibits theft but also allows the owner to locate where they last parked their bike. BitLock (Bitlock) and Skylock (Skylock) are examples of a smart keyless lock for a bicycle. Another stand-alone bicycle add-on that has been invented is the Bluetooth and GPS-enabled Helios (Helios bars), a set of aluminum handlebars that integrates headlight and blinker system for bicycles.

12.5 Potential Opportunities on the Internet of Bicycles

Without any ambiguity, the goal of IoB is the enablement of seamless integration of any potential bicycle sensor device into the Internet, thus creating a platform for new forms of machine-to-machine communication and machine-to-human communication. The implementation of IoB is expected to face some challenges as it may require substantial remodification of the protocol stack from the physical layer up to data representation and service composition.

12.5.1 Long Range, Low-Rate, Low-Power, Low-Cost Internet of Bicycle Wireless Backhaul

The three wireless network topologies capable of simulating the IoT concept are the cellular, IEEE 802.15 standards, and Low-Power Wide-Area Networks (LPWANs). At the moment, tested IoT applications are based on multihop short-range IEEE 802.15.4 standards that include: ZigBee and Bluetooth (Gubbi, Buyya, Marusic, & Palaniswami, 2013). Fortunately, this is same technology driving the BWSN. These technologies offered the needed degrees of freedom to implement many IoT user applications as they operate in the license-exempt radio spectrum, thrives on very low power consumption, and suitable for miniaturized devices, which are fundamental requirements for wireless sensor-driven IoT devices. However, ZigBee and Bluetooth suffer from poor coverage in the range of 10 to 100 m line-of-sight that limits their application especially in scenarios that require urban-wide coverage similar to the IoB concept.

On the other hand, the traditional cellular technology specifically, Third-Generation Partnership Project, which drives the universal mobile telecommunications system (UMTS) and long term evolution (LTE) wireless standards, was not originally designed to support IoT because of the huge number of devices requiring connectivity (Centenaro et al., 2015). The IoT traffic will be sporadic involving sending short packages and requires lots of signaling overhead. To illustrate this point, take for instance, when bicycles are fitted with road intersection accident warning systems, packets are transmitted only at the road intersection and when there is an impending danger. In essence, BWSN cannot be supported by cellular topology. Nonetheless, there are concerted efforts to revamp the 2G/GSM to cater for the expected IoT traffic through Cellular IoT architecture (Centenaro et al., 2015).

A temporary solution may lie in the so-called LPWANs. The LPWANs networks operate in the sub-GHz, unlicensed frequency bands using mesh or star topologies. A unique feature of LPWANs is the presence of collector node referred as the gateway to other collocated LPWANs. There are several LPWANs standards and proprietary technologies such as TV White Space technology IEEE 802.22 wireless rural area network (WRAN)

(Fitch et al., 2011), LoRa (*Long Range*) (Centenaro et al., 2015), SIGFOX™ and Ingenu™. The actual description of these technologies is beyond the scope of this chapter. In summary, for low requirement of IoT data rates, the wireless backbone for IoB should meet the listed criteria:

- *Operating frequency bands*: 784, 868, 915 MHz
- *Transmission range*: 10 to 15 km in rural areas, and 2 to 5 km in urban areas
- *Topology support*: For large mesh networking or star clusters
- Quality of service (QoS) Support for low latency devices

12.5.2 High-Rate 5G Backhaul

Industry experts believe that 5G will hit the market by the beginning of 2020. However to date, the decision on 5G spectrum band license has yet to be allocated. One of the goals of 5G is to cater for the rising aggregate spectrum demand driven by the IoT concept. Some IoT applications will require an aggregated data rate in order of Gbps that can only be met by the proposed 5G networks attributes (Mitola et al., 2014):

- Combines multiple physical layer (PHY), medium-access control, radio access network (RAN), mobility management IP core technologies involving multiple bands with different propagation characteristics, and access rules.
- Heterogeneous coexistence involving networks with transmission power, access scheme, cell topology consisting picocell, femtocell, microcell, and macrocell.
- Interference mitigation and management techniques with QoS constraints.

12.5.3 Energy-Efficient Transmission Algorithm

In the IoB concept, hundreds or even thousands of sensing nodes might be involved using data fusion rules. Hence, there is a need to exploit large numbers of these sensing nodes to create a cheap and energy-efficient but high-quality network. Therefore, protocols must be designed capable of achieving fault tolerance when confronted with node failure while minimizing energy consumption. Inspired by these problems, several efficient routing algorithms have been proposed such as Low-Energy Adaptive Clustering Hierarchy (Arora et al., 2016). The protocol provides localized coordination and control of cluster setup and operation, randomized rotation of cluster head selection, and data compression capabilities. Although performance wise, this protocol is efficient, however, no specific standard was attached to this protocol.

Exploiting this limitation, several energy-efficient algorithms have been proposed for bicycle sensor transmission. Baechler (2015) has proposed an energy-efficient low-power algorithm based on either ANT/ANT⁺, ZigBee, or Bluetooth standard. The proposal includes an adaptive power-reduction algorithm to reduce the power consumption at the speed/cadence sensor bicycle node. Furthermore, the algorithm is capable of switching between redundancy and converged data algorithm and the sleep/wake algorithm for torque sensor node power consumption as presented in Section 12.3. Several other works have utilized XBee and ANT wireless protocol (Dementyev et al., 2013). Both wireless protocols considered the low power consumption, low cost and weight, small size with acceptable communication range, and above all, they operate in the license-exempt 2.4 GHz industrial, scientific, and medical (ISM) band. Of recent, an improved version of the

Bluetooth technology known as Bluetooth Smart or Bluetooth Low Energy has emerged (Khssibi et al., 2013). Though several energy-efficient algorithms have been proposed for bicycle sensor, there is a need to explore the following in the context of IoB:

- Energy-efficient algorithm based on distance and traffic
- Low complexity cluster head selection
- Optimal cluster size
- Admission control policies

12.5.4 Energy-Harvesting Radios

Energy-harvesting techniques are an integral component of the modern sensor network to prolong the wireless sensor transmitter lifetime. The details on recent research on energy-harvesting radios have been presented in Section 12.3.5. However, the question remains on the most effective and optimal approach among the listed approaches to power the IoB radios.

The OpenBike (OpenBike) project mentioned in Section 12.5.5 is an example of energy-harvesting effort by converting pedal power to electric power and subsequently, charging the battery. Nonetheless, some research gaps still exist on energy-harvesting protocol design:

- Designing a protocol that exploits intermittent and random energy arrival
- Optimal energy-scheduling algorithm taking cognizance of energy harvest constraint
- Specific algorithm categorizing source of the energy power source
- Adaptive and response algorithm capable of selecting/switching to different sources of energy

12.5.5 Big Data Analytics

Each of the IoB drivers discussed in the previous section has the capacity to generate small data which will snowball into big data. Big data is simply making intelligence out of structured and unstructured dataset and has been increasingly adopted in the many sectors such as: health, sciences, engineering, education, and the society at large (Hashem et al., 2015). Not only the big data techniques possess the capacity to assist vulnerable road users in automating and computerizing their schedules, but they also provide reduced infrastructure maintenance cost, efficient management, and user access. There are many advantages of exploring the big data techniques in IoB concept, which include: extrapolating hidden/latent information as well as pattern recognition using any of the statistical tools. IoB is expected to fit into the four Vs of big data, which are volume, variety, velocity, and value (Ding et al., 2014). Wireless sensor lacks the memory and computing power to analyze data and makes an informed decision. Therefore, IoB will be inclined toward cloud computing. The advantages of using cloud computing schemes are the elimination of expensive hardware, a complex algorithm, dedicated hardware space, and a reduction in CAPital EXpenditure (CAPEX) and OPerating EXpenditure (OPEX). However, before these performances gain can be achieved, there is a need to address some of the basic challenges of big data in the context of IoB data capture, storage, search, analysis, and virtualization.

12.5.6 Network Security Challenge

As with other ICT-based system, security challenges must be overcome for IoB concept to be accepted. Security measures ensure the interactions between bicycle to bicycle or human operators to bicycle over the wireless link are conducted in a secured platform devoid of foreseeable security challenges. Several security mechanisms that must be integrated in IoT in general and adopted by the IoB have already been enumerated by the research community (Heer et al., 2011; Roman et al., 2013) such as:

- Protocol and network security characterized by the cryptographic algorithm for secured end-to-end communication.
- A simple and yet complex identity management scheme embedded in a highly integrated authentication framework driven by a user-friendly authorization module.
- Privacy policy to address concerns bordering on profiling, stealing, and loss of control must be designed. Anonymity and transparency are the key attributes of the privacy modules that must be upheld.
- Robust and fault tolerance are essential for IoB concept as a cyclist cannot afford to have a system that is not resistance to a fault.
- Trust and governance framework encompass machine learning and human control in the IoB concept. Although trust creates a platform for entities to deal with future actions of collaborating devices, governance invokes the human mechanism ability to punish, exclude, and readmit users.

Highlighting the possible security framework is important, it is equally important to expose some of the tactics that will be deployed to execute the security threats stated below:

- *Denial of service (DoS)*: This is the most widely anticipated techniques to launch security attack in a wireless system. DoS can target IoB system by jamming their communication channels. Thus, rendering the system inaccessible.
- *Physical damage*: Although DoS is technical involving the use of highly sophisticated tools, physical damage denotes a condition in which the attacker has direct access to the IoB hardware.
- *Eavesdropper*: Denotes incubation stage in which the attacker in a passive mode is focusing on monitoring, analyzing, and observing the data flow chain in the system.
- *Node capture*: This is the advanced form of eavesdropping in which the attacker decides to wirelessly and remotely control the IoB system.

12.6 Conclusion

IoBs is expected to be one of the *things* to capitalize from the IoT. The benefit of IoB has improved rider safety, greener transportation option to the urban community, improve health management of the cyclist, and new economies of scale opportunities. The IoB technologies compose of wireless communication and sensor technology, backhaul network, cloud computing, and big data processing. Energy has been identified as one of the most critical resources toward enabling IoB due to the requirement of low bicycle weight, mobility support, low cost, and low system complexity. However, battery technology still has not reached maturity. One of the grand challenges toward IoT is a lifetime of the sensors. An alternative way is to exploit another form of surrounding energy, known as energy harvesting. The taxonomy view on energy harvesting techniques has been presented. Ultimately, the future aim is to achieve battery-free supply. Besides that, several energy-efficient approaches have been identified. A prototype has been developed in this work as an example of a case study based on sleep/wake strategy and radio optimization. Finally, latest development and potential research topics in IoB have also been highlighted in this chapter.

References

Abbasi, A. Z., Islam, N., & Shaikh, Z. A. (2014). A review of wireless sensors and networks' applications in agriculture. *Computer Standards & Interfaces, 36*(2), 263–270.

Adibi, S. (2012). Link technologies and BlackBerry mobile health (mHealth) solutions: A review. *IEEE Transactions on Information Technology in Biomedicine, 16*(4), 586–597.

Amzucu, D. M., Li, H., & Fledderus, E. (2014). Indoor radio propagation and interference in 2.4 GHz wireless sensor networks: Measurements and analysis. *Wireless Personal Communications, 76*(2), 245–269.

Anchora, L., Capone, A., Mighali, V., Patrono, L., & Simone, F. (2014). A novel MAC scheduler to minimize the energy consumption in a Wireless Sensor Network. *Ad Hoc Networks, 16,* 88–104.

Arora, V. K., Sharma, V., & Sachdeva, M. (2016). A survey on LEACH and other's routing protocols in wireless sensor network. *Optik-International Journal for Light and Electron Optics, 127*(16), 6590–6600.

Ba, H., Demirkol, I., & Heinzelman, W. (2013). Passive wake-up radios: From devices to applications. *Ad Hoc Networks, 11*(8), 2605–2621.

Baca, A., Kornfeind, P., Preuschl, E., Bichler, S., Tampier, M., & Novatchkov, H. (2010). A server-based mobile coaching system. *Sensors, 10*(12), 10640–10662.

Baechler, H. (2015). Sensor apparatus and method for determining pedalling cadence and travelling speed of a bicycle: Google Patents number: US9075076B2, BMC Trading AG: Switzerland.

Balbinot, A., Milani, C., & Nascimento, J. S. B. (2014). A new crank arm-based load cell for the 3D analysis of the force applied by a cyclist. *Sensors, 14*(12), 22921–22939.

Bini, R. R., Hume, P. A., & Cerviri, A. (2011). A comparison of cycling SRM crank and strain gauge instrumented pedal measures of peak torque, crank angle at peak torque and power output. *Procedia Engineering, 13,* 56–61.

Bitlock. Smart keyless bike lock. Retrieved from https://bitlock.co.

Buratti, C., Conti, A., Dardari, D., & Verdone, R. (2009). An overview on wireless sensor networks technology and evolution. *Sensors, 9*(9), 6869–6896.

Caballero, I., Sáez, J. V., & Zapirain, B. G. (2011, June). *Review and new proposals for zigbee applications in healthcare and home automation.* Paper presented at the International Workshop on Ambient Assisted Living, Torremolinos-Málaga, Spain.

Carrano, R. C., Passos, D., Magalhaes, L. C., & Albuquerque, C. V. (2014). Survey and taxonomy of duty cycling mechanisms in wireless sensor networks. *IEEE Communications Surveys & Tutorials, 16*(1), 181–194.

Casamassima, F., Farella, E., & Benini, L. (2013). *Power saving policies for multipurpose WBAN.* Paper presented at the 23rd International Workshop on Power and Timing Modeling, Optimization and Simulation (PATMOS), Karlsruhe, Germany p. 83–90.

Casas, O. V., Dalazen, R., & Balbinot, A. (2016). 3D load cell for measure force in a bicycle crank. *Measurement, 93*, 189–201.

Castagnetti, A., Pegatoquet, A., Le, T. N., & Auguin, M. (2014). A joint duty-cycle and transmission power management for energy harvesting WSN. *IEEE Transactions On Industrial Informatics, 10*(2), 928–936.

Cavallari, R., Martelli, F., Rosini, R., Buratti, C., & Verdone, R. (2014). A survey on wireless body area networks: Technologies and design challenges. *IEEE Communications Surveys & Tutorials, 16*(3), 1635–1657. doi:10.1109/SURV.2014.012214.00007

Centenaro, M., Vangelista, L., Zanella, A., & Zorzi, M. (2015). Long-range communications in unlicensed bands: The rising stars in the IoT and smart city scenarios. *arXiv preprint arXiv:1510.00620. 23*(5): p. 60–67.

Chen, K., Ma, M., Cheng, E., Yuan, F., & Su, W. (2014). A survey on MAC protocols for underwater wireless sensor networks. *IEEE Communications Surveys & Tutorials, 16*(3), 1433–1447.

Citi Bike NYC The Bike-Share Planning Guide. Retrieved from www.citibikenyc.com.

Cortés-Sánchez, J., Velázquez-Ramírez, A., Lucas-Bravo, A., Rivero-Angeles, M. E., & Salinas-Reyes, V. A. (2014). On the use of electromagnetic waves as means of power supply in wireless sensor networks. *EURASIP Journal on Wireless Communications and Networking, 2014*(1), 1–10.

Cotuk, H., Bicakci, K., Tavli, B., & Uzun, E. (2014). The impact of transmission power control strategies on lifetime of wireless sensor networks. *IEEE Transactions on Computers, 63*(11), 2866–2879.

Dai, G., Qiu, J., Liu, P., Lin, B., & Zhang, S. (2012). Remaining energy-level-based transmission power control for energy-harvesting WSNs. *International Journal of Distributed Sensor Networks, 2012*, 1–12.

de Paz Alberola, R., & Pesch, D. (2012). Duty cycle learning algorithm (DCLA) for IEEE 802.15.4 beacon-enabled wireless sensor networks. *Ad Hoc Networks, 10*(4), 664–679. doi:http://dx.doi.org/10.1016/j.adhoc.2011.06.006.

Decuir, J. (2014). Introducing Bluetooth smart: Part II: Applications and updates. *IEEE Consumer Electronics Magazine, 3*(2), 25–29. doi:10.1109/MCE.2013.2297617.

Dementyev, A., Hodges, S., Taylor, S., & Smith, J. (2013, April). *Power consumption analysis of Bluetooth Low Energy, ZigBee and ANT sensor nodes in a cyclic sleep scenario.* Paper presented at the IEEE International on Wireless Symposium (IWS), Beijing, China.

Dhasian, H. R., & Balasubramanian, P. (2013). Survey of data aggregation techniques using soft computing in wireless sensor networks. *IET Information Security, 7*(4), 336–342.

Dill, J., Monsere, C. M., & McNeil, N. (2012). Evaluation of bike boxes at signalized intersections. *Accident Analysis & Prevention, 44*(1), 126–134.

Ding, G., Wu, Q., Wang, J., & Yao, Y.-D. (2014). Big spectrum data: The new resource for cognitive wireless networking. *arXiv preprint arXiv:1404.6508*, p. 1–13.

Dong, L., Tao, H., Doherty, W., & Young, M. (2015). A sleep scheduling mechanism with PSO collaborative evolution for wireless sensor networks. *International Journal of Distributed Sensor Networks, 2015*, 12, 517250.

Farris, I., Militano, L., Iera, A., Molinaro, A., & Spinella, S. C. (2016). Tag-based cooperative data gathering and energy recharging in wide area RFID sensor networks. *Ad Hoc Networks, 36*, 214–228.

Fischer, E. L., Rousseau, G. K., Turner, S. M., Blais, E. J., Engelhart, C. L., Henderson, D. R., ... Tobias, P. A. (2010). *Pedestrian and bicyclist safety and mobility in Europe.* Transportation Research Board (TRB): Washington, DC.

Fitch, M., Nekovee, M., Kawade, S., Briggs, K., & MacKenzie, R. (2011). Wireless service provision in TV white space with cognitive radio technology: A telecom operator's perspective and experience. *IEEE Communications Magazine, 49*(3), 64–73.

Flusche, D. (2012). Bicycling means business: The economic benefits of bicycle infrastructure. *The National Academies of Sciences, Engineering, and Medicine, 17*, 28.

Garcia, M., Sendra, S., Lloret, J., & Canovas, A. (2013). Saving energy and improving communications using cooperative group-based wireless sensor networks. *Telecommunication Systems, 52*(4), 2489–2502.

Gasparrini, S., Gambi, E., & Spinsante, S. (2013, October). *Evaluation and possible improvements of the ANT protocol for home heart monitoring applications.* Paper presented at the IEEE International Workshop on Measurements and Networking Proceedings (M&N), Naples, Italy.

Georgakakis, E., Nikolidakis, S., Vergados, D., & Douligeris, C. (2011). An analysis of Bluetooth, zigbee and Bluetooth low energy and their use in WBANs. In J. Lin & K. Nikita (Eds.), *Wireless Mobile Communication and Healthcare* (Vol. 55, pp. 168–175). Berlin, Germany: Springer.

Gharghan, S. K., Nordin, R., & Ismail, M. (2014). Energy-efficient ZigBee-based wireless sensor network for track bicycle performance monitoring. *Sensors, 14*(8), 15573–15592.

Gharghan, S. K., Nordin, R., & Ismail, M. (2015). An ultra-low power wireless sensor network for bicycle torque performance measurements. *Sensors, 15*(5), 11741.

Gharghan, S. K., Nordin, R., & Ismail, M. (2016). Energy efficiency of ultra-low-power bicycle wireless sensor networks based on a combination of power reduction techniques. *Journal of Sensors, 2016*, 21. doi:10.1155/2016/7314207.

Goldhammer, M., Strigel, E., Meissner, D., Brunsmann, U., Doll, K., & Dietmayer, K. (2012, September). *Cooperative multi sensor network for traffic safety applications at intersections.* Paper presented at the 15th International IEEE Conference on Intelligent Transportation Systems, Anchorage, AK.

Griffin, B., & Detweiler, C. (2012, May). *Resonant wireless power transfer to ground sensors from a UAV.* Paper presented at the IEEE International Conference on Robotics and Automation (ICRA), Saint Paul, MN.

Gu, L., & Stankovic, J. A. (2005). Radio-triggered wake-up for wireless sensor networks. *Real-Time Systems, 29*(2–3), 157–182.

Gubbi, J., Buyya, R., Marusic, S., & Palaniswami, M. (2013). Internet of Things (IoT): A vision, architectural elements, and future directions. *Future Generation Computer Systems, 29*(7), 1645–1660.

Hashem, I. A. T., Yaqoob, I., Anuar, N. B., Mokhtar, S., Gani, A., & Khan, S. U. (2015). The rise of "big data" on cloud computing: Review and open research issues. *Information Systems, 47*, 98–115.

Hassan, M. A. A. (2012, June). *A review of wireless technology usage for mobile robot controller.* Paper presented at the Proceeding of the International Conference on System Engineering and Modeling (ICSEM'12), Singapore.

Hayashi, H., Kagami, O., & Harada, M. (2012, June). *Results of field trials with wide-area ubiquitous network.* Paper presented at the IEEE MTT-S International Microwave Symposium Digest (MTT), Montreal, QC.

Hayes, T., & Ali, F. (2016). Robust ad-hoc sensor routing (RASeR) protocol for mobile wireless sensor networks. *Ad Hoc Networks, 50*, 128–144.

Heer, T., et al. *Security Challenges in the IP-based Internet of Things. Wireless Personal Communications,* 2011. 61(3): p. 527–542.

Helios bars. Transform any bike into a smart bike. Retrieved from www.kickstarter.com/projects/kennygibbs/helios-bars-transform-any-bike-into-a-smart-bike.

Hövding. Airbag for urban cyclists. Retrieved from www.hovding.com.

Incebacak, D., Zilan, R., Tavli, B., Barcelo-Ordinas, J. M., & Garcia-Vidal, J. (2015). Optimal data compression for lifetime maximization in wireless sensor networks operating in stealth mode. *Ad Hoc Networks, 24*, 134–147.

Jung, J. W., Wang, W., & Ingram, M. A. (2011, March). *Cooperative transmission range extension for duty cycle-limited wireless sensor networks.* Paper presented at the IEEE 2nd International Conference on Wireless Communication, Vehicular Technology, Information Theory and Aerospace & Electronic Systems Technology (Wireless VITAE), Chennai, India.

Karapistoli, E., Gragopoulos, I., Tsetsinas, I., & Pavlidou, F.-N. (2009). A MAC protocol for low-rate UWB wireless sensor networks using directional antennas. *Computer Networks, 53*(7), 961–972.

Karasabun, E., Korpeoglu, I., & Aykanat, C. (2013). Active node determination for correlated data gathering in wireless sensor networks. *Computer Networks, 57*(5), 1124–1138.

Kaseva, V. A., Kohvakka, M., Kuorilehto, M., Hännikäinen, M., & Hämäläinen, T. D. (2008). A wireless sensor network for RF-based indoor localization. *EURASIP Journal on Advances in Signal Processing, 2008*(1), 1–27.

Khanafer, M., Guennoun, M., & Mouftah, H. T. (2014). A survey of beacon-enabled IEEE 802.15.4 MAC protocols in wireless sensor networks. *IEEE Communications Surveys & Tutorials, 16*(2), 856–876.

Khatiwada, B., & Moh, S. (2015). A novel multi-channel MAC protocol for directional antennas in ad hoc networks. *Wireless Personal Communications, 80*(3), 1095–1112.

Khssibi, S., Idoudi, H., Van Den Bossche, A., Val, T., & Saidane, L. A. (2013). Presentation and analysis of a new technology for low-power wireless sensor network. *International Journal of Digital Information and Wireless Communications, 3*(1), 75–86.

Kohvakka, M., Suhonen, J., Hämäläinen, T. D., & Hännikäinen, M. (2010). Energy-efficient reservation-based medium access control protocol for wireless sensor networks. *EURASIP Journal on Wireless Communications and Networking, 2010*(1), 1–22.

Kohvakka, M., Suhonen, J., Kuorilehto, M., Kaseva, V., Hännikäinen, M., & Hämäläinen, T. D. (2009). Energy-efficient neighbor discovery protocol for mobile wireless sensor networks. *Ad Hoc Networks, 7*(1), 24–41.

Kong, P., & Ku, H. (2016). Efficiency optimising scheme for wireless power transfer system with two transmitters. *Electronics Letters, 52*(4), 310–312.

Kuhn, T., Jaitner, T., & Gotzhein, R. (2008). Online-monitoring of multiple track cyclists during training and competition (p81) *The Engineering of Sport, 7*, 405–412.

Langhammer, N., & Kays, R. (2012, August). *Enhanced frequency hopping for reliable interconnection of low power smart home devices.* Paper presented at the 8th International Wireless Communications and Mobile Computing Conference (IWCMC), Limassol, Cyprus.

Lanzisera, S., Mehta, A. M., & Pister, K. S. (2009, June). *Reducing average power in wireless sensor networks through data rate adaptation.* Paper presented at the IEEE International Conference on Communications, Dresden, Germany.

Li, H., Li, J., Wang, K., Chen, W., & Yang, X. (2015). A maximum efficiency point tracking control scheme for wireless power transfer systems using magnetic resonant coupling. *IEEE Transactions on Power Electronics, 30*(7), 3998–4008.

Li, M., & Zhuang, M. (2012, August). *An overview of Physical layers on wireless body area network.* Paper presented at the International Conference on Anti-Counterfeiting, Security and Identification (ASID), Taipei, Taiwan.

Liu, J., C. Chen, and Y. Ma (2012). Modeling and performance analysis of device discovery in bluetooth low energy networks, 1538-1543. Paper presented at the 6th *IEEE Global Communications Conference* (GLOBECOM) Anaheim, CA.

Liu, J., Li, M., Yuan, B., & Liu, W. (2015). A novel energy efficient mac protocol for wireless body area network. *China Communications, 12*(2), 11–20.

Liu, J., Sun, G., Zhao, D., Yao, X., & Zhang, Y. (2012). MCU-Controlling based Bluetooth data transferring system. *Procedia Engineering, 29*, 2109–2115. doi:10.1016/j.proeng.2012.01.271.

Maharjan, B. K., Witkowski, U., & Zandian, R. (2014, September). *Tree network based on Bluetooth 4.0 for wireless sensor network applications.* Paper presented at the 6th IEEE European Embedded Design in Education and Research Conference (EDERC), Milano, Italy.

Mansoor, N., Islam, A. M., Zareei, M., Baharun, S., Wakabayashi, T., & Komaki, S. (2015). Cognitive radio ad-hoc network architectures: A survey. *Wireless Personal Communications, 81*(3), 1117–1142.

Marin-Perianu, R., Marin-Perianu, M., Havinga, P., Taylor, S., Begg, R., Palaniswami, M., & Rouffet, D. (2013). A performance analysis of a wireless body-area network monitoring system for professional cycling. *Personal and Ubiquitous Computing, 17*(1), 197–209.

Mayne, K. (2016a). Smarter cycling series: What's the future for the bike industry in the world of driverless cars? Retrieved from https:ecf.com/news-and-events/news/smarter-cycling-series-what's-future-bike-industry-world-driverless-cars.

Mayne, K. (2016b). Cyclists and public bike sharing—The best kept secret in Smart City data collection? European Cyclists' Federation. Retrieved from https://ecf.com/news-and-events/news/cyclists-and-public-bike-sharing.

Mesin, L., Aram, S., & Pasero, E. (2014). A neural data-driven algorithm for smart sampling in wireless sensor networks. *EURASIP Journal on Wireless Communications and Networking, 2014*(1), 1–8.

Mitola, J., Guerci, J., Reed, J., Yao, Y.-D., Chen, Y., Clancy, T. C., … McGwier, R. (2014). Accelerating 5G QoE via public-private spectrum sharing. *IEEE Communications Magazine, 52*(5), 77–85.

Mraz, L., Cervenka, V., Komosny, D., & Simek, M. (2013). Comprehensive performance analysis of zigBee technology based on real measurements. *Wireless Personal Communications, 71*(4), 2783–2803. doi:10.1007/s11277-012-0971-1.

Naeem, M., Illanko, K., Karmokar, A., Anpalagan, A., & Jaseemuddin, M. (2013). Energy-efficient cognitive radio sensor networks: Parametric and convex transformations. *Sensors, 13*(8), 11032–11050.

Nagarajan, R., & Dhanasekaran, R. (2013, April). *Implementation of wireless data transmission in monitoring and control.* Paper presented at the International Conference on Communications and Signal Processing (ICCSP), Melmaruvathur, India.

Nechibvute, A., Chawanda, A., & Luhanga, P. (2012). Piezoelectric energy harvesting devices: An alternative energy source for wireless sensors. *Smart Materials Research, 2012*, 13.

Nintanavongsa, P., Naderi, M. Y., & Chowdhury, K. R. (2013, April). *Medium access control protocol design for sensors powered by wireless energy transfer.* Paper presented at the Proceedings of IEEE INFOCOM, Turin, Italy.

Olieman, M., Marin-Perianu, R., & Marin-Perianu, M. (2012). Measurement of dynamic comfort in cycling using wireless acceleration sensors. *Procedia Engineering, 34*, 568–573.

Padgette, J., Scarfone, K., & Chen, L. (2012). Guide to Bluetooth security. *NIST Special Publication, 800*, 121.

Qin, T., Meng, Y., Li, L., Wan, H., & Zhang, D. (2014). An energy-saving algorithm for wireless sensor networks based on network coding and compressed sensing. *China Communications, 11*(13), 171–178.

Ramakrishnan, S., & Krishna, B. T. (2014, February). *Closed loop fuzzy logic based transmission power control for energy efficiency in wireless sensor networks.* Paper presented at the International Conference on Computer Communication and Systems, Chennai, India.

Rasouli, H., Kavian, Y. S., & Rashvand, H. F. (2014). ADCA: Adaptive Duty Cycle Algorithm for energy efficient IEEE 802.15.4 beacon-enabled wireless sensor networks. *IEEE Sensors Journal, 14*(11), 3893–3902. doi:10.1109/JSEN.2014.2353574.

Rault, T., Bouabdallah, A., & Challal, Y. (2014). Energy efficiency in wireless sensor networks: A top-down survey. *Computer Networks, 67*, 104–122.

Ravindranath, M. (2014). Cisco CEO at CES 2014: Internet of things is a \$19 trillion opportunity. *The Washington Post.* (accessed on January 8, 2014). Retrieved from http://www.washingtonpost.com/business/on-it/cisco-ceo-at-ces-2014-internet-of-things-isa-19-trillion-opportunity/2014/01/08/8d456fba-789b-11e3-8963-b4b654bcc9b2_story.html

Rawat, P., Singh, K. D., Chaouchi, H., & Bonnin, J. M. (2014). Wireless sensor networks: A survey on recent developments and potential synergies. *The Journal of Supercomputing, 68*(1), 1–48.

Roman, R., J. Zhou, and J. Lopez, *On the features and challenges of security and privacy in distributed internet of things. Computer Networks*, 2013. 57(10): p. 2266-2279.

Rosas, F., & Oberli, C. (2012, May). *Modulation optimization for achieving energy efficient communications over fading channels.* Paper presented at the IEEE 75th Vehicular Technology Conference (VTC Spring), Yokohama, Japan.

Rout, R. R., & Ghosh, S. K. (2013). Enhancement of lifetime using duty cycle and network coding in wireless sensor networks. *IEEE Transactions on Wireless Communications, 12*(2), 656–667.

Ruiz-Garcia, L., Lunadei, L., Barreiro, P., & Robla, I. (2009). A review of wireless sensor technologies and applications in agriculture and food industry: State of the art and current trends. *Sensors, 9*(6), 4728–4750.

Sendra, S., Lloret, J., García, M., & Toledo, J. F. (2011). Power saving and energy optimization techniques for wireless sensor networks. *Journal of Communications, 6*, 439–459.

Setiawan, M. A., Shahnia, F., Rajakaruna, S., & Ghosh, A. (2015). ZigBee-based communication system for data transfer within future microgrids. *IEEE Transactions on Smart Grid, 6*(5), 2343–2355.

Shin, H.-Y., Un, F.-L., & Huang, K.-W. (2013, July). *A sensor-based tracking system for cyclist group.* Paper presented at the Seventh International Conference on Complex, Intelligent, and Software Intensive Systems (CISIS), Taichung, Taiwan.

Skylock. The smartest bike lock. Retrieved from www.skylock.cc.

Smart Pedal. The connected bike enabling solution. Retrieved from connectedcycle.com/pedal.html.

Smith, P. (2011). *US Patent CS-213199-AN.* Comparisons between low power wireless technologies. A&V Elettronica: Italty.

Soini, M., Nummela, J., Oksa, P., Ukkonen, L., & Sydänheimo, L. (2008). Wireless body area network for hip rehabilitation system. *Ubiquitous Computing and Communication Journal, 3*(5), 42–48.

Solos. Performance eyewear. Retrieved from www.solos-wearables.com.

Srisooksai, T., Keamarungsi, K., Lamsrichan, P., & Araki, K. (2012). Practical data compression in wireless sensor networks: A survey. *Journal of Network and Computer Applications, 35*(1), 37–59.

SRM. Retrieved from http://www.srm.de.

Sudevalayam, S., & Kulkarni, P. (2011). Energy harvesting sensor nodes: Survey and implications. *IEEE Communications Surveys & Tutorials, 13*(3), 443–461.

Tabish, R., Ben Mnaouer, A., Touati, F., & Ghaleb, A. M. (2013, November). *A comparative analysis of ble and 6lowpan for u-healthcare applications.* Paper presented at the 7th IEEE GCC Conference and Exhibition (GCC), Doha, Qatar.

Talele, A. K., Patil, S. G., & Chopade, N. B. (2015, January). *A survey on data routing and aggregation techniques for wireless sensor networks.* Paper presented at the International Conference on Pervasive Computing (ICPC), Pune, India.

Thielen, D., Lorenz, T., Hannibal, M., Köster, F., & Plättner, J. (2012, September). *A feasibility study on a cooperative safety application for cyclists crossing intersections.* Paper presented at the 15th International IEEE Conference on Intelligent Transportation Systems, Anchorage, AK.

USDOT. (2010). The national bicycling and walking study: 15-year status report., FHWA-PD-94-023. Washington, DC: Federal Highway Administration.

VeloCittà. Better use of bicycle share systems. Retrieved from http://velo-citta.eu/.

Walker, W., Aroul, A. P., & Bhatia, D. (2009, September). *Mobile health monitoring systems.* Paper presented at the 2009 Annual International Conference of the IEEE Engineering in Medicine and Biology Society, Minneapolis, MN.

Wang, J. (2013). Zigbee light link and its applications. *IEEE Wireless Communications, 20*(4), 6–7.

Wegman, F., Zhang, F., & Dijkstra, A. (2012). How to make more cycling good for road safety? *Accident Analysis & Prevention, 44*(1), 19–29.

WHO. (2015). Global status report on road safety, NLM classification: WA 275. Geneva, Switzerland: World Health Organization.

Wong, A. W., Dawkins, M., Devita, G., Kasparidis, N., Katsiamis, A., King, O., … Burdett, A. J. (2013). A 1 V 5 mA multimode IEEE 802.15.6/Bluetooth low-energy WBAN transceiver for biotelemetry applications. *IEEE Journal of Solid-State Circuits, 48*(1), 186–198. doi:10.1109/JSSC.2012.2221215.

Xie, L., Shi, Y., Hou, Y. T., & Lou, A. (2013). Wireless power transfer and applications to sensor networks. *IEEE Wireless Communications, 20*(4), 140–145.

Yan, R., Sun, H., & Qian, Y. (2013). Energy-aware sensor node design with its application in wireless sensor networks. *IEEE Transactions on Instrumentation and Measurement, 62*(5), 1183–1191.

Yang, C.-M., Wu, C.-C., Chou, C.-M., & Yang, C.-W. (2009, November). *Textile-based monitoring system for biker.* Paper presented at the 9th International Conference on Information Technology and Applications in Biomedicine, Larnaca, Cyprus.

Yang, Y., Yeo, J., & Priya, S. (2012). Harvesting energy from the counterbalancing (weaving) movement in bicycle riding. *Sensors, 12*(8), 10248–10258.

Yongtai, H., Lihui, L., & Yanqiu, L. (2012). Design of solar photovoltaic micro-power supply for application of wireless sensor nodes in complex illumination environments. *IET Wireless Sensor Systems, 2*(1), 16–21.

Yoshida, S., Noji, T., Fukuda, G., Kobayashi, Y., & Kawasaki, S. (2013). Experimental demonstration of coexistence of microwave wireless communication and power transfer technologies for battery-free sensor network systems. *International Journal of Antennas and Propagation, 2013*, 10.

Yu, K.-M., Zhou, J., Yu, C.-Y., Liu, J.-Y., Lee, C.-C., Chang, H.-W., & Hsieh, H.-N. (2009, December). *An event-based wireless navigation and healthcare system for group recreational cycling.* Paper presented at the 5th International Conference on Mobile Ad-hoc and Sensor Networks, Fujian, China.

Zebbane, B., Chenait, M., & Badache, N. (2015). A group-based energy-saving algorithm for sleep/wake scheduling and topology control in wireless sensor networks. *Wireless Personal Communications, 84*(2), 959–983.

Zhan, B., & Yu, X. (2013). Wireless node design in smart home system. In Y. Yang & M. Ma (Eds.), *Proceedings of the 2nd International Conference on Green Communications and Networks (GCN 2012)* (Vol. 3, pp. 545–552). Berlin, Germany: Springer.

Zhang, P., Sun, L., Zhang, P., Hou, R., Tian, G., & Liu, X. (2013, November). *Wireless network design and implementation in smart home.* Paper presented at the 6th International Conference on Intelligent Networks and Intelligent Systems (ICINIS), Shenyang, China.

13

Self-Adaptive Cyber-City System

Iping Supriana, Kridanto Surendro, Aradea Dipaloka, and Edvin Ramadhan

CONTENTS

13.1 Introduction: Cyber-City Systems and Seven Pillars of Life

The growth of urban population in the world nowadays encourages the growth of the level of needs which is increasingly out of control. This is based on the fact that the level of knowledge in the population continues to increase, so (ITU-T, 2014) it led to the social, economic, and environmental issues becoming connected to each other. This fact shows that the characteristics of the system to control a city has the diversity of interrelated elements and will lead to a variety of challenges and problems in the provision of public services (PSn).

Therefore, the city management needs to have an intelligent solution and creates a sustainable environment to (ITU-T, 2014) manage various infrastructure resources, environmental resource, monitoring the activities in the city, and more needs. All of this management will relate to the system requirements of the city-management system, such as transportation management, energy management, water management, waste management, municipal administration management, health services management, and other PSn management.

Internet of Things (IoT) is a concept that can answer the challenges. Nowadays, there are several definitions of it, but basically, they have the same goal to extend the benefits of Internet connectivity that can connect a variety of real-world objects as physical and virtual representation continuously. One of the implementation of IoT for the needs of city-management system is how to utilizing the functions of information and communication technology is developed on a concept called the cyber city or smart city or intelligent city or some other term, that basically aimed to meet the requirement of city management development through the systematic formulation. The proposed system is packed with various facilities, such as the efficiency and effectiveness of implementation, speed of access, accuracy of services, and others. But this various advantages will not be able to make a significant contribution if only temporary. This is related to the detail and completeness of the concept to meet the characteristic of the growth in a city.

In this discussion, we will describe a model about how to establish a system that can accommodate a variety of changes in the application environment, how to manage the change, and how to make the system has the capability to adapt and fit the changes. So that this model can be an inspiration in developing a scenario of systems that needed to manage a city. The model that has been designed inspired by the seven pillars of life (taklif) (Nabulsi, 2010), which is composed of the universe, law, reason, nature, lust, freedom of choice, and time. These seven pillars are represented through an agent-oriented approach BDI (belief, desire, intention). So the points of the inspiration are, we need to consider which part is associated with the domain model and which part is associated with control model in building the management software system, in this case, the city-management system. It is expected to create a computational model that can be used as guidance in developing a system of cyber-city systems and other IoT software systems in general, with the self-adaptive ability which can handle the issue of change and growth of systems.

13.2 The Basis of System Development

Self-adaptive systems (SAS) is presented as an alternative solution to the problems associated with the complexity of the IoT system, including the demands of autonomy, automation, adaptability, flexibility, scalability, reliability, speed, and others (Supriana and Aradea, 2015). SAS is a system that can automatically take appropriate action based on the knowledge that the system has about what is happening in the system itself, guided by the goal, and assisted by the users who are given the access (Ganek and Corbi, 2015). SAS can modify the system behavior in response to the changes in the system itself or the changes in its environment (Cheng et al., 2014). This definition shows that the system has the knowledge, which can be used to achieve its objectives through several means, and is able to make the adaptation in behavior based on events in the environment.

The characteristics of each element in the environment of an IoT-management system require a mapping and alignment with the system that will be the ultimate driving machine in the achievement of a city manager. So (Aradea et al., 2014), all of the activity that occurs in the environment of this IoT system requires a mechanism that can represent the behavior of the system in real-world change. SAS concept was developed as a solution to this problem. The constructed cyber-city system must have the capability to reflect the requirements of each element of a city, to the adaptability of changes in the system and its environment.

13.2.1 Representation of System

Basically, the universe was created for man to be given the mandate as the leader on earth, who is able to organize and manage all the resources of the earth. This practice is called the taklif (Nabulsi, 2010). The process of taklif comprises seven pillars (Nabulsi, 2010), namely (ITU-T, 2014) (1) universe, (2) reasonable, (3) nature, (4) religion or laws, (5) lust, (6) the freedom to choose, and (7) time. The explanation of all seven pillars of this taklif as can be seen in Table 13.1.

TABLE 13.1

Seven Pillars of the Taklif System

Pillars	Explanation
The Universe	All content of the earth, including the behavior of the components of the universe, can be grouped into two kinds of roles in the taklif system, that is, as the knowledge and as utilization. Human reasoning works on the principles of harmony with the universe and the principles of law. There are three principles of human reasoning, namely the principle of *cause and effect* (something happens for a reason), the principle of *goal* (something exists must be the goal), and the principle of *anticontradictory* (two things can apply when they are opposites).
Reasonable	The human mind will process two facts, namely the fact sensuous (reality) and logical facts (explanation). The logical fact has a higher position than the sensuous fact. Humans have the potential to think broadly but limited by the information provided by the senses. There is a lot of information about the past and the future that cannot be reached by the senses. If we relate this thinking to the concept of religion, we can state that religion is the only thing that can be delimiters in the laws or reason.
Nature	Naturally, humans were created with moral instincts. This moral instinct that defines the rules of human life. But there are some human attitudes which are contrary to this rule, this attitude appears because of character flaws and physical to the human. This weakness makes the natural rule to be violated, distorted, and even erased. This weakness is a challenge for humans, this rule can be restored to its natural state through learning or management that is able to organize and disenchant the people to respect the rule of nature.
Religion or Laws	Religion is a concept of how people should behave toward the creator, fellow human beings, and its natural surroundings. The truth of human thought and the natural order must be calibrated with the rules of religion; truth in religion is generally considered an absolute, whereas reason and nature can be distorted by lust. To determine the truth as the basis of the rule, the truth must meet four requirements, which is carried by religion or laws, approved by human reasoning, in accordance with nature, and confirmed by objective reality.
Lust	Lust is the desire to achieve the goal, with their desire for a need for pleasure and love of something inherent in man, the lust will make him it will strive to achieve it. But lust is neutral. Lust is controlled by the choice of reason and nature, whether just as satisfying a need or as a means of achievement of objectives as stated in the rules of religion or laws.
Freedom of Choice	Freedom of choice is central to the taklif system. In this case, the religious rules known as the belief which acts as the basis of the freedom to choose. In this world, there are many choices and each choice has its own consequences. Religious rules and nature that will be a reference for these choices.
Time	All interactions of the six pillars of the taklif system which have been described previously run in the dimension of space and time. Its space is the earth and the universe, whereas the time is a series of events recorded since humans are born until his death.

Source: Nabulsi, R. M. *7 Pilar kehidupan: Alam semesta, akal, fitrah, syariat, syahwat, kebebasan memilih, dan waktu,* Gema Insani, Jakarta, 2010; Agustin, R. D., Supriana, I., Model Komputasi Pada Manusia dengan Pendekatan Agent, Kolaborasi BDI Model dan Tujuh Pilar Kehidupan sebagai Inspirasi untuk Mengembangkan Enacted Serious Game, *Konferensi Nasional Sistem Informasi (KNSI),* ITB, 2012.

13.2.2 Construction of Model

Based on the model of the representation of system in Section 13.2.1, the construction model of developed software in the IoT concept, referring to the seven pillars of the taklif system, is mapped to the BDI Model (Bratman, 1987; Mora, 1999; Russel, 2003). A more detailed description can be found in the paper (Agustin and Supriana, 2012). The BDI Model is a theory of reasoning from a practical standpoint (practical reasoning) and were adopted in an agent-oriented architecture.

There are two main processes in practical reasoning, namely deliberate and means-end reasoning. Deliberate is the process of deciding specific circumstances to be achieved, called the intention. The means-end reasoning is the process to draw up a plan to achieve that intention. The deliberate process begins with understanding the stimulus from the environment using the belief, that the perception of the agent (which is assumed to be true) on the environment. The outcome of this process is the captured problem classification by the agent at the detected time. Based on these results, the system will search for the best reaction to solve the given problem. In the desires of the BDI model, these solutions are emerging as an alternative option which generated sporadically, so the various alternative solutions appear; but because it is generated sporadically, there may be a solution that is unrealistic or inconsistent.

Before adding a goal to the state of intention, all alternative choices in the state for this goal needs to be verified and validated, to be realistic with the condition of the agent, and consistent with the objectives set (Mora, 1999). Furthermore, in the state of intention, the system has obtained some options, to be realized as an action. In the end, the mental state will generate an action into some subactions to be executed by the effector. The mapping of these BDI models of the seven pillars of the taklif system is shown in Figure 13.1.

1. Pillars of components in the universe are the facts of each entity and its environment. The behavior of the components in the universe in the form of events can be captured by the information-model context. In this case, there are two kinds of roles, namely the introduction of knowledge and utilization, which means how to recognize and utilize the facts.

2. The next component is the representation of the pillars of sense. Pillars of sense mean to know the nature and to work based on the principles that correspond to the goal, nature, and religion or law (rules of how to behave/delimiters sense). This component evaluates the condition by processing the facts of reality and logical facts based on occurred events. In this evaluation phase, the facts will be processed, diagnosed, and classified based on the problem, so we can determine how to choose understanding.

3. The processing of facts would trigger some response that will be taken by the action component in the process of adaptation. There will be some choices in response to the prevailing condition in accordance with the rule of lust (functional requirements and nonfunctional requirements), and to generate the response options, it will be processed by the system itself without being given the freedom to choose.

4. Components of sense will choose several response options in accordance with the model of rules in religion or law (as described). In this process, selection and filtering are performed between some alternatives of priority response, so that the results are just a few alternatives with the highest priority.

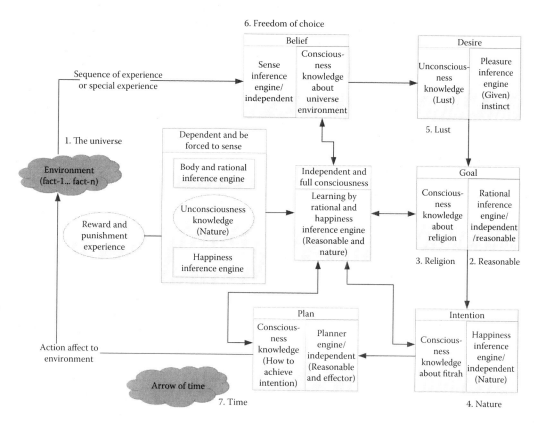

FIGURE 13.1
The mapping of the seven pillars in the taklif system into the BDI model.

5. Representation of the qolbu components will filter out some of the alternate options again from remaining responses, in reference to the understanding of nature's rule (instinctive understanding of the religion or the law of nature) until the obtained result is the best response that will be implemented.

6. Based on the selected response, some planning for the adjustment process can be arranged, and the effector of the body part system to execute the selected response can be determined in a certain order. The result of the execution of the effector is called as action.

7. The actions taken will get a reward or punishment based on the accuracy of the selected response with religious rules and nature should be. This stage will be a learning process for the system and will become a trigger for updating knowledge so that the system can determine the best response action in the future.

The characteristics of the seven pillars of the taklif system in conducting adaptation response are shown in Figure 13.1. The pillars are represented by 4 major rules in 3 BDI states, which is *desires* (R1), *goal* (R2 and R3), and the *intention* (R4); the states will determine the understanding of *belief*. The development strategy for each of these rules are as follows:

1. *Rule-1 (generic structures)*: Represent the real world in the form of abstract classes to model the goal.
2. *Rule 2 (conceptual rule)*: Identifying the generic structure that meets the basic specifications of fact; this rule will also guide the system in determining an alternative initial option based on the understanding of belief to environmental stimuli.
3. *Rule-3 (configuration rule)*: Define the configuration of the components, verify and validate an alternative option of conceptual results of operation rule.
4. *Rule 4 (specific rule)*: A special rule that is used if needed or if operating results do not meet certain criteria to obtain one or more options that are most relevant actions.

13.3 Self-Adaptive Model

This section will discuss the proposed model to build software systems with self-adaptive capabilities in the IoT concept. The framework was developed as shown in Figure 13.2, consisting of two main parts, as follows:

1. The domain model is represented as goal model. The domain model is a component that provides the IoT concept's basic functions and application logic. This model was developed based on goal-oriented approach, which can exploit the human-oriented abstractions such as agents and some other concepts, so it can represent real-world conditions.
2. The control model is represented as inference engine that controls or manages the target system through the adaptation logic. These models apply some patterns of action of the agent through the transition rules. The control model considers context-aware scenarios to represent a control strategy to meet the requirements of self-adaptation.

The domain model is a requirement of the IoT-goal system. Domain models are mapped to the BDI models for defining goals, planning, and other elements. The results of the mapping are represented by agent definition file and the plan specification. The concept of this agent represents the context of the IoT system as a fact in belief base and environment class for the purposes of monitoring functions (Supriana and Aradea, 2016). This approach quite provides the variability at run-time, but we extend it through the mapping into the BDI models, which is based on seven-pillar taklif with rule-based, which are constructed in a more general and more flexible, so it can provide a better way of analyzing the variability and can help in detailing the behavior of the system to meet the goals and adaptation action.

The control model is the setting of the behavior of the IoT system, tasked to monitor the environment and adjust the system if necessary, for example, reconfigure when there is a change goal, optimize themselves when the operation changes, and able

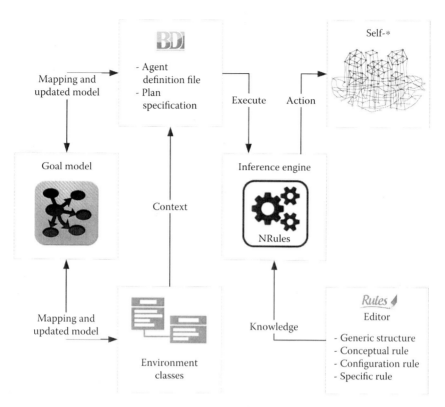

FIGURE 13.2
Framework for self-adaptive systems.

to handle certain types of errors. The pattern applied is an extension of the architecture event-condition-action (ECA) which in one or more *event* refers to the state of the current system. The models will change the application or IoT-system environment. Meanwhile, one or more *condition* refers to the time when a particular event occurs and the action rule is activated. So one or more *action* can be activated under certain conditions through the operation of rules in determining the behavior of the adaptive system.

13.3.1 Rule Representation

Any changes occurred in the IoT system is seen as a model of automata and the relationship of the possible state. Those changes can occur from an initial state until it reaches the final state. So the description of automata can be expressed as the model of rules which describes the relationship between the two states.

The representation of rules of these designed automata models, automatically, can detect the changes in the states and determine the necessary actions for adaptations. Scenarios are developed through the formation of the rule to identify similarities in the properties of each list of knowledge. In general, the rule-based system consists of 11 classes (Wu, 2004) as follows: rule, left-hand side (LHS), expression, fact, slot, composite expression, unary expression, binary expression, right-hand side (RHS), action, and function, as shown in Figure 13.3.

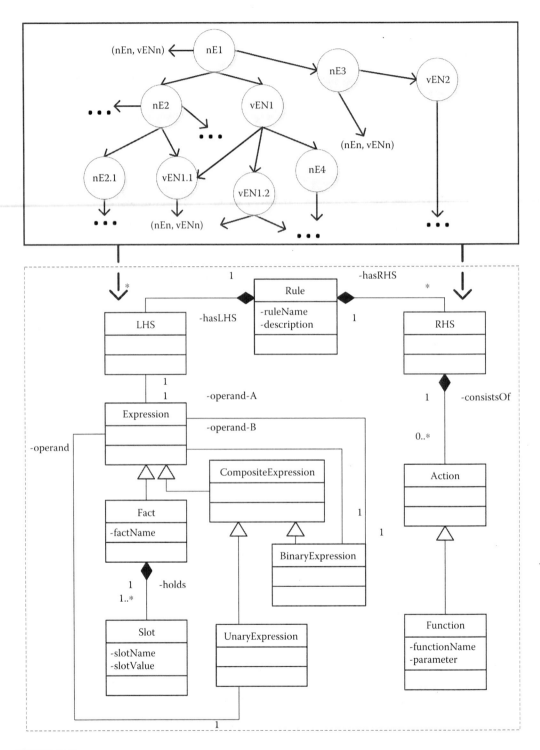

FIGURE 13.3
Knowledge tree and rule-based systems.

This rule-based system has a production rule in the form *if condition > then action >*. Action > on the RHS is concrete. This production rule is a practical consequence of a particular condition, whereas the other rules in the rule base are known as derivation rule. Conclusion > on the RHS is more abstract and becomes the logical consequence of certain conditions. However, a production rule can apply derivation rule using certain actions such as *assert* that expressed knowledge.

An action > defined on the RHS is determined by the expression on the LHS, which is conformity between rules and facts. Therefore, to create a self-adaptive ability, the rule-based system should be expanded as discussed in Section 13.3.2. The main objective is to equate the LHS in accordance with the formulation of the property slot, so there is a set of inner properties that become the source of facts, such as variables, constants, tuples, and so on. The basis of the mechanism is comparing the two or more collections of properties with a certain structure based on the criteria required.

In Figure 13.3 (bottom), we can see that the LHS is an expression that can be a fact, a single expression (pattern) which is characterized by a name and a collection of slots, or may be a composite expression with conditional elements (and, or, not). The conditional element is also used to connect a single expression (facts) or a composite expression. Class of fact has a containment reference in the class slot with the attribute "slotName" and "slotValue." The changes in this slot attribute will be detected based on any of the elements contained in the facts. For example, in the structure of the following basic rule:

$$\text{if } \langle(\text{fact-1}) \ (\text{fact-2})...(\text{fact-}N)\rangle \text{ then}\langle(\text{action-1})...(\text{action-}N)\rangle \tag{13.1}$$

Information models of facts have variations on slotName and slotValue based on the defined information context. The information models are defined to be a relation between entities in the system and the environment. Information models are also represented as an interconnected class and the facts in working memory. Thus, a system entity would have a slotName such as "nameEntity" (nE1, nE2, nE3 ... nEn) and has a view component for the environment as slot "viewEnvironment" (vEN) and can be composed of other facts, for example (vEN1, vEN2, vEN3,... vENn,), and each slot is possible to have more specific (vEN1.1, vEN1.2, ... vEN1.n), and the specification is also very possible to have other new slot. Representation of fact is a list of knowledge owned by each entity, and we define the list of this knowledge in a graph form or tree structure of knowledge, as we can see in Figure 13.3.

The tree structure of knowledge will add, delete, update, and customize every vertex automatically in response to the occurred state changes in accordance with the facts. The structure of property description of the knowledge tree can be managed by mapping the patterns of the seven pillars of taklif into the BDI model. So the representation of rules will be a reference for developed adaptation mechanisms to make a rule expansion of the ECA model (Daniele, 2006; Pires et al., 2008). This model can handle a wider scope of the solution space, and flexible, by using the following basic structure:

WHEN <event>; one or more transition state

IF <condition>; conditions that must be met to trigger action

THEN <action>; one or more actions when the event takes place

VALID TIME <time period>; period of adaptation actions

(13.2)

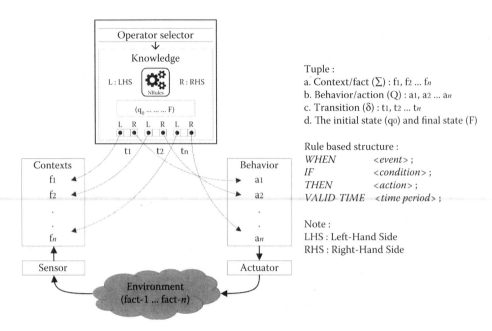

FIGURE 13.4
Model of rule representation.

Based on these descriptions, to define the needs of models, we set a tuple of the system, which are as follows: (a) Facts context information (Σ: f1 … fn) as the set of inputs, (b) Action of system behavior (Q: a1 … an) as a set of finite status, (c) Transition function or operator (δ: t1 … tn) as a function of the change, (d) Preliminary data (q0) as initial status, and (e) information targets status as the set of final status.

Figure 13.4 describes the relationship between variables and functions of the developed model. Information context is a set of inputs to be monitored as an event of the facts from any context information. The condition will be evaluated with reference to a specific event that occurs within the set of states, and it will trigger the adaptation action through the transition function of the prevailing condition; so the status of the target can be achieved by the action of the expected behavior.

Descriptions of the model are as follows:

1. Σ: Context information in the form of a set of facts (f1 … fn) is captured from the event environment and can determine the change in status.

2. Q: Behavioral actions (a1 … an) form the set of behaviors status of the system, starting from the initial state (q0) to the status of which can be targeted (F).

3. δ: Transition function (t1 … tn) as time of action adaptation based on the evaluation of the condition, to change the initial status (q0) to the status of the target (F).

13.3.2 Structure of Knowledge Base

With reference to the representation of prepared rules, any change in the IoT system state can be expressed as series of a node changes (Supriana et al., 1989); each situation will be modeled into a node, and changes of state will be described as the arc which connecting two nodes. So the used structure to handle the dependencies between the elements of knowledge is a directed graph.

The motivation for this development of the knowledge base structure is the reasoning that the approach of rule-based systems in most applications only focused on the rules of behavior of the software that is designed specifically, while the problems that may arise relate to the variability that can be served. This problem triggers the idea to formulate a model of the system structure to be more common and flexible, can be applicable to a wide scope of variability in the system and its environment, and expanded flexibly. The development of this model is targeted to meet the needs of the system to capture the variability in context and behavior of the system itself and its environment. This model can also set the criteria of assurance for the system management and its adaptation mechanisms.

In the design of the developed structures, each node in the graph has some slot categories, and each category has a number of attributes that can be increased, as shown in Figure 13.5.

The basic mechanism for knowledge management is prepared to meet the scope of a life cycle system: birth (create), use (use), updated (update), and deleted (delete). Developed-management module consists of initiation slot, expansion slots, adjustment slot, and inspection slot. The searching mechanism in a knowledge space is performed by using the access to the slot module, control of the trajectory, and the search delay. This mechanism can handle input consisting of context information, change and the output as behavioral adaptation actions of the sensors and actuator module. Overview model development can be seen in Figure 13.6.

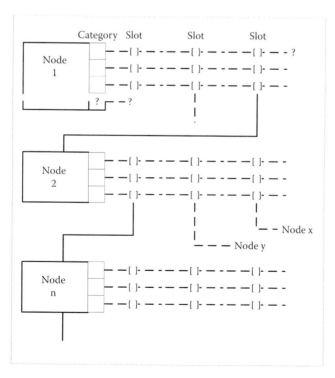

FIGURE 13.5
Structure of knowledge base.

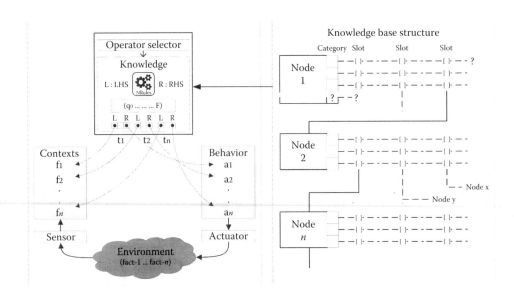

FIGURE 13.6
Representation and structure of knowledge base.

13.3.3 Reconfiguration Strategy

The main objective of this strategy is to define the domain model and determine the most relevant adaptation response options. The notation used to construct this algorithm consists of the following:

1. A model of goal (G) consists of connected nodes by its property attributes (P); each node also consists of a number of states (S_n), which may have a contribution to a soft goal.

2. Some states (S_n) may consist of the initial state (q0) and the status of the target (F); this state is influenced by several processing of facts (Σ: f_n) on the LHS, which will determine the behavior of the action (Q: a_n) and on the RHS, through the transition function (δ: t_n)

3. The process of recognition performed by observing Σ: f_n as a trigger for S_n in each G, until S can be set as a reference of plan preparation to realize a number of Q: a_n, including its mapping to the component (C).

To realize the construction of the system and the need for changes, we define four basic operations (create, read, update, and delete). This operation is mapped to the operation of the components (Kramer & Magee, 2007) that can be seen in Chart 13.1.

In addition, we define the rule that represents each element in the components of goal systems. This rule is compiled based on the rules of goal decomposition (Nakagawa et al., 2008); if the goal decomposition is AND-decomposition, then the parent goal will require multiple attributes of the relation (port) for each child goal with a one-to-one degree of relationship. But if the goal decomposition is OR-decomposition, then the parent goal will provide the conditional attributes of the relation (port) to each goal, with the one-to-many degree of relation. In this activity, we also need to set the properties for each goal (functional) and a soft goal (nonfunctional). Chart 13.2 illustrates the algorithm to define a primitive component based on the goal model.

Chart 13.1 Configuration Algorithm

Components Configuration

G ← (goal, soft goal)
C ← (components)
for all N in goalModel do
    G ← C: configuration components for operation
    for each (Σ, Q) ≠ ø do
        (create, read, update, delete) ← DiagnoseNode (G)
        create ← create component instance C from type G
        read ← connect or disconnect C1 to C2
        update ← set mode of C
        delete ← delete component instance C
        end for
    goalModel m ← reconfiguration(create, read, update, delete)
    enactModel(m)
end for

Chart 13.2 Diagnosis Algorithm

Goal Diagnosis

for all G in m do
    G ← addProperty(P)
    if G decomposition = AND-decomposition then
        G parent ← add multiplePort // sum of G childrenPort
        for all G in G children do
            port ← add providedPort
            C configuration ← DiagnoseNode
        else if G.decomposition = OR-decomposition then
            G parent ← add conditionalPort
            for all G in G children do
                port ← add conditional providedPort
                C configuration ← DiagnoseNode
            end for
        end if
        end for
    end if
end for

The development-control strategy to observe and regulate every component of the system that has been defined, by using the design pattern (Abuseta and Swesi, 2015) that was inspired by the model of MAPE-K (IBM, 2005), and modified in accordance with the previous system requirements. Chart 13.3 shows the algorithm to monitor and analyze the needs of the adaptation plan.

Chart 13.3   Observation Algorithm

States Observation

for all G in m do
    $m \leftarrow (\sum f_n)$ *// at run-time*
    $G \leftarrow getValue(P)$ *// time triggered or event triggered*
    for each value(S) in P do
        $S \leftarrow$ *combining internal and external value(P)*
        if S in S.target(F) ≠ P.threshold then
            systemState ← new S.system(S.instance) and
            systemStateLog ← save(S.system) and
            send information(S.system) to analyzerManager
        end if
        for each S.system in analyzerManager do
            analyzer ← update(logs) actual S.system
            search(S.system) in symptomRepository
            if symptom ≠ ø then
                create(adaptationRequest) and
                update(adaptationRequest) for plan specification
            else
                addSymptom to symptomRepository and
                create(adaptationRequest) and
                send information(adaptationRequest) for plan specification
            end if
        end for
    end for
end for

There are a number of properties on the model of goals that must be read and measured (observation) in concurrency. In the Java programming language, multithreading techniques are relevant to meet this need. This activity represents the state system at run-time by using time-triggered or event-triggered; this process is performed in response to any request or event. The state of the system at run-time is represented by a combination of the internal and external property value; the desired state is directed by a goal and a soft goal. Violations of the state detected by the threshold level of each goal property and a new run-time system state are stored in the system state logs for analysis.

Violation of the goals and requirements of the system is analyzed based on the symptom repository. The symptom repository is a collection of symptoms that has been set for the system to avoid and heal itself. This component is part of the knowledge base. The symptom repository is equipped with facilities to add new symptoms that occur during runtime analysis through operating add symptom. The symptom class consisting of an associative array is used to store any symptoms, which represent each event of symptoms and its value represents the condition relating to the event. Some programming techniques for this class and interfaces are (Abuseta and Swesi, 2015) map (Java), dictionary (Python), and associative arrays (PHP). If the analysis detects the presence of symptoms, then the plan component will receive a demand signal for adaptation and reconfigure the system based on the engine policy. Chart 13.4 shows the algorithm reconfiguration plan.

The policy engine provides the high-level goals that control the operation and functions of related systems. The general form is ECA rules; these rules are used to determine the action when the event occurred and meet certain conditions. The policy engine is represented as a knowledge base, which provides an interface for system administrators to determine and change the system policy. In our version, the engine is being expanded with the model of rule editor, and also additions or changes in the specification can be performed by editing the knowledge base directly, or reset back into the system.

Chart 13.4 Reconfiguration Algorithm

Reconfiguration Plan

for each adaptationRequest(S.system) do
 init ← set work (Σ, Q)
 while δ {t_n (f_n, a_n) | $n \neq \emptyset$} do
  δ ← find that the LHS of the operator match with work say it found
  if found is only one then
   RHS ← set work
   else if work is equivalent with target then
   stop succeed
  end if
  if found is more than one then
   found ← set work one of the found
   backs ← put rest of found
  end if
  if found is empty then
   else if backs is empty then
   stop failed
   else
    backs ← set work one of backs
   end if
  end if
 for all $\delta(t_n)$ is found do
  a ← construct correctiveAction(addAction)
  changePlan ← newChangePlan(a_n)
  send changePlan to one or more executors
  for each a in executor do
   actuator ← update(a_n) // one or more actuators
   S.system ← reconfiguration m with actuator
   // set new value for C (DiagnoseNode)
   systemStateLog ← saveState(S.system)
  end for
 end for
end for

Each request of adaptation will be represented as a state in the system, which detected based on the occurred events. A set of δ can be expressed as δ {*tn* (*fn*, *an*) | $n \neq \emptyset$}, with "*fn*" is the fact of context (the property of goal) and "*an*" is the behavior of the action that expected to a specific contextual *n*. The quality of the inference engine depends on the selection of state for the adaptive action on the RHS class. This quality is also determined by the expression in the LHS class. The expression is a match between the rule and the facts. For example, comparing two or more collections of properties with certain structures, based on the required criteria. The used key strategies are the forward strategy, reusing (reuse) the existing fundamental component, and matching the required specifications.

The changing plan contains the set of action for the execution component to perform the adaptation action. This action should be carried out in some specific order, for example, sequentially or concurrently or both. The execution component will use a number of actuators to set new values of the property in the target system and its environment. The corrective action of the execution component on a number of the actuator will bring the system back to the desired state or an acceptable state, and then this state will be saved in the log of the system state by the actuator.

13.3.4 Component-Based Software

In this section, a description of the reconfiguration strategy that has been discussed in Section 13.3.3 will be implemented in the component-based software. The component is a specific function unit which interacts with using the interface through the delivery of a service. The component-based system is a collection of components which are connected to each other, so it has the desired functions. In general, the interface between the components is a standard form, so the components can be connected and have a good interoperability. Based on the variety of this interface, the components can be divided into several types, namely (a) the input component, (b) the process components, (c) output component, and (d) the delivery components.

The input component is a component that only has an interface that generates information. The process component is defined as a component which has a pair or more connections between the input interface and an output interface. In general, information for the output interface in the components of the process have a functional relationship with the information on the input interface. Some of the functions can be owned by any other process components, including structural, preparation, grouping, and counting function.

Output component is a component that only has the interface that receives the output information from the processing component. And the delivery component is a supporting component that mediates the delivery between the input component and output component. By using this delivery component, the system can become more compatible to be accessed in the variety of platforms; it is mostly needed in a system with the concept of IoT.

Each component can provide independent services, connected or collaborated services, and free active services. This independent service allows a component to perform its function itself, actively, without requiring the presence of other components. The components with the connected services or collaborated services are the component with the interaction of two or more components connected. In particular, this process can take place sequentially or in parallel, and the component of the free active services is a component that is active and running but has not performed its function until there are triggers that appear from any other active component.

One of the approaches that can be adopted to formulate the architecture of the software components, for the needs of a cyber-city system in the concept of IoT, is to utilize the architectural description languages, which was developed in the Darwin model (Magee et al., 1996; Magee et al., 1996; Hirsch et al., 2006). This model is a declarative component-based architectural description languages that supports a hierarchical model and graphical modeling, easy to use in determining the software components, including interconnection and structure by using formal modeling notation. The specification of this software component architecture will be incorporated into the requirements of the self-adaptive cyber-city system in Section 13.4, as a motivation of the case to provide a more concrete.

13.4 Case Study of Cyber-City System

This section discusses the case about the development of the cyber-city system in the IoT concept; the goal of this case is to make the cyber-city system with the ability to capture requirements of each element of a city and has a good adaptability to changes and have the capability to handle the growth of the system in its environment.

13.4.1 Modeling of System

The development of the cyber-city system can be initiated through the domain modeling in the IoT concept; the system will identify each entity until the processes or resources can be defined, and the relation between each other can be explained. Each of these processes is connected to each interface that serves as executor of the system. This activity can be done with the goal-based modeling approach. Furthermore, each defined element in the system is mapped into the software components, until the operational mechanism of the system can be determined, whether the components are set to dynamically or statically.

The stages are performed to identify and define the IoT context of adaptation mechanisms; so in every identified context, we can build a knowledge tree, as shown in Figure 13.7. As an example, according to the context of the user interface identification mechanism, the structure of knowledge can anticipate the growth of user requirements, both of internal or external stakeholders. In another example, with the basic mechanism of service-context identification, the knowledge structure can anticipate the changes and growth of requirements on the activity and the process of converting data into system services. In addition, it is possible to add an identification mechanism of other contexts if necessary, based on the changes of facts or the growth of the system.

The illustration in Figure 13.7 shows that the structure of actor tree and data tree will be automatically added, deleted, updated, and adapted to every vertex, in response to changing circumstances and growth occurs. As an illustration, every actor tree in the cyber-city system will be linked to the requirements and changes of actors in government, private, personal, and so on, as a user of the system. Data tree will be associated with the process of change and growth activity of each actor in the service system, whether it is in the process of trade, education, transportation, health, tourism, entertainment, and so on. Based on the formulation of this knowledge-tree structure, we can define the requirements of PSn in the form of the service catalog as shown in Table 13.2.

The service catalog describes details of the service for each actor. Duration is the service availability of time, for example, 2×24 hours, 8:00 a.m. to 4:00 p.m., and so on. Options are a class of selected service based on the level of interest or influence to the goal, for example, level 1: A response (1-2), level 2: B response (2-3), and level 3: C response (3-4). Moreover, the

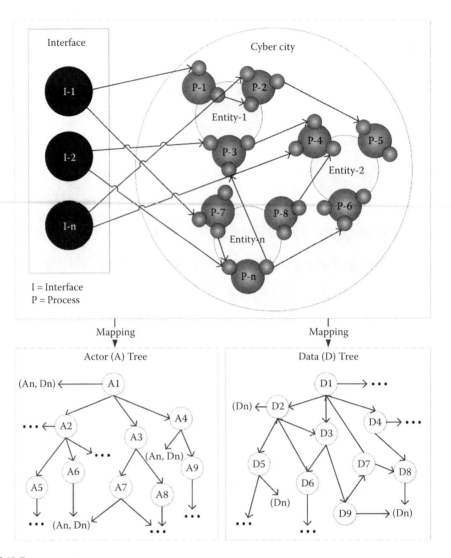

FIGURE 13.7
The mapping of elements in a cyber-city system into the knowledge tree.

TABLE 13.2

Service Catalog

| Service Catalog | Description | | | User | | | |
	Duration	Option	Response	U_1	U_2	U_3	U_n
Service-1.1	24×7	2	2	√	–	√	√
Service-1.2	24×7	2	1	√	–	√	–
Service-1.n	24×7	1	1	–	√	–	√
Service-2.1	24×7	1	2	√	√	√	–
Service-2.2	08–16	3	4	–	√	√	√
Service-2.n	08–16	1	1	√	√	–	–
Service-n.m	24×7	3	3	–	–	√	√

response is the incident recovery activity of the service class, which is classified based on the level of seriousness of the incident, for example, level 1: 10 min, level 2: 30 min, level 3: 1 hour, and level 4: 2 hours, depending on the impacts of incidents.

The changes that can lead to incidents in the service catalog can be categorized as the facts of (fn) changes, for example:

1. f1: The political policy of the city can be associated with the changes in infrastructure and the structure of the city departments.

2. f2: The growth of activity, regarding changes to business processes and growth of PSn.

3. f3: Service providers, changes legalization, are caused by the demand for a new form, features, and algorithms, in the application of external services.

The category of service changes can be considered to set the service catalog for strategies, management, and recovery. Request for a fast response to make changes in the system can be made by performing a matching process between the occurred facts against the facts in the structure of the knowledge tree. The service catalog consists of business-process service which contains the entire PSn and technical service (Tsn). Tsn relates to all technical specifications of the technology. Figure 13.8 shows the adaptation mechanisms from capturing events by the agent from the fact (f1, f2, ..., fn) in the system environment. The agents also capture the context of information from the service catalog (PSn, Tsn) until the action implementation.

An event is the information of any context which is connected with the specification of the new facts inside environmental and service catalog (Σ = f1, f2 ... fn; PSn, TSn). The created condition will evaluate the situation in accordance with the occurred specific events, including the captured characteristics of the service and the changes in every context (Q = fn $\longleftrightarrow$ PSn, fn $\longleftrightarrow$ TSn). So the evaluation of the service level agreement (SLA) is performed to select the most appropriate action behavior.

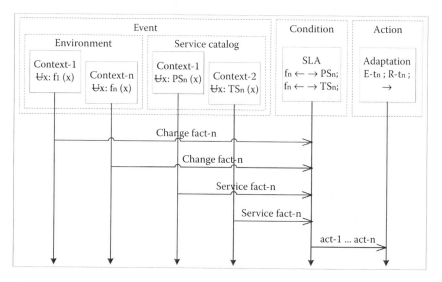

FIGURE 13.8
The dynamic behavior of the system.

Based on the prevailing condition, the action of adaptation is performed ($\delta = E\text{–}tn; R\text{–}tn$). The reconfiguration process of evolution ($E\text{–}t1, E\text{–}t2, \ldots E\text{–}tn$) at the time ($t$) made upon the consideration of the SLA attainment targets evaluation, although the action of reconfiguration ($R\text{–}t1, R\text{–}t2, \ldots R\text{–}tn$) or the handling changes that have not been stipulated in the SLA. Basically, this PSn adaptation action deals with the authorization of the service portfolio, for example, when the service should be updated, replaced, maintained, refactored, dismissed, and rationalized. All of these services represented adaptation actions within the control of changes to the structure of the tree of knowledge.

13.4.2 System Configuration

The mechanisms are developed to manage the service catalog, as shown in Figure 13.9. The modeling process uses the TAOM4E tools (TAOM4E n. d.). A goal of the cyber-city system services is managed using three subgoal activities such as the user interface, service application, and service providers. Each goal can have subgoals and also plans that contribute to the soft goal.

The modeling services through the goals element, as shown in Figure 13.9, are mapped into the software components as shown in Figure 13.10. Each goal and plan will form a primitive component so that there are four component groups—3 groups serve to detect the context and 1 group as the reconfiguration components. The three groups are "detect context-1" for the user interface, "detect context-2" for the service application, and "detect context-3" for the service provider. All three groups of these components require a group of components that has the function to perform the reconfiguration; the

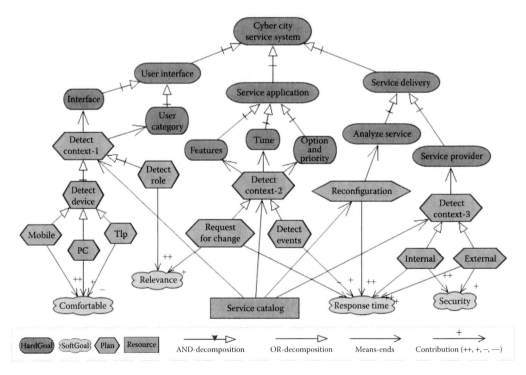

FIGURE 13.9
The service catalog management for a cyber-city system.

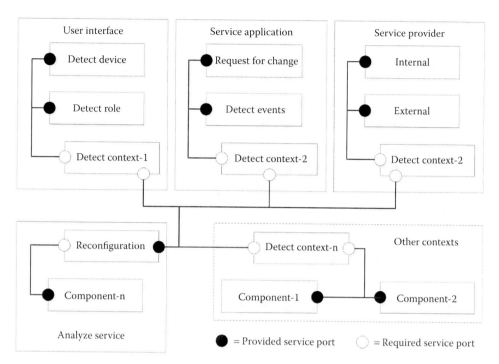

FIGURE 13.10
Software components configuration.

reconfiguration components will determine the adaptation strategies for each component that needs it, so the presence of this component is generic. That means if at run-time systems require the addition of new components, then automatically, these components can be added as shown in the dashed line in Figure 13.10. It is an example of adding a new component to detect new context. Likewise, if a component is not needed or need to be changed or need to be reused, it can be executed at run-time.

The request for change component is primitive for the component that requires reconfiguration. The model is initialized by the variable the component dynamic, which is determined by conditions at run-time. Achievement of the goal-service application consists of features, time, options, and priority. This achievement is the target of the component. This target is activated based on the soft-goal criteria and constraints. The soft goal is a consideration of the contribution, against the nonfunctional system, whereas constraints are the domain assumptions, which are set based on the strategy reconfiguration as described in Section 13.3. The components of "Detect Context-2" is a composite component that describes the primitive component interconnect and needs with the component reconfiguration through the type of OR relations in accordance with the modeling goal.

Based on the configuration of developed components, each component will establish a knowledge according to the needs of their respective functions. Figure 13.11 will illustrate the knowledge tree from the results of software components operations. For example, the user interface and its goal plan "detect context-1" will form a tree of actors or users, whereas the goal service application with the plan "detect context-2" will form a tree of data or services, so does the other goals and plans.

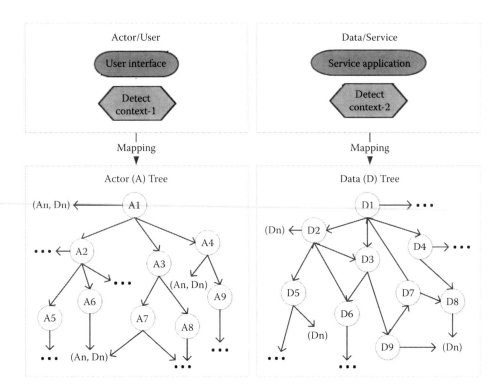

FIGURE 13.11
The mapping of components into the knowledge tree.

13.5 Discussion and Conclusion

13.5.1 Discussion

SAS is a research area that has a broad spectrum because it deals with a variety of factors and various conditions that must be recognized. Discussion performed by limiting the scope of the basic goal-oriented in requirements engineering approach.

In this article, we introduced an approach that is based on the goal-oriented in requirements engineering approach by inserting the additional elements through the primitive construction of the system as discussed in Sections 13.3 and 13.4. Figure 13.12 shows the overall architecture models, which can be used as a guide for developers in constructing a SAS in general, and self-adaptive cyber-city system in the IoT concept in particular.

Description of the system-development model in Figure 13.12 is as follows:

1. The development activities begin with the goal modeling to represent the IoT domain model. The available tools at this stage can be adjusted with an adopted goal-model approach, for example:
   a. TAOM4E (TAOM4E n. d.) to model Tropos (Bresciani et al., 2004)
   b. Objectiver (http://www.objectiver.com/) to model KAOS (Dardenne et al., 1993)

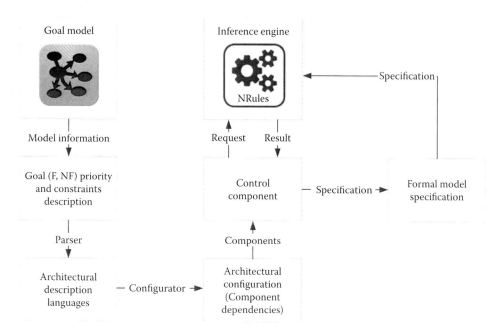

FIGURE 13.12
Development model of a system.

    c. Open OME (http://www.cs.toronto.edu/km/openome/)

    d. STS-ml (http://istar.rwth-aachen.de/tiki-index.php?page=i*+Tools)

    e. OmniGraffle (http://www.omnigroup.com/omnigraffle), and others

      Or we can arrange a description of this goal by developing its metamodels, for example, using EMF (http://www.eclipse.org/modeling/emf/updates/ or http://www.eclipse.org/modeling/emf/downloads/).

2. By the time of preparing the description of these goals, we need to ensure that any functional (F) requirement should be represented as a goal and nonfunctional (NF) requirements as soft-goals that can reflect the quality criteria of the system. How to determine the parameters of the priorities and constraints are discussed in Section 13.3.

3. After mapping the description of the goals that have been made in the architecture description language, we configure it into a software component. This is accomplished by using a software to generate the required components.

4. The preparation of the software component architecture can refer to Section 13.3.4. The supporting code generator tools such as java emitter templates (JET) Template (http://www.eclipse.org/modeling/m2t/downloads/) can be used to transform the eclipse modeling framework (EMF) meta-model. Or some other tools mentioned in point a, which has been equipped with the code generation facility.

5. The component control will coordinate all the components to give the inference function. This component generates formal model specifications to perform the selected reasoning tasks.

6. The components of inference engine can be developed by drafting rules, as discussed in Section 13.3.3. This rule-based system can also be modeled using the EMF meta-model tools. Some of the tools that can be used as an engine rule are CLIPS (http://www.ghg.net/clips/CLIPS.html), JESS (http://www.jessrules.com/jess/download.shtml or http://herzberg.ca.sandia.gov/jess/), jDREW (http://www.jdrew.org/jDREWebsite/jDREW.html), Mandarax (http://mandarax.sourceforge.net/), or using the Datalog Rules and DLV inference engine (Leone et al., 2006), and others; they can be adapted to the description of the requirements of the goal.

The selection of tools that will be used in this IoT system development must be definitely determined by studying the alignment between the description of goals and the needs of the inference engine by considering any capabilities of tools to cover all requirements from the developed concept.

13.5.2 Conclusion

Characteristics of the cyber-city system have much diversity of elements and are very dynamic; it is caused by the very quick growth of the environment. The proposed solution in this article is formulated through two approaches. First, importance to understand and capture the variability in context (IoT concept) and behavior of the system. This is realized through the domain modeling context with the logic-based approach. Second, the availability of the knowledge structure and the quality of the inference engine that can handle the scope of more broad and flexible solution space. This is realized through the modeling of the inference context with the rule-based approach.

The main highlight of discussion in this article is the provision of the SAS to meet the scope of a system life cycle in the IoT concept; this relates to the control of automation which is the part of the main concept of IoT. The given idea is an opportunity that can be followed up in the control of automation of the future of the IoT. So the concept of IoT system is expected to become more aware and adaptive; all orders can be performed automatically, organized, intelligent, acting independently in context and situation, or event environment.

The adaptation strategies as a mechanism for reconfiguration are proposed to overcome this problem; a series of modifications were developed to represent the problem space. In addition, this strategy also prepares the system to be able to conduct a search of the solution space as an appropriate alternative solution automatically in the implementation of the control strategy. The developed concept was inspired by the seven pillars of taklif systems which are the representations of life in the universe. These pillars are formulated into the rules of systems with self-adaptive capabilities. We believe this proposed concept will contribute to the provision of software systems in the context of a dynamic environment, and especially in the construction of the cyber-city systems in the IoT concept. We expect this concept to accommodate any issues of diversity and dynamism inside the system and its elements.

References

Abuseta, Y. and K. Swesi. Design patterns for self adaptive systems engineering, *International Journal of Software Engineering & Applications* (*IJSEA*), 6(4), pp. 11–28, 2015.

Agustin, R. D. and I. Supriana. Model Komputasi Pada Manusia dengan Pendekatan Agent, Kolaborasi BDI Model dan Tujuh Pilar Kehidupan sebagai Inspirasi untuk Mengembangkan Enacted Serious Game, *Konferensi Nasional Sistem Informasi* (*KNSI*), ITB, 2012.

Aradea, I. Supriana, and K. Surendro. An overview of multi agent system approach in knowledge management model, *International Conference on Information Technology Systems and Innovation* (*ICITSI*), IEEE, School of Electrical Engineering and Informatics, ITB, 2014.

Bratman, E. M. *Intentions, Plans, and Practical Reason*, CSLI Publications, Stanford, CA, 1987.

Bresciani, P., A. Perini, P. Giorgini, F. Giunchiglia, and J. Mylopoulos. TROPOS: An agent-oriented software development methodology, *Journal of Autonomous Agents and Multi-Agent Systems*, 8(3): 203–236, 2004.

Cheng, B. H. C., K. I. Eder, M. Gogolla, L. Grunske, M. Litoiu, H. A. Muller, et al. Using models at runtime to address assurance for self-adaptive, In *Model@run.time Foundations, Applications, and Roadmaps, Lecture Notes in Computer Science* (*LNCS*), Vol. 8378, pp. 101–136, Springer, 2014.

Daniele, L. M. Towards a rule-based approach for context-aware applications, Master Thesis, University of Cagliari, Italy, 2006.

Dardenne, A., A. van Lamsweerde, and S. Fickas. Goal directed requirements acquisition, In *Selected Papers of the Sixth International Workshop on Software Specification and Design*, pp. 3–50, Elsevier Science Publishers B.V., Amsterdam, 1993.

Ganek, A. G. and T. A. Corbi. The dawning of the autonomic computing era, *IBM Systems Journal*, 42(1): 5–18, 2003.

Hirsch, D., J. Kramer, J. Magee, and S. Uchitel. Modes for software architectures. In V. Gruhn and F. Oquendo (Eds.), *EWSA 2006, Lecture Notes in Computer Science* (*LNCS*), Vol. 4344, pp. 113–126, Springer-Verlag, Berlin, Germany, 2006.

IBM, *An Architectural Blueprint for Autonomic Computing*, IBM, Hawthorne, NY, 2005.

ITU-T, Overview and role of ICT in smart sustainable cities Telecommunication Standardization Sector, 2014.

Kramer, J. and J. Magee. Self-managed systems: An architectural challenge, *Future of Software Engineering, FOSE'07*, pp. 259–268, The ACM Special Interest Group on Software Engineering (SIGSOFT), IEEE Computer Society, Washington, DC, May 2007.

Leone, N., G. Pfeifer, W. Faber, T. Eiter, G. Gottlob, S. Perri, and F. Scarcello. The DLV system for knowledge representation and reasoning. *ACM Transactions on Computational Logic* (*TOCL*), 7(3): 499–562, 2006.

Magee, J., N. Dulay, S. Eisenbach, and J. Kramer. Specifying distributed software architectures. In *Fifth European Software Engineering Conference* (ESEC95), Barcelona, September 1995.

Magee, J., J. Kramer, and D. Giannakopoulou. Analysing the behaviour of distributed software architectures: A case study. In *5th IEEE Workshop on Future Trends of Distributed Computing Systems*, pp. 240–245, 1996.

Mora, M. C. BDI models and systems: Reducing the Gap. In J. P. Muller et al. (Eds.), ATAL'98, LNAI 1555, pp. 11–27, 1999. Springer-Verlag, Berlin, Germany, 1999.

Nabulsi, Dr, R. M. *7 Pilar kehidupan: Alam semesta, akal, fitrah, syariat, syahwat, kebebasan memilih, dan waktu*, Gema Insani, Jakarta, 2010.

Nakagawa, H., A. Ohsuga, and S. Honiden. Constructing self-adaptive systems using a KAOS model, In *Proceedings of the SASOW*, pp. 132–137, IEEE, 2008.

Pires, L. F., N. Maatjes, M. van Sinderen, and P. D. Costa. Model-driven approach to the implementation of context-aware applications using rule engines, In M. Mühlhäuser, A. Ferscha, and E. Aitenbichler (Eds.), *Constructing Ambient Intelligence. AmI Workshops. Communications in Computer and Information Science*, vol. 11, pp. 104–112, Springer, Berlin, Germany, 2008.

Russell, S. and P. Norvig. *Artificial intelligence, a Modern Approach*, 2nd Prentice Hall, Upper Saddle River, NJ, 2003.

Supriana, I. and D. Aradea. Model self-adaptive sebagai landasan sistem untuk menunjang penumbuhan komunitas, *Keynote Paper Seminar Nasional Teknologi Informasi dan Komunikasi (SENTIKA)*, Vol. 6, Yogyakarta, Maret 18–19, 2016.

Supriana, I. and D. Aradea. Automatically relation modeling on spatial relationship as self-adaptation ability, *Internationl Conference on Advanced Informatics: Concept, Theory and Application (ICAICTA)*, Vol. 2, IEEE, Bangkok, Thailand, 2015.

Supriana, I., S. Wahyudin, and A. Mulyanto. Pengembangan motor inferensi untuk aplikasi sistem pakar dalam model diagnosa, Technical Report, School of Electrical Engineering and Informatics, Bandung Institute of Technology, Kota Bandung, Indonesia, 1989.

Tool for Agent-Oriented visual Modelling for Eclipse (TAOM4E) and its plugin t2x (Tropos4AS to Jadex), developed by the Software Engineering group at Fondazione Bruno Kessler (FBK), Trento, available, including the extensions, at http://selab.fbk.eu/taom.

Wu, C. G. Modeling rule-based systems with EMF, Eclipse Corner Article, Copyright (c) 2004 Chaur G. Wu. All rights reserved, 2004.

14

The Role of Big Data, Cloud Computing, and Mobile Technologies in the Development of IoT Ecosystems

Janusz Wielki

CONTENTS

14.1 Introduction

The dynamic development of information technology moves it into the next phase of its evolution at a rapid pace. This is the phase in which the first computers are able to receive data from virtually any type of physical object (Bisson et al. 2013). The situation results from the fact that components such as sensors, processors, and software are becoming much more a part of the product, and at the same time, have the capability to connect to the accompanying outer infrastructure layer due to their connectivity. Moreover, connecting literally everything to the Internet becomes economically justified (World Economic Forum 2015). As a result, there were considerably more capabilities to create cyber-physical systems, that is, systems integrating the physical and virtual worlds (Industrie 4.0 Working Group 2013; Schwab 2016). The concept of the Internet of Things (IoT) has emerged in this context (Barbier et al. 2013; Gartner 2014; Ericsson 2015). Simultaneously, it is not possible to create the technological infrastructure of the IoT system without the use of three other leading technologies, which in itself carry an enormous transformation potential. This applies to cloud computing, advanced big data analysis tools, and mobile solutions (Burkitt 2014; Aharon et al. 2015; Heppelmann and Porter 2014, 2015; ITU 2015). In this context, the main motivation connected with this chapter is to broaden knowledge about the role of three technologies (big data, cloud computing, and mobile technologies) in a process of creation solutions based on the IoT concept. Its basic goal is an attempt to identify the role and place which these three complementing technologies play in building IoT ecosystems and their functioning, and what kind of mutual relations among them exists.

14.2 The Concept of IoT and the Most Important Factors Stimulating Its Development

The analysis of the literature on the subject shows that there is no single universal definition of the IoT, and the authors emphasize issues differently. And so (Tapscott and Tapscott 2016), they use the term IoT in the context of *installing intelligence into existing infrastructure*, whereas Keen refers to the *intelligent devices on the network* (Keen 2015). Heppelmann and Porter propose a more extensive definition. According to them, the phrase *Internet of Things* was created to reflect a growing number of smart, connected products and highlight the new opportunities they can represent (Heppelmann and Porter 2014). In this context, Ericsson clarifies that a connected device is "a physical object that has an IP stack, enabling two-way communication over a network interface" (Ericsson 2015). On the other hand, a considerable group of definitions applies the concept of IoT to the use of sensors, actuators, and communication technologies embedded in physical objects and the opportunities that they bring about. And so, according to Bauer et al. (2014), "the Internet of Things refers to the networking of physical objects through the use of embedded sensors, actuators, and other devices that can collect or transmit information about the objects." On the other hand, according to Bisson et al. (2013), "the Internet of Things refers to the use of sensors, actuators, and data communications technology built into physical objects—from roadways to pacemakers—that enable those objects to be tracked, coordinated, or controlled across a data network or the Internet." While Aharon et al. (2015) define the IoT as "sensors and actuators connected by networks to computing systems. These systems can monitor or manage the health and actions of connected objects and machines. Connected sensors can also monitor the natural world, people, and animals." Similarly, Dobbs et al. (2015) define IoT. According to them, the concept of the IoT is related to the "embedded sensors and actuators in machines and other physical objects that are being adopted for data collection, remote monitoring, decision making and process optimization in everything from manufacturing to infrastructure to health care." On the other hand, the IoT according to the European Commission is a term referring to "devices of all sorts (…) equipped with sensors and actuators, connected to the Internet, allowing them to monitor their status or the environment, to receive orders or even to take autonomous action based on available information" (European Commission 2015). However, the definition given by The White House (2014) emphasizes the IoT as the capability of devices to communicate with other devices. According to this definition, the IoT is "a term used to describe the ability of devices to communicate with each other using embedded sensors that are linked through wired and wireless networks." In turn, the consulting company PwC defines IoT as an ecosystem created by the equipment, tools, and facilities. They define the IoT as the "ecosystem of Internet-connected devices, operational tools and facilities" (PWC 2016). For Schwab, the IoT is a relationship between things and humans that can be realized through the use of various technological solutions. As he writes, IoT "in its simplest form, it can be described a relationship between things (products, services, places, etc.) and people that is made possible by connected technologies and various platforms" (Schwab 2016). Rifkin sees IoT in a similar spirit. As he writes, "the Internet of Things will connect everything with everyone in an integrated global network" (Rifkin 2014). In turn, Cisco proposes a completely different definition of the IoT. For them, "IoT is simply the point in time when more *things or objects* were connected to the Internet than people" (Evans 2011). There are a number of factors stimulating the development of the IoT. Undoubtedly, the continuous technological progress is a key factor. It caused the critical building blocks of computing, needed to create IoT ecosystems and growing exponentially for years reached such a level of technical maturity, and correspondingly low price that the universal implementation of solutions based on this concept became possible (Brynjolfsson and McAfee 2014). This applies to such aspects that are a key to its implementation as capabilities and prices of sensors,* computing power, disk space, network connections, both wired and wireless networks, or progress in the area of creating low-cost power systems. Actions taken to eliminate the limitations of connectedness between the components of the IoT ecosystems have become an important element stimulating the development of the IoT. An example is the introduction of IPv6 solving the problem of exhausting the available IP addresses, as a result, removing one of the major barriers to the development of IoT (Barbier et al. 2013).

14.3 The Most Important Opportunities and Possibilities Related to the Use of IoT by Enterprises

When it comes to the senior management view on the key benefits of using the concept of the IoT, one can indicate top five categories according to a study conducted by KPMG in 2015. The key ones are the improvement of productivity (20%) and the acceleration of innovation cycle in organizations (20%). Further important aspects are the possibilities of greater diversity of products or services offered by the company (16%), cost reduction (13%), and an increase in the level of profitability (10%) (Figure 14.1).

At the same time, the senior management indicated the areas in the same study in which they see the greatest potential for monetization for their companies resulting from their use of solutions from the realm of the IoT. Those related to consumers and consumer markets (22%) are by far dominant. The next such areas are as follows: technology (13%), aerospace, space and defense (10%), education (10%), and automotive/transportation (9%) (KPMG 2015). However, it is worth noting that, as McKinsey's forecasts indicate, the target value generated by IoT solutions for B2B will be twice the size of those created for the consumer market (Aharon et al. 2015) (Figure 14.2).

When it comes to the possibilities related to the use of smart-connected products, which allow the creation of value in IoT solutions, result from four basic functionalities (Figure 14.3).

The first ones are sensors built into smart products that enable to monitor the following:

- Its health, operation, and use
- External environment

---

* For example, sales of sensors being a key component of each IoT ecosystem have been increasing at a rate of 70% per annum since 2010 (Bisson et al., 2013).

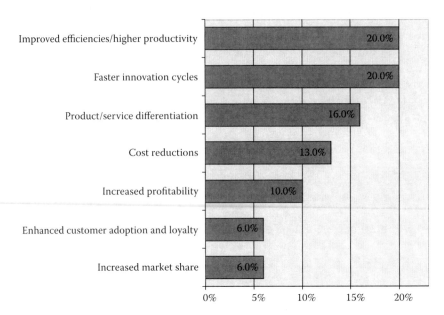

FIGURE 14.1
The key benefits of using the concept of the Internet of things.

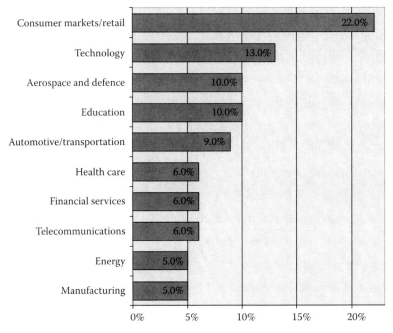

FIGURE 14.2
IoT monetization opportunities.

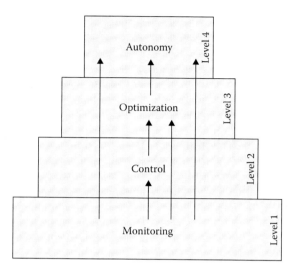

FIGURE 14.3
Four types of opportunities offered by smart, connected products.

In turn, software contained both in the product itself and the infrastructure layer offers even more possibilities; namely, it enables to remotely control the product and its functions and personalization of its operation on a scale that previously was not possible to achieve. Monitoring capabilities and the resulting wide-data stream combined with the control opportunities offered by smart products allow the organizations to optimize their performance in an extremely wide range. This applies to aspects such as the significant improvement in the operation of the product itself or its predictive diagnostics and repair (Heppelmann and Porter 2014). The three functionalities discussed above allow the smart products to achieve an unprecedented degree of autonomy. Moreover, different levels of autonomy can be achieved, which are as follows (Heppelmann and Porter 2014):

- Independent operation
- Independent coordination of operation with other products or systems
- Autonomous improvement of the product and its personalization
- Autonomous self-diagnosis and support service

These four functionalities offered by smart-combined products provide business organizations with two basic types of opportunities. They are related to the following (Aharon et al. 2015):

- Reconstruction of business processes
- Creation of new business models

When it comes to the reconstruction of business processes, there are a number of related opportunities that differ depending on the settings to which they apply. According to McKinsey, one can identify nine such areas with the greatest economic impact (Figure 14.4).

These are, for example, changes in the design of domestic appliances in the home setting; namely the process is based on the analysis of usage-based design. In the retail setting, the key application areas of the IoT solutions are as follows: checkout automation and goods layout optimization or individualization of promotional activities in stores.

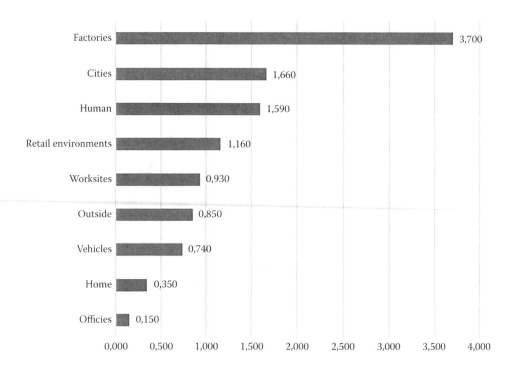

FIGURE 14.4
Potential economic impact of IoT in 2025 (in trillions USD).

On the other hand, in the office setting, the IoT applications include organizational redesign and the monitoring of employees or the use of augmented reality for training purposes. Regarding the transformation of business processes in the factories, the key issues include optimization of operations and the related improvement in productivity, optimization of the use of equipment and supplies, predictive maintenance, the maintenance of equipment, and the occupational safety and health. Another setting that enables a deep reconstruction of business processes is related to nonstandard production worksites, such as oil and gas or construction sites. In this case, the most important capabilities are similar to the previous case, with additional IoT-enabled R&D activities. When it comes to the setting associated with the various types of vehicles, the key areas of IoT-based solutions include repair and condition-based maintenance, equipment design based on an analysis of their usage or presales analytics. Vehicles are also related to the capabilities of reconstruction of logistics processes. This applies to things such as real-time routing, the use of connected navigations, or the use of transport monitoring systems (Aharon et al. 2015). Regarding the opportunities associated with the implementation of new business models, we can identify ten basic types in this context. These include the following:

1. Business models based on *anything-as-a-service* concept

2. Business models based on the use of new forms of outsourcing

3. Business models based solely on the data and their use

4. Business models based on additional services related to the physical product offered to customers

5. Business models based on smart products that are sources of added value for the customer

6. Business models based on behavioral profiling
7. Hybrid business models
8. Business models based on offering IoT platforms
9. Business models based on offering comprehensive IoT infrastructure solutions
10. Business models based on offering extended services

The model being the product-as-a-service is the *mainstream* in the first group. Its development is related to much more widely observed processes of migration from the customer buying the product to the one in which the manufacturer retains ownership of it, and the customer uses it, paying for its real use. The development of functionalities related to intelligent, connected products provides big opportunities in this area. Rolls Royce, which offers their engines to airlines in the *power-by-the-hour* model, is one of the companies pioneering in this field. In this model, the airlines pay for the real engine use time, instead of incurring a one-off cost of its purchase and additional costs of maintenance and repair. Xerox also employs such a model by monitoring the actual use of their photocopiers via the installed sensors (Heppelmann and Porter 2015). The development of smart systems also allows for the implementation of business models based on offering new forms of outsourcing. Another such example is the Pacific Control company operating in Dubai. It offers the remote monitoring of buildings, airports, and hotels based on the IoT (The Economist 2010).

For the third category, the development of smart-connected devices enables one to collect vast amounts of different types of data that can be used to create business models based on their usage. Skyhook Wireless company, which offers specific information acquired based on geolocation data they collect, is an example of this approach. They can include information such as which local bars will be the most popular on a specific day and time, how many people will go near the billboard at a given date and a specific time, or what is the density of people in a specific urban area on a given day and time. The company uses anonymous geolocation data collected from mobile users of its services in every major American city during the past 24 months to carry out this analysis (The Economist 2010; Mims 2010).

The development of the IoT also enables one to implement business models based on providing customers with additional services related to the physical product they purchased and use. Caterpillar company is one example of this approach. Specialized teams advise customers on how to optimize the deployment of equipment, when a smaller number of machines suffice and how to achieve better fuel efficiency through the stock of machines based on analysis of data collected from each machine used on the construction site (Heppelmann and Porter 2015). Heidelberger Druckmaschinen, a manufacturer of printing presses, offers a similar type of service based on over 1000 sensors installed in them (The Economist 2010). Another group is business models based on providing customers with smart products that are sources of additional benefits to them. Play Pure Drive is an example of such a solution. In this case, Babolat company has transformed a traditional product into a smart one that provides players with the opportunity to improve their technique by the use of a dedicated application, a tennis racket equipped with appropriate sensors, and a system enabling the connection to the smartphone. Clothing manufacturer Ralph Lauren made a similar move by offering smart PoloTech Shirt. It collects all parameters including pulse, the intensity of the movement, calories burned, and many others with built-in sensors in real-time during exercise and transmits them to a smartphone or smartwatch (Heppelmann and Porter 2014; Ralph Lauren 2016). Another group of business models are those based on behavioral profiling.

The system for establishing insurance rates based on monitoring of the driving behavior through the suitable telemetry device mounted in the vehicle is an example of this type of solution. The American insurance company Progressive offers this solution under the name of Snapshot (Burkitt 2014; Progressive 2016). Coverbox uses a similar system on the British market (The Economist 2010). Hybrid business models are a compromise between the models of product-as-a-service and traditional purchase of products by customers. They connect sales with, for example, different types of service contracts based on the monitoring of the device operations. Another group is business models based on offering IoT platforms to the users. Apple HomeKit is an example of such a solution. It controls various home devices from different manufacturers through the smartphone application. HealthKit platform, which enables the integration of devices for monitoring people's health and activity, is another example of a solution by the same company (Burkitt 2014). The next group of business models is those based on providing comprehensive IoT infrastructure solutions. ThingWorx platform is one of well-known examples of this approach. It provides comprehensive services for the creation of IoT solutions (ThingWorx 2016). The last group is business models based on the provision of extended services. This forward-looking category includes solutions based on the use of data and information collected by the providers of various IoT services and providing their own services based on them. Operations of insurance companies working on solutions which include creating their own portfolio based on cooperation with companies offering various types of IoT systems designed to monitor health and physical activity are an example of this approach (Burkitt 2014).

14.4 The IoT Ecosystem and Its Most Important Components

The appropriate technological infrastructure is essential to achieve the functionalities described earlier and the related new business models. Figure 14.5 shows the key elements.

Product hardware consists of sensors, processors, and elements providing communication (antenna ports, etc.) embedded in the product and supporting its traditional, mechanical, and electrical components. In turn, the product software is the operating system, on-board applications, user interface, and product control components embedded in the product. At the communication level, suitable communication protocols such as MQTT, CoAp, AMQP (Dwyer 2015; Perera 2015) providing communication between the product and product cloud are the necessary element of the system. The latter element of the IoT system is composed of four layers. The first one (smart product application) are applications that run on remote servers, which manage the monitoring, control, optimization, and autonomous operations of product features. Another one (product data database) is a system that allows the aggregation, normalization, and data management both in real-time and on historical data acquired from the product. Another layer of product cloud (rules/analytics engine) is related to the rules, business logic, and big data analytical capabilities providing algorithms used during the product operation and revealing the related new product insights. The last layer (application platform) is an application development and execution environment, which allows quick creation of smart networked business applications that use data access, visualization, and run-time tools. On the contrary, tools for managing system access and authentification of users and product security, connectivity, and individual layers of the product cloud are an extremely important component

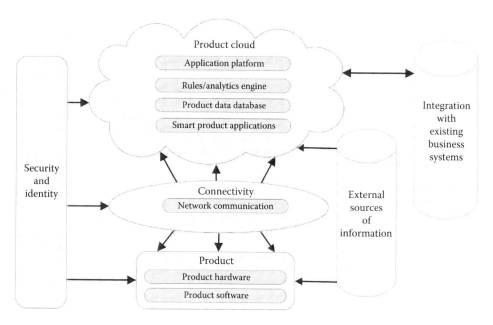

FIGURE 14.5
IoT system technological infrastructure.

of the IoT system. The entire infrastructure also has to have access to external sources of information (weather, traffic, energy prices, social media, etc.) necessary from the point of view of the capabilities offered by the product. As the technological infrastructure of IoT is part of the overall information technology (IT) infrastructure in an organization, the tools that integrate data from smart-connected products into key IT systems of a company, such as enterprise resource planning (ERP), customer relationship management (CRM), or product lifecycle management (PLM) are necessary (Heppelmann and Porter 2014).

14.5 The Analysis of the Role and Relationships between Advanced Big Data Business Analytics, Cloud-Based Solutions, and Mobile Technologies in the Context of Creating the Architecture and Operations of the IoT

The appropriate infrastructure layer is necessary to allow the operations of the IoT ecosystem. The solutions offered as a part of it must provide the following (Burkitt 2014; Heppelmann and Porter 2014; Deichmann et al. 2015):

- Continuous connection of ecosystem elements to the Internet at all levels and through various transmission channels (persistent connections, Wi-Fi, cellular, or Bluetooth)

- Adequate computing power and disk space necessary to collect and process vast amounts of data supplied from numerous endpoints of the system

- Software running on external servers, management monitoring, control, optimization, and autonomous operations of functionality offered by the product
- Connection to social networks
- Services related to data management and analysis
- Opportunities to develop the necessary software

It is not possible to create such an infrastructure of the IoT system without the use of three other leading technologies, which in itself carry an enormous transformation potential (Figure 14.6).

It is necessary to ensure reliable operation of the entire IoT ecosystem in the communication layer at the basic level to think about creating solutions, IoT at all. Hence, ubiquitous connectivity is seen as one of the key enablers that stimulate the development of the IoT (Aharon et al. 2015). Generally speaking, one can distinguish two market segments related to IoT connections, that is, massive IoT connections and critical IoT connections. The first one is characterized by high connection volumes, low cost, requirements on low-energy consumption, and small data traffic volumes. The characteristics of the other one are ultrareliability and availability and low latency (Ericsson 2016). IoT systems use different kinds of technical solutions, such as fixed connections, Wi-Fi, cellular, or Bluetooth connections to ensure adequate connectivity. According to the report by Ericsson, the compounded annual growth rate of the number of IoT connections is to be 23% between 2015 and 2021. At the same time, the highest level of growth is expected in cellular IoT and is projected at 1.5 billion of various types of connected devices (Ericsson 2016). These forecasts clearly show that mobile technologies grow into a key solution when it comes to providing connectivity in IoT systems. At the same time, the development of 5G network, characterized by greater capacity and lower energy requirements, make them very important due to these qualities from the point of view of stimulating the growth of IoT (Ericsson 2016). At the same time, the role of *cloud* technologies is crucial due to the number of devices operating in the IoT ecosystem and the amount of obtained, collected, and processed data. It would be difficult to imagine the creation of IoT without these solutions, both for technical (e.g., scalability) and economical reasons. The progress in this

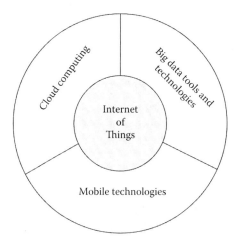

FIGURE 14.6
Technologies supporting development of IoT ecosystems.

area taking place in recent years has become a driver of IoT market development, and they are related to such aspects as follows (Dwyer 2015; RightScale 2016; Wielki 2015b):

- Increasing throughput and reliability of Internet connections
- The development of web services and their increasing availability
- The processes of commoditization of information technology
- Decreasing consumer concerns related to the security of cloud solutions

There are several key factors that determine the usefulness of *cloud* technologies in the creation of IoT solutions. Undoubtedly, the use of a *cloud* infrastructure provides a whole range of opportunities to reduce costs. They apply to both physical and software-related elements. As for the first one, it concerns such aspects as follows (Wielki 2015a):

- Reduction or elimination of waste related to the low level of hardware utilization
- Reduction of costs connected with hardware maintenance
- Lowering costs related to energy consumption
- Possibilities for the permanent analysis of costs and the selection of the optimal service level

In the case of software, it is related to the following (Wielki 2015a):

- Purchase and installation of software, its maintenance, and upgrade
- Purchase of wrongly selected software
- Low level of software usage
- Developing and testing of applications

In the case of critical IoT systems, it is also a question of cloud infrastructure's scalability. It makes it possible to add the necessary computational capacity, disk space, bandwidth, and new services or applications in a way that is not only cost effective, but also much faster (Bisson et al. 2013; Clutch 2016). All three major *cloud*-service models are useful in the development of the IoT infrastructure, that is, software-as-a-service (SaaS), infrastructure-as-a-service (IaaS), and platform-as-a-service (PaaS). The first two are necessary for the operation of three layers of product cloud, that is (Figure 14.4), smart-product application, product-data database, and rulet/analytics engine. The third one is very important from the point of view of the application platform layer. At the same time, while creating the IoT infrastructure in the *cloud* model, organizations can take advantage of each of the three basic types of cloud, that is, private, public, and hybrid. Research shows that private and hybrid clouds record the highest growth among businesses (RightScale 2016). However, the connection of devices and providing the infrastructure for their proper operation is just the first step toward building a competitive advantage based on the IoT systems. The real value is when organizations are able to transform vast amounts of collected data, often in real time, into real business intelligence that can be taken advantage of, for example, in the creation of new business models. Here comes the crucial role of advanced-analytical systems, that is, big data systems, also known as big data and advanced analytics (Magnin 2016). Such solutions are, according to IDC, "a new generation of technologies and architectures, designed to economically extract value from very large volumes of a wide variety of data by enabling high-velocity capture, discovery,

TABLE 14.1

The Differences between the Big Data Analytical Systems and Traditional Ones

	Big Data Analytical Systems	Traditional Analytics
Data type	Unstructured data	Data formatted in rows and columns
Volume of data	100 TBs to PBs	Tens of TBs or less
Flow of data	Continual data flow	Static pool of data
Utilized analysis methods	Machine learning	Hypothesis-based
Primary purpose of system usage	Products based on data	Internal decision support and services

and/or analysis" (Gantz and Reinsel 2012). Table 14.1 shows the comparison of big data analytical systems and the traditional ones.

The amount of data from IoT systems is growing and will grow exponentially (Brown et al. 2011). According to IDC, the global amount of data will increase tenfold from 4.4 ZB to 44 ZB in 2014–2020. In 2020, ten percent of all the world's data will come from the IoT systems. At the same time, we should emphasize that what is important is not the amount of data, but the amount of payload. At the moment, not all of the data collected are used, and those that are used in many cases are not fully utilized (e.g., in the case of massive amounts of data coming from devices operating within IoT systems, their use in the prediction and optimization is critical) (Aharon et al. 2015). However, this situation will change according to IDC. According to these forecast, the number of useful data shall increase in 2013–2020 from 22% to 35% (Adshead 2014). In these circumstances related to the development of IoT and rapid growth of data available to companies from different devices, the growing interest in tools that provide them with the capability to conduct advanced business analytics should not be unusual. According to the forecast by ABI Research, the market for this type of tools that enable to integrate, analyze, and create reports based on data from IoT systems is expected to reach the value of $30 billion in 2021 (Stackpole 2016). At the same time, the implementation, maintenance, and development of advanced analytical solutions, such as business intelligence (BI), require the use of an appropriate but costly infrastructure or having the staff with appropriate qualifications (Olszak 2014). When they are based on cloud computing technologies, and with the implementation of the analytics-as-a-service (AaaS) model, there is the possibility of reducing those costs, use the latest solutions available on the market, and at the same time, eliminating potential staffing issues (Thompson and Van der Walt 2010; Intel IT Center 2014). The situation is similar with regard to big data projects. In the case of projects of this type, known as cloud-based big data analytics (Intel IT Center 2014), organizations can build their own analysis ecosystem based on big data technology and the abovementioned three basic *cloud* service models (Figure 14.7).

Solutions available in the IaaS model enable the organizations to build infrastructural foundations for the whole big data ecosystem and implement the advanced analytical services available in the SaaS model. Examples of services available in the first one are solutions that are as follows:

- Amazon Web Services
- Windows Azure
- Citrix CloudPlatform

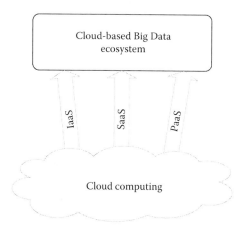

FIGURE 14.7
Big data ecosystem based on cloud technologies.

- Vmare Cloud Suite
- Rackspace

When it comes to big data analytics services available in the SaaS model, the examples of these solutions are as follows:

- Amazon Elastic MapReduce
- Google BigQuery
- Rackspace Hadoop
- Windows Azure HDInsight
- FICO Big Data Analyzer

At the same time, the organization can combine the use of different analytical solutions for their various business purposes, such as customer sentiment analysis, risk management, and asset management. Services available in the PaaS model, which can be used by organizations as a platform to develop their advanced analytical tools, are also a *complementary* part for the big data ecosystem (Intel IT Center 2014). Examples of the services offered in the model PaaS include solutions such as

- Google App Engine
- Windows Azure
- Force.com
- Red Hat OpenShift
- VMare Cloud Foundary

At the same time, there are attempts to develop standards for creation of advanced big data business analytics based on *cloud* solutions (ITU 2015).

14.6 Conclusion

Business organizations are on the verge of profound changes in their operations. They result from the dynamic developments in IT, which enable to increasingly use the Internet infrastructure to receive data from virtually any type of physical object. This gives the organizations a whole range of unprecedented opportunities in terms of value-creation processes based on the implementation of new business models and reconstruction of those used thus far. At the same time, it is necessary to build a reliable, adequately efficient, scalable, and cost-effective infrastructure layer to allow the use of the emerging opportunities related to the IoT and build the competitive advantage based on it. With regard to these issues, the role of the three leading technologies, which dynamically develop in the last few years and significantly change today's economic reality, that is, mobile technologies, cloud computing, and advanced analytical tools for big data, is very important. The implementation of new business models outlined in this chapter or deep transformation of processes in various social and economic areas would not be possible without them, and without them reaching the appropriate level of maturity. It should also be noted that the abovementioned technologies bring a whole range of various challenges (Wielki 2015c, 2015d), which can overlap and intensify in certain circumstances. This applies to, for example, such important aspects as data security or privacy issues. In this case, each organization must evaluate and weigh the potential benefits and risks, building the IoT ecosystem which is optimal from their point of view and which integrates solutions which are cheaper, but with a lesser degree of control (i.e., operating in the public cloud), with more expensive solutions having the higher degree of control (i.e., operating in the private cloud). Undoubtedly, the creation of adequately reliable IoT system is not a simple task, but nevertheless the continuous progress in three technologies discussed in this chapter significantly facilitates this task.

References

Adshead, A. 2014. Data set to grow 10-fold by 2020 as Internet of things takes off. http://www.computerweekly.com/news/2240217788/Data-set-to-grow-10-fold-by-2020-as-internet-of-things-takes-off (accessed April 10, 2014).

Aharon, D., Bisson, P., Bughin, J. et al. 2015. The Internet of Things: Mapping the value beyond the hype. http://www.mckinsey.com/~/media/McKinsey/dotcom/Insights/Business%20Technology/Unlocking%20the%20potential%20of%20the%20Internet%20of%20Things/Unlocking_the_potential_of_the_Internet_of_Things_Full_report.ashx (accessed July 1, 2015).

Barbier, J., Bradley, J., Handler, D. 2013. Embracing the Internet of everything to capture your share of $14.4 Trillion. https://www.cisco.com/web/about/ac79/docs/innov/IoE_Economy.pdf (accessed September 6, 2014).

Bauer, H., Patel, M., Veira, J. 2014. The Internet of things: Sizing up the opportunity. http://www.mckinsey.com/industries/high-tech/our-insights/the-internet-of-things-sizing-up-the-opportunity (accessed December 23, 2014).

Bisson, P., Bughin, J., Chui, M. et al. 2013. Disruptive technologies. http://www.mckinsey.com/~/media/McKinsey/dotcom/Insights%20and%20pubs/MGI/Research/Technology%20and%20Innovation/Disruptive%20technologies/MGI_Disruptive_technologies_Full_report_May2013.ashx (accessed June 1, 2013).

Brown, B., Bughin, J., Byers, A., Chui, M., Dobbs, R., Manyika, J. 2011. Big data: The next frontier for innovation, competition, and productivity, http://www.mckinsey.com/mgi/publications/big_data/pdfs/MGI_big_data_full_report.pdf (accessed May 30, 2011).

Brynjolfsson, E. and McAfee, A. 2014. *The Second Machine Age*. New York: W. W. Norton & Company.

Burkitt, F. 2014. A strategist's guide to the Internet of things. strategy+business, Issue 77 (November). http://www.strategy-business.com/article/00294?tid=27782251&pg=all (accessed November 19, 2014).

Clutch. 2016. Enterprise cloud computing survey. https://clutch.co/cloud#survey (accessed February 11, 2016).

Davenport, T. 2014. *Big data@work*. Boston, MA: Harvard Business School Press.

Deichmann, J., Roggendorf, M., Wee, D. 2015. Preparing IT systems and organizations for the Internet of Things. http://www.mckinsey.com/industries/high-tech/our-insights/preparing-it-systems-and-organizations-for-the-internet-of-things (accessed November 30, 2015).

Dobbs, R., Manyika, J., Woetzel, J. 2015. *No Ordinary Disruption*. New York: PublicAffairs.

Dwyer, I. 2015. Internet of things. http://cdn2.hubspot.net/hubfs/553779/PDFs/Ironio-IoT-Whitepaper-Jun2015_4.pdf?t=1455213785378 (accessed July 9, 2015).

Ericsson. 2015. Ericsson mobility report. http://www.ericsson.com/res/docs/2015/mobility-report/ericsson-mobility-report-nov-2015.pdf (accessed November 30, 2015).

Ericsson. 2016. Ericsson mobility report. http://www.ericsson.com/res/docs/2016/ericsson-mobility-report-2016.pdf (accessed June 29, 2016).

European Commission. 2015. Monitoring the digital economy & society 2016–2021. http://ec.europa.eu/newsroom/dae/document.cfm?doc_id=13706 (accessed December 30, 2015).

Evans, D. 2011. The Internet of things. http://www.cisco.com/web/about/ac79/docs/innov/IoT_IBSG_0411FINAL.pdf (accessed April 30, 2011).

Gantz, J. and Reinsel, D. 2012. The digital universe in 2020: Big data bigger digital shadows, and biggest growth in the far east. http://www.emc.com/collateral/analyst-reports/idc-the-digital-universe-in-2020.pdf (accessed December 29, 2012).

Gartner. 2014. Gartner says 4.9 Billion connected "Things" will be in use in 2015. http://www.gartner.com/newsroom/id/2905717 (accessed November 20, 2014).

Heppelmann, J. and Porter, M. 2014. How smart, connected products are transforming competition, *Harvard Business Review*. November, pp. 64–88.

Heppelmann, J. and Porter, M. 2015. How smart, connected products are transforming companies, *Harvard Business Review*. October, pp. 96–114.

Industrie 4.0 Working Group. 2013. Recommendations for implementing the strategic initiative Industrie. 4.0. http://www.acatech.de/fileadmin/user_upload/Baumstruktur_nach_Website/Acatech/root/de/Material_fuer_Sonderseiten/Industrie_4.0/Final_report__Industrie_4.0_accessible.pdf (accessed July 30, 2014).

Intel IT Center. 2014. Big data in the cloud. http://www.intel.com/content/dam/www/public/us/en/documents/product-briefs/Big Data-cloud-technologies-brief.pdf (accessed April 29, 2014).

ITU. 2015. Big data—cloud computing based requirements and capabilities. https://www.itu.int/rec/dologin_pub.asp?lang=e&id=T-REC-Y.3600-201511-I!!PDF-E&type=items (accessed December 23, 2015).

Keen, A. 2015. *The Internet is Not the Answer*. London: Atlantic Monthly Press.

KPMG. 2015. The changing landscape of disruptive technologies. https://techinnovation.kpmg.chaordix.com/static/docs/TechInnovation2015-Part2.pdf (accessed December 26, 2015).

Magnin, C. 2016. How big data will revolutionize the global food chain. http://www.mckinsey.com/business-functions/digital-mckinsey/our-insights/how-Big Data-will-revolutionize-the-global-food-chain (accessed August 29, 2016).

Mims, C. 2010. Follow the Smart Phones. https://www.technologyreview.com/s/418451/follow-the-smart-phones/ (accessed April 23, 2011).

Olszak, C. 2014. Business intelligence in cloud. *Polish Journal of Management Studies*, 10(2): 115–125.

Perera, S. 2015. IoT analytics. http://resources.idgenterprise.com/original/AST-0163543_wso2_whitepaper_iot-analytics-using-Big Data-to-architect-iot-solutions.pdf (accessed December 29, 2015).

Progressive. 2016. The fair way to pay for car insurance. https://www.progressive.com/auto/snapshot/ (accessed February 9, 2016).

PWC. 2016. Industry 4.0: Building the digital enterprise. https://www.pwc.com/gx/en/industries/industries-4.0/landing-page/industry-4.0-building-your-digital-enterprise-april-2016.pdf (accessed May 1, 2016).

Ralph Lauren. 2016. The polo tech shirt. http://www.ralphlauren.com/product/index.jsp?productId=69917696 (accessed April 30, 2016).

Rifkin, J. 2014. *The Zero Marginal Cost Society*. New York: Palgrave Macmillan.

RightScale. 2016. State of the cloud report. http://assets.rightscale.com/uploads/pdfs/RightScale-2016-State-of-the-Cloud-Report.pdf (accessed July 30, 2016).

Schwab, K. 2016. *The Fourth Industrial Revolution*. Cologny/Geneva: World Economic Forum.

Stackpole, B. 2016. Making IoT magic with analytics. http://www.computerworld.com/article/3078754/internet-of-things/making-iot-magic.html#tk.drr_mlt (accessed June 3, 2016).

Tapscott, A. and D. Tapscott. 2016. *Blockchain Revolution*. New York: Penguine Random House.

The Economist. 2010. Augmented business. http://www.economist.com/node/17388392/ (accessed November 29, 2010).

The White House. 2014. Big data: Seizing opportunities, preserving values. https://www.whitehouse.gov/sites/default/files/docs/big_data_privacy_report_may_1_2014.pdf (accessed May 26, 2014).

ThingWorx. 2016. One platform—Limitless possibilities. http://www.thingworx.com/ (accessed April 16, 2016).

Thompson, W. and J. Van der Walt. 2010. Business intelligence in the cloud. *SA Journal of Information Management*, 12(1). http://www.sajim.co.za/index.php/SAJIM/rt/printerFriendly/445/444&.

Wielki, J. 2015a. An analysis of the opportunities and challenges connected with utilization of the cloud computing model and the most important aspects of the migration strategy. In *Annals of Computer Science and Information Systems, Vol. 5. Proceedings of the 2015 Federated Conference on Computer Science and Information Systems*. Ed. M. Ganzha, L. Maciaszek, M. Paprzycki, pp. 1569–1574. New York City/Warszawa: Institute of Electrical and Electronics Engineers/Polskie Towarzystwo Informatyczne.

Wielki, J. 2015b. Analiza możliwości wykorzystania modelu cloud computing w kontekście redukcji kosztów związanych z funkcjonowaniem infrastruktury IT współczesnych organizacji. *Problemy Zarządzania*, 13(2) (52): 204–216.

Wielki, J. 2015c. The opportunities, impediments and challenges connected with the utilization of the cloud computing model by business organizations. In *Information Management in Practice*. Ed. B. Kubiak, and J. Maślankowski, pp. 11–26. Gdańsk: Faculty of Management University of Gdańsk.

Wielki, J. 2015d. The social and ethical challenges connected with the big data phenomenon. *Polish Journal of Management Studies*, 11(2): 192–202.

World Economic Forum. 2015. Deep shift: Technology tipping points and societal impact. http://www3.weforum.org/docs/WEF_GAC15_Technological_Tipping_Points_report_2015.pdf (accessed September 30, 2015).

15

Data Processing in IoT Using an Enterprise Service Bus

Siddharth S. Prasad, Robin Singh Bhadoria, and Mohit Mittal

CONTENTS

15.1 Introduction

The Internet of Things (IoT) is a collection of several physical devices or embedded gadgets that are interconnected through well-known network such as *internet*. Such devices need to be followed the same border of guidelines while accessing and facilitating their client. There have been several challenges that need to be addressed in related to interoperability between different gadgets, central coordination for handling data, and utilization of resources. For example, developers are working with applications that create massive volumes of new and rapidly changing data types. These data types could be in the form of structured, semistructured, unstructured, and polymorphic (Miorandi et al., 2012; Swan, 2012). The processing of all these different types of data is a challenging task. If an application requires the processing of just anyone of that, it creates lot of ambiguity in recognition of such data type and formats. ESB is software framework that provides multiple service integration over common and shared platform. ESB also support multiple data format and type that helps any service to recognize its format.

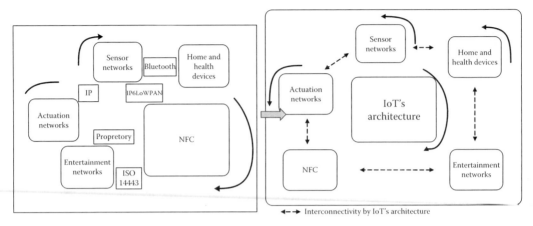

FIGURE 15.1
From different technologies to Internet of Things.

It also enabled the system with known set of pattern, namely, Enterprise Integration Pattern (EIP). When there are many types of data that need to be processed, current solutions could be implemented using ESB.

In the past few years, there has been a massive increase in the interest of developing such as NFC, Sensor network, and many more as shown in Figure 15.1. With the increase in the popularity of IoT in today's world, efficient processing of the data sent from different device can be merged with the IoT platform. The major task that needs to be dealt with is to provide integrating platform which could help in handling data and its format (Dastjerdi & Buyya, 2016). As in many applications, even the slightest of latency could be problematic. Making this transmission of data more efficient would have a great impact on such applications. Such impaction could efficiently be done by using ESB.

15.2 Enterprise Service Bus

An Enterprise Service Bus (ESB) fundamentally is architecture. It is a set of principles and rules for the integration of many applications together over an infrastructure that is bus-like. There are various ESB products. They all enable users to build architecture of this type but differ in the way they do it and capabilities that are offered by them. ESB architecture's core concept is that you integrate various applications by inserting a communication bus between them and then enable all applications to talk to the bus (Bhadoria, 2015; Bhadoria et al., 2017). This reduces coupling between systems so that they can communicate without dependency on or knowledge of each other on the bus. The ESB concept was created as a result of the need to do away with point-to-point integration, which becomes very hard to manage and brittle over time. Point-to-point integration leads to custom integration code being distributed among applications with no central way to track or troubleshoot. *Spaghetti code*, is the popular term used to describe this, and it does not scale as it creates close dependencies between applications (Aggarwal et al., 2013).

An ESB's primary duties:

- Control and monitor routing of the exchange of messages between services
- Resolve contention between the components of communicating service
- Control versioning and deployment of services
- Marshal usage of services that are redundant
- Cater for services such as message and event queuing and sequencing, security or exception handling, event handling, data transformation and mapping, enforcing good quality of communication service and protocol conversion

15.3 Integration Core Principles

Mapping ESB architecture to five core principles of integration:

- *Service integration*: Composing many existing fine-grained components into an individual higher order composite service to achieve appropriate *granularity* of services and promote manageability and reuse of the basic components.
- *Protocol transformation*: Data transformation between standard data formats and particular data formats required by each connector of the ESB. Standard (or Canoncial) data formats greatly simplify the requirements for transformation associated with implementation of a large ESB in which there are lots of providers and consumers, each with their own data formats and definitions. For example, conversion of formats between CSV, Cobol to either JSON or SOAP/XML, and vice-versa.
- *Incorporate transportation*: Transport protocol negotiation between multiple formats (such as JMS, JDBC, and HTTP).
- *Mediation*: It interprets different interfaces for the following reasons: (a) incorporate different service versions to provide compatibility with existing services, (b) support implementation for multiple channels with underlying service component. It may involve multiple interfaces simultaneously to handle same component with standards formats such as SOAP/XML.
- *Nonfunctional consistency*: It includes the consistency associated with security and monitoring policies that are applied in handling service. Moreover, scalability and availability could be achieved through multiple service instances that are plugged-in to ESB. It provides with improved throughput (in terms of scalability) and eradicate single-points-of-failure (SPOFs), which is also covers the primary objective availability.

15.4 Service Modeling and Realization

15.4.1 How Modeling Improves Service-Oriented Architecture

Service-Oriented Architecture's power is in its ability to enable agility of business through reuse and integration of business process (Amsden, 2007). This is achieved by SOA in two ways: (a) By encouragement of solutions organized around services that are reusable and encapsulate functional capabilities separated from their implementations and (b) by providing methods to manage coupling between functional capabilities.

Modeling can be used to lean the gap between today's business demands and service deployed based solutions. SOA models the level of abstraction in handling data in service so that one can focus on business goals to achieve high reusability and availability. Such development approaches can be used in implementing SOA features such as scalability, reusability, and availability with platforms of Java (J2EE) or IBM CICS. This helps in meeting the agile goals of business functional and nonfunctional objectives.

The term service-oriented architecture (SOA) has several overtones. SOA is used commonly by practitioners to both define a style of architecture and describe a common infrastructure that enables IT systems that are built using that style of architecture to operate. These are useful perspectives that are technology focused, but they are not enough by themselves.

To reach its potential, an SOA-based IT infrastructure (or SOA) needs to be relevant to business, thus business-driven and implemented to support the business. SOA solutions need to be designed in ways that are connected to the requirements of the business that they fulfill. This is difficult to accomplish if the requirements of the business are captured in several XML documents that describe collection of web services.

What do we need? A way to formalize the requirements of business and increase the level of abstraction such that SOA can resemble business services more closely and how those services might meet the goals and objectives of the business. The deployed solution is thus tied to its intended business value. Simultaneously, we need a method to isolate business concerns from the SOA platforms that support them.

These goals can be achieved with the help of Modeling and model-driven development (or MDD). Models allow us to abstract the implementation details away and focus on the issues that drive architectural choices. To some degree, the approach we will depict applies one of the fundamental principles of SOA to the development of SOA solutions: loose coupling, and separation of concerns. Here, we neatly separate the responsibilities and tasks of business analysts from those of IT staff.

Generally, business analysts will be centered on business operational and organizational requirements necessary to meet business objectives and goals that accomplish some business vision. Frequently, they are not related to (nor skilled enough to deal with) IT concerns, such as cohesion and coupling, reuse, security, distribution, data integrity, persistence, failure recovery, concurrency, and others. Further, tools for business process modeling often do not have the necessary capabilities to address these concerns, and, if they did, they most probably would not be effective tools for business analysts.

15.4.2 Service Realization

The first step in the design of service implementation is to provide the services. That is, to figure out what service capabilities will be provided by which service providers. This is an integral part of designing an SOA, as the choice of providers establishes the relationships between service providers and consumers. Therefore, this establishes both the coupling between the parts of a system and its capabilities.

All the operations could be put into one service and have a straightforward solution. But all clients would depend on that single service, which would lead to very high degree of coupling. Any change in the provider would lead to a possible change in all consumers. This was a typical issue with module libraries in the old days of programming in C.

A separate service could also be created for each functional capability, but this would result in a system that is very complex that would not reflect good cohesion and encapsulation. The communication between different clients would be difficult in implementing different interfaces for different data format and type. The reason for such mismatch in interoperability is the support for different protocols by different clients in interaction. The prominent solution for such interoperability could be ESB. It not only organizes data and services together but also mediates the system with global security of Denial of Service (DoS) attack.

In the end, choosing the participants of the service is something that takes a bit of skill and can be affected by lots of compromises. Distribution can play a key role in handling data from different interfaces that could be managed by means of different connectors. Designing SOA-based solution should be independent of service clients and provider's location, but that is not very practical solution in general. Services that are deployed to global solution must not affect the overall performance, security, and availability of SOA system. A system that ignores all these features in its solution may lead to unacceptable implementations of service that decreases the service proliferation.

15.5 Enterprise Integration Patterns for Data Handling in IoT

15.5.1 What Is the Need for Integration?

Business applications today rarely live in isolation. Instant access is what is expected by users to all business functions that can be offered by an enterprise, regardless of which system the functionality may reside in. This requires disparate applications to be linked with integrated solution that is based on middleware. This gap is lean with the aid of ESB that provides the *plumbing* such as routing, data transport, and data transformation.

15.5.2 What Makes Integration So Hard?

It is a complex task to handle multiple service integrations for different data patterns together to build a solution for specific application. There are so many other conflicts that are related to different drivers that support data functionality. Middleware architecture plays an important role in handling services more efficiently and provides inevitable additions to overall monitoring for system architecture (Bonomi et al., 2012). Commonly, different service vendors provide best practices and methodologies for handling service through their well-defined interface, but these interfaces are inefficient in recognizing different data format and its type. This gives a rise to conflict for software counterfeit in controlling data and its associated format.

15.5.3 Asynchronous Data Architecture

The best strategy for service integration has been supported by architectures that handle data asynchronously. It allows the loose coupling of services that overcomes the limitations of remote execution. That is why asynchronous messaging that exchange data from multiple services are enabled by Enterprise Integration Patterns (EIPs). Unfortunately, asynchronous messaging has its pitfalls. A lot of assumptions are valid when developing single, synchronous applications but do not support dynamic environment. Service providers neutrality make the design of system independent of these pitfalls so that it enhance the system performance and robustness for data architectures (Atzori et al., 2010).

15.5.4 How Can Patterns Help?

Patterns are a proven approach to capturing knowledge of experts, especially in those fields, where there are no simple universal (*one size fits all*) answers, such as object-oriented design, application architecture, or message-oriented integration. A specific problem is tackled by each pattern by the discussion of design considerations and presentation of an elegant solution that balances the forces. Many a time, the first approach that strikes us is not the solution, but an approach that, through actual use, has evolved over a period of time. Consequently, each pattern incorporates into it gets the experience that past architects have gained by building solutions frequently and learning from their shortcomings. Thus, no pattern is *invented*. Rather, they are observed and discovered from actual practice (Kulkarni et al., 2016).

Integration of services on common platform of IoT can be handled with the help of potential end-points. These end-points provide interface to interact with other services from different domains. This also facilitate to exchange data presented in different format and type. ESB play a vital in handling and organizes services (Sarma & Girão, 2009). This scenario is well depicted in Figure 15.2.

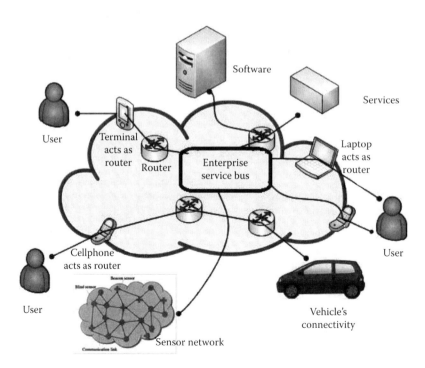

FIGURE 15.2
End-point communications in IoT.

15.6 Challenges for IoT Data

Devices having network interface can be categorized into three depending on resources it possess:

- Devices with multiple processors of high speed with large memory (gigabytes) and multiple networking options, such as PC, laptops, tablets, and smartphones.
- Devices with few megabytes of memory used in different industrial or home automation.
- Devices with very small processing power, memory and usually having a single network interface. These devices might operate for a longer period on long-lasting batteries. This class includes sensor nodes (such as 8-bit microcontrollers with 32-kB RAM and 512-kB flash storage).

The first two classes of devices mentioned above have many frameworks and protocols that are confirmed to be efficient enough already, whereas devices specified above in last case make it tough to figure an interoperable system with aid of same protocols. Henceforth, there is limited applicability of such techniques that are used to integrate systems. Such devices should be synchronized with real-time requirements (Cooper & James, 2009).

Cloud infrastructure is frequently used with IoT, and it provides solution based on devices that are producing data. Figure 15.3 represents the service flow for IoT that gives an idea of actual stream of dataflows. Initially, some data are created from IoT device software development kit (SDK) that is responsible for set of client libraries

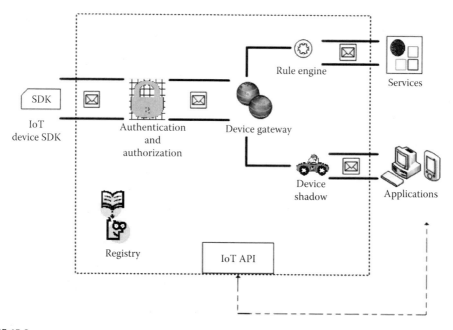

FIGURE 15.3
Service flow control in IoT.

and exchange of information. When this information is validated through predefined technique authentication and authorization, this step is secured with encryption. The device gateway channelizes this secure message to various services that could receive/ send this secure message to further service with the help of rule engines. This rule-based engine initially transforms the message based on certain logic that follows some business information exchange. Applications that receive this secure message through device shadows would keep the device state during entire service transaction and maintains this connection. For each information stream, registry is maintained, which assigns unique identity to each device.

The main power of the IoT paradigm lies in its ability to provide real-time data from numerous different distributed sources to other smart entities, machines, and people for many different types of services. One major issue is that the underlying data from the various resources are immensely heterogeneous. This can be very noisy and is usually very distributed and of large scale (Zorzi et al., 2010).

It is therefore hard for other entities to use the data productively, without a clear definition of what is accessible for processing. To enable productive use of this very distributed and heterogeneous data, frameworks are required to describe the data in a way that is intuitive enough, so that it becomes more easy to use, that is, the issue of semantic interoperability gets addressed. This leads to challenges both in terms of providing scalable, high quality, and real-time analytics and also in terms of describing intuitively to users information about what sort of services and data are available in a variety of scenarios (Ali et al., 2016).

Thus, techniques to analyze, query, manage, and clean the data in the distributed way are required. The cleaning is performed usually at the time of data collection and is often built into the middleware that interfaces with the sensor devices. So, the research on the cleanliness of data is often studied in regards to things-oriented vision. The issues of implementing standardized descriptions and data access for smart services are studied in the context of standardized web protocols and interfaces, and querying/description frameworks as offered by semantic web technology. The idea is basically to reuse the already existing web infrastructure intuitively, so the distributed nature and heterogeneity of the various sources of data can be integrated seamlessly with the different services (Bandyopadhyay & Sen, 2011).

These problems are examined usually within the web of things' context and the semantic web visions. Therefore, the end-to-end data management of IoT technology requires the collaboration and unification between the various aspects of the development of these technologies in order to provide an infrastructure that is seamless and effective (Negash et al., 2015).

15.7 Conclusion

The data and objects in IoT are heterogeneous, as opposed to the World Wide Web of documents wherein objects are described themselves in terms of a natural lexicon. Also, the objects and data in the IoT paradigm may not be naturally available in a way that is descriptive enough to be searchable, unless, of course, an effort is made to create descriptions that are standardized for these objects in terms of their properties. The cleaning of data in IoT may be required for different reasons that are concluded as follows:

- When data are obtained from conventional sensor devices, it may be incomplete, noisy or may require probabilistic uncertain modeling.
- RFID data are duplicate, redundant, noisy, and even incomplete with regards to sensor readers.
- The method of privacy-preservation may require a reduction of data quality intentionally, in which case methods are required for processing of privacy-sensitive data.

References

Aggarwal, C. C., Ashish, N., & Sheth, A. (2013). The internet of things: A survey from the data-centric perspective. In C. C. Aggarwal (Ed.), *Managing and Mining Sensor Data* (pp. 383–428). NY: Springer.

Ali, M. I., Ono, N., Kaysar, M., Shamszaman, Z. U., Pham, T. L., Gao, F., … Mileo, A. (2016). Real-time data analytics and event detection for IoT-enabled communication systems. *Web Semantics: Science, Services and Agents on the World Wide Web, 42,* 19–37.

Amsden, J. (2007). *Modeling SOA, Part 1. Service Identification.* IBM Developer Works. Retrieve from: https://www.ibm.com/developerworks/rational/library/07/1002_amsden/index.html?S_TACT=105AGX15&S_CMP=LP.

Atzori, L., Iera, A., & Morabito, G. (2010). The internet of things: A survey. *Computer Networks, 54*(15), 2787–2805.

Bandyopadhyay, D., & Sen, J. (2011). Internet of things: Applications and challenges in technology and standardization. *Wireless Personal Communications, 58*(1), 49–69.

Bhadoria, R. S. (2015). Performance analysis for enterprise service bus in SOA system. *International Journal of IT Business Strategy Management, 1*(2015), 9–16.

Bhadoria, R. S., Chaudhari, N. S., & Tomar, G. S. (2016). The performance metric for enterprise service bus (ESB) in SOA system: Theoretical underpinnings and empirical illustrations for information processing. *Information Systems, 65*(2017), 158–171.

Bonomi, F., Milito, R., Zhu, J., & Addepalli, S. (2012, August). Fog computing and its role in the internet of things. In *Proceedings of the first edition of the MCC workshop on Mobile cloud computing* (pp. 13–16). Helsinki, Finland: ACM.

Cooper, J., & James, A. (2009). Challenges for database management in the internet of things. *IETE Technical Review, 26*(5), 320–329.

Dastjerdi, A. V., & Buyya, R. (2016). Fog computing: Helping the internet of things realize its potential. *Computer, 49*(8), 112–116.

Kulkarni, P. H., Kute, P. D., & More, V. N. (2016). IoT based data processing for automated industrial meter reader using Raspberry Pi. In *Internet of Things and Applications (IOTA), International Conference on* (pp. 107–111). IEEE.

Miorandi, D., Sicari, S., De Pellegrini, F., & Chlamtac, I. (2012). Internet of things: Vision, applications and research challenges. *Ad Hoc Networks, 10*(7), 1497–1516.

Negash, B., Rahmani, A. M., Westerlund, T., Liljeberg, P., & Tenhunen, H. (2015). LISA: Lightweight internet of things service bus architecture. *Procedia Computer Science, 52,* 436–443.

Sarma, A. C., & Girão, J. (2009). Identities in the future internet of things. *Wireless Personal Communications, 49*(3), 353–363.

Swan, M. (2012). Sensor mania! the internet of things, wearable computing, objective metrics, and the quantified self 2.0. *Journal of Sensor and Actuator Networks, 1*(3), 217–253.

Zorzi, M., Gluhak, A., Lange, S., & Bassi, A. (2010). From today's intranet of things to a future internet of things: a wireless-and mobility-related view. *IEEE Wireless Communications, 17*(6), 44–51.

Section IV

Future Research, Scope, and Case Studies of the Internet of Things and Ubiquitous Computing

16

Infection Tracing in i-Hospital

Mimonah Al Qathrady, Ahmed Helmy, and Khalid Lmuzaini

CONTENTS

16.1 Introduction and Background

Infection transmission risks are present in all hospital settings as well as other communities. According to the CDC (2007), humans—patients, healthcare personnel, or visitors—constitute the primary source of infectious agents, along with other objects such as contaminated equipment. Direct contact and indirect contact with individuals or objects are the primary methods of transmission of infectious agents. Also, Middle East Respiratory Syndrome has an outbreak in South Korea was traced to a patient who has been admitted to a hospital (Tina, 2016), where the virus is transmitted to 82 other people, mostly visitors and fellow patients. Moreover, according to a survey that was based on a large sample of U.S. acute care hospitals, about 1 in 25 hospital patients have at least one health-care-associated infection (Magill et al., 2014). These, as well as cases of H1N1 Swine Flu and Ebola, in which more than 1,300 patients were admitted in a Singapore Hospital because of H1N1 (Subramony et al., 2010). Also, there are more than 4K reported deaths in West Africa in the first nine months of the Ebola epidemic (Team, 2014). These cases provide concrete examples that motivate our work of infection tracing.

When there is a case of epidemic disease, it is vital to identify the original sources to control and prevent further spread of infection by tracing the sources and then tracing forward from the sources and identifying the population with higher risk including the nodes that contacted the sources and might be infected as well.

As many smart hospitals and communities are equipped with mobile and sensing devices to track the objects' movement and their encounters, we propose to enhance these technologies and act immediately whenever a case of infection is detected. Our target is an intelligent hospital that deploys Internet of Things and able to track the encounter between individuals and objects.

Using a mobile application to collect encounter information has been used in the past different studies for different purposes. In iTrust (Kumar & Helmy, 2012), the encounter information is filtered and ranked for social reasons to identify trusted encounters. It ranks the encountered nodes based on the duration and frequency of encounters.

Other researchers have worked on studying the infection transmission and dynamics in different populations. They captured the encounters between peoples using sensing devices. A wireless sensor network technology was used to obtain close proximity data between individuals during a typical day at an American high school; then, the data are used to construct the social network relevant for infectious disease transmission (Salathé et al., 2010). Also, face-to-face interactions between the attendees at a conference were captured using radio-frequency identification (RFID), the spread of epidemics along these interactions was simulated (Stehlé et al., 2011).

Mobile phone devices and Bluetooth have been used in other works as well to collect the encounters and study or map the epidemic disease. Mobile phones are used to collect human contact data and record information such as locality, user symptoms for flu or cold, and human interactions. Then, the data were used to develop mathematical models for the spread of infectious diseases (Yoneki & Crowcroft, 2014). The sensed interaction by mobile phone and Bluetooth data is proved to be suitable for modeling the spread of disease and increase the predictive power of epidemic models (Farrahi et al., 2015). Also, GPS in mobile phones was used for tracking the epidemic on the map; after the patient is diagnosed with Dengue fever, the patient's location is used to draw the epidemic map (Reddy et al., 2015)

The previous researches have focused on modeling the disease transmission, which is only one component of the i-hospital infection-tracing framework.

More intelligent hospitals in the future will deploy Internet of Things which consists of heterogeneous devices. As a result, the encounter collection framework in this chapter is designed to be used with different heterogeneous devices as it will be the case with the Internet of Things. The sensing devices will gather the encounter data. Thus, the applications that use the framework are intended to be technology independent. They could potentially work with data collected from various technologies, for example, RFID or mobile phones.

Another important part of our infection-tracing framework is tracing back to infection sources. The concept of tracing back has been usually used as a defense mechanism in computer networks to identify the source of attacks (Peng et al., 2007). Kim and Helmy (2010) introduced a protocol framework to trace back the attackers in mobile multihop networks; the problem is complicated by mobility.

Tracing back the attackers in a computer network is less complicated than tracing the disease infection for several reasons: the main reasons are related to the nature of the agents' spread methods and ranges. There may be a different range (and method) for the disease spread, sensing, and communication. Such differences must be taken into consideration during the design, simulation, and analysis. Also, to make the attack happen in computer networks, there must be some kind of communication between nodes, so information about the communicating nodes can be saved. On the other hand, when the disease spreads from one person to another, it is impossible to record the exact moment of the infected nodes during the transmission process.

Trace back and forward for a source of problem and subjects that have been affected by a problem were used in the foodborne outbreak to identify the source of a product that was implicated in the foodborne outbreak (Weiser et al., 2013). Tracing back the supply chain is different than tracing the disease. This is because the supply chain database has certain data, whereas the certain data when the infection transmitted between two individuals are not available.

Several researchers have tried to solve the problem of identifying a source of different kind of infection like rumors when it spread in the society. For example, Shah and Zaman (2011) obtained an estimator for the rumor source based on the infected nodes and the underlying network structure. Others considered the problem of identifying an infection source based on an observed set of infected nodes in a network (Luo & Tay, 2013). The works of finding the source out of the infected population usually have knowledge about the group of infected nodes and attempt to figure out which one is the source.

The idea of infection tracing in this chapter has more challenging scope where the knowledge about infected nodes is not available, and the system attempts to find the sources and population at risk after knowing only one case.

Contact tracing is an important mean of controlling infectious diseases. Armbruster and Brandeau (2007a) developed a simulation model for contact tracing and used it to explore the effectiveness of different contact-tracing policies in a budget-constrained setting. A simulation model of contact tracing is used to evaluate the cost and effectiveness of different levels of contact tracing (Armbruster & Brandeau, 2007b).

This chapter will explain the infection tracing framework that exploits the Internet of Things in the intelligent hospital that collects encounter information. Also, it will explain how this encounter data can be used to trace automatically the source of infection and population at risks as soon as one case of infection is detected.

An i-hospital system that traces back from infected node to identify the original sources of infection will control and prevent further spread of infection and then guide the trace to identify infected population even before the patients report their cases.

The first part of the chapter is about an i-hospital high-level framework that gathers the encounter data from different heterogeneous devices of the Internet of Things (IoT), processes and utilizes them for various applications. It focuses on the encounter collection and processing.

The second part is about the infection-tracing framework. It shows how the encounter data that collected from IoT can be utilized to tackle the tracing problem in the intelligent hospital. Then, the disease infection source trace back problem is defined. Selection methods that work on traced to identify the more likely infected nodes are described. Then, simulation and evaluation results that use real wireless data are presented.

16.2 i-Hospital Nodes Tracking and Processing High-Level Framework

The i-hospital nodes tracking and processing high-level framework consists of three main components: encounter sensing and data collection side, server side, and application side. First, the encounter data are collected, aggregated, processed, and sent to the server, where it is processed further. The data will serve a wide variety of applications. One of the applications is infection tracing. The framework components interact with each other, by sending data, control messages, or parameters between each other. This framework assumes the availability of the infrastructure; other architectures will be described in the next section where the hospitals lack or have limited infrastructures.

16.2.1 Encounter Sensing and Collection Side

This side is represented by the block A in Figure 16.1. The encounter sensing and data collection side consists of three main components: the sensing devices in the IoT that capture their encounters, the network architectures, and the filtration and processing of encounter data before sending them to the server.

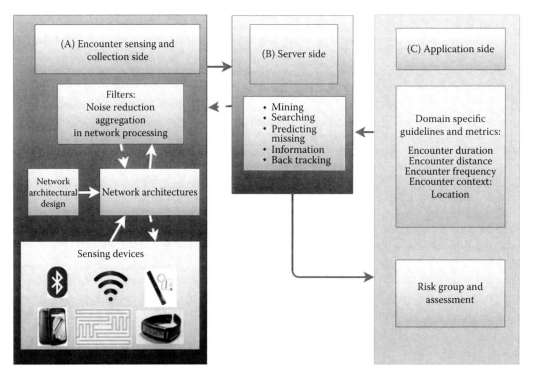

FIGURE 16.1
i-Hospital high-level framework for collecting encounters and processing them. Block (A) shows the encounter collection side. Block (B) shows the server side that is used for storing and running the applications. Block (C) represents the application sides, where various applications with different parameters can use the collected encounter data.

16.2.1.1 Sensing Devices

Smart hospitals project used an RFID as the main device to track objects and people (Fuhrer & Guinard, 2006). However, our i-hospital framework works with different heterogeneous sensing devices in IoT to collect their encounter information. For example, RFID tags, mobile and Bluetooth devices, access points, IoT-based wristbands, and IoT-enabled beds and equipment. These devices have different ranges when sensing encounters, and the i-hospital framework takes the sensing ranges into consideration.

RFID could be passive or active tags attached to human or objects. Its range depends on its type. Tags could record its encounter with other tags as in active tags, or the readers track tags. RFID-passive tag has a microcircuit and an antenna. It receives its power from the energizing electromagnetic field of an RFID reader (or interrogator). The energy coupled from the electromagnetic field undergoes voltage multiplication and rectification to power the passive tag's microelectronics. The tags should be placed within the range of an RFID reader. Passive RFID tags are different in how they receive the power from readers or how to communicate data to readers. Active teas used batteries and can serve longer ranges than passive tags. Cisco (2014) gives more explanation for the differences between active and passive tags can be found in the Cisco (2014).

The read range of passive tags (tags without batteries) depends on many factors: the power of the reader, interference from metal objects, or other RF devices and the frequency of operation. In general, low-frequency tags are read from a foot or less. High-frequency tags are read from about three feet and ultra-high frequency (UHF) tags are read from 10 to 20 feet. Active tags use batteries to boost read ranges to 300 feet or more (RFID journal, 2016).

Regular mobile device with Bluetooth capability can use Bluetooth to record encounters. Also, some devices running Bluetooth can be attached to objects to record encounters. Bluetooth record encounters within 10 to 100 m.

Another example of sensing devices is access points, in which they record the MAC address of associated devices. Two devices are more likely to be encountered if they associated with the same access point at overlapped intervals of time.

Each sensing device has its capabilities and limitation. For example, access points have a wider range, but it registers encounters that might not be captured by RFID. Thus, incorporate different sensing devices and combining their encounter records will potentially help increase the accuracy of encounter tracking data. For example, data from different sources could identify more encounters than what could be reported by only one sensing device. They also could identify that the probability of two subjects did not encounter each other even if they have been reported as being encountered by one sensing device.

In fact, human or objects in the intelligent hospital will be more likely be equipped with heterogeneous devices that cooperatively perform the sensing and tracking functions. Other tracking devices may be placed in static places like RFID readers. Other sensing devices are expected to be carried by mobile users like mobile devices. Mobile devices may not only capture its encounter with each other by using Bluetooth, for example, but also they enhance the functionality of RFID technology. For example, RFID readers may miss reading tags when they are out of its range, but some mobile device may have the ability to read tags and could supplement the tag information to the server.

Other devices in the environment are special nodes and have more capabilities than the rest such as filters. They process the data before sending them to the server.

When talking about the heterogeneity of sensing technology, we have two aspects: the heterogeneity of the device's capabilities and the heterogeneity of nodes that are equipped with the sensing devices.

Not only RFID passive and active have different capabilities, but also the mobile phones themselves have different capabilities. For example, sensing between mobile phone is not bidirectional where one device has more power and can read the other device while the other one cannot read it. This device heterogeneity should be addressed. Their effect on the collected data must be measured and predicted.

Nodes are not only heterogeneous because of the sensing devices that sense their encounters, but they are also different based on their roles. For example, some nodes are mobile and move from one place to another, whereas other nodes are static. Nodes similar to equipment are different based on their frequency of use, will they be used only for one time like a bed sheet or could be used multiple times like a wheelchair or an expensive X-ray device. If the nodes are human, their role can be defined as if they are patient, visitor, or hospital staff. The node's mobility can be based on their role too, by taking into consideration the area they are allowed to move to or restricted from.

16.2.1.2 Network Architectures

The high-level framework assumes the availability of an infrastructure that receives the encounter data. When the devices send their data to the server, there are different possible architectures of collecting the data. One approach is that the devices send the data directly to the server, and then the data are processed and stored in the server. The drawback of this method is contention on the server side. Another approach is that the devices send their data to other dedicated devices in the networks, where these dedicated devices process the data, and then send a cleaner version of the data to the server. Having dedicated devices in the network to clean the data before sending it to the server will reduce the overhead on the server, as only limited devices connected to the server, the dedicated ones, and not all the devices in the area.

This concept of collecting data is different than network architecture concept when there is a lack of infrastructure or servers, which will be explained in the next section.

16.2.1.3 Filters and Aggregation

The purpose of the filters is to reduce the overhead on the server by removing noise, aggregate data, and remove redundancy. They process the data before sending them to the server. Filters reduce the pressure on the server by reducing the amount of data flood to it, especially when they remove the redundant data. They are also reducing the overhead on the server processing as they are cleaning the raw data, filtering the noisy and redundant raw data, and cross-correlation of reported information from different devices before sending them. There are many levels of aggregation performed by the filters; the level depends on the granularity of the system. The basic level is an aggregate of the result reported by a device or multiple devices. For example, if the mobile device keeps sensing the same equipment, and the equipment does not change its position, then instead of sending the same information to the server, the report could be aggregated into one report indicating that the time period and that equipment is still in its position. They may also detect missing information and run some algorithm to fill the gap. Detecting missing encounter information and inferring them is also run on the server side as it needs much information from multiple sources similar to layout planning and rich encounter data.

Filter nodes could be special ones with storage and processing capabilities, in which the mobile devices send the collected information to them similar to dedicated devices described earlier in the network architecture; The other option is installing some filter functionality within all or selected groups of mobile devices.

16.2.1.4 Sending the Tracking and Sensing Information to the Server

Sending encounter information can be done either by pulling or pushing the data. When nodes push their data, they do it constantly or periodically without waiting to be asked. On the other side, if the data will be taken by pulling, the nodes wait until it being queried about their data or asked to send them. In this problem, push and pull methods may be used at the same time, with different spectra. Nodes could push information after a specific amount of time and after processing and aggregating them.

Also, it may depend on kind of the tracked objects; for instance, information about expensive objects is constantly pushed. Another way is to recognize the frequency of query about different kinds of objects; if the object is frequently queried, then information about it is constantly pushed; other information is pushed when the node, for example, is leaving the facility; otherwise, its encounter data are sent when they are requested.

16.2.2 Server Side

The server plays an important role in this framework, in which it receives the mobile and encounter data, processes them further, and runs the application—mining the massive amount of data such as classifying nodes or clustering them into groups.

After receiving the encounter data, the server will put the encounter data into further processing, such as detecting the gap and filling the missing information. Detecting the presence of missing encounter information is very crucial in an environment similar to a hospital.

When the server detects missing data, it is expected to infer the missing data.

In some cases, when there is something abnormal in the data it receives, it will notify the administration about the case. These notifications might send to a human being who is in charge of interacting with the system. Then, the human will answer the server if the situation is normal or not. For example, the situation is normal if the crowd is not available in the facility for that period of time for some reason, and eventually there is no sensing information. In this case, the system may not go further to fill the gap and adjust the model.

The presence of missing information can be detected by either direct inferring or more complex model that relies on analysis of historical information.

16.2.3 Application Side

Various applications will utilize the tracking and sensing information that has already been collected. The applications supply the server with their domain-specific parameters. Using an appropriate algorithm with the application parameters, the servers will return back a suitable solution or recommendations for each problem.

An example of problems that will highly benefit from knowledge about encounters between objects is improving the hospital layout. In a previous study (Brown et al., 2014), the impact of building space on social interactions was measured by using encounter data that were collected by wearable sensing devices.

Another significant problem, which is the core of this chapter, is infection tracing. Examples of application parameters are duration, frequency, distance, or encounter definition. Duration is the amount of time two nodes encounter each other, whereas frequency is the number of times two nodes have encountered each other regardless of the encounter duration. Distance can be defined as how many meters two nodes were far apart. The distance is usually a probabilistic distance, and it depends on the sensing device that records the encounters of two nodes. The encounter is used to be defined as two nodes have been in specific distance from each other at the same time. However, an application similar to infection tracing may have various definitions for encounters. It may define at the nodes that have been in the same place within T time, as in indirect encounter, which will be described in the next section.

16.3 Infection Tracing: The Infection-Tracing Framework

Infection tracing is an important application that utilizes IoT in the intelligent hospital and the collected encounter data. The application goal is to trace infection by identifying the sources of infection and population that might be at risk of being infected.

The framework that is used for tracing is shown in Figure 16.2. It consists of three main parts: encounter, tracing, and system evaluation side. Encounter side focuses on the encounter definition and parameters while dealing with infection. Tracing side describes infection tracing algorithms. Finally, the evaluation side defines the metrics that will be used to measure the infection tracking system.

16.3.1 Encounters Side

This side concerns about the encounter collection and processing for tracing infection. Block A in Figure 16.2 shows its parts. The high-level framework described in Section 16.3 has already explained the general parts of the encounter data collection. The encounter side in this section will point to encounter specifications that should be taken into consideration when we are dealing with infection tracing.

Different architectures of encounter collection will be explained briefly. Then, we will point to other encounter issues such as encounter ranges, direct and indirect encounter, and the subject of encounter: human or objects.

16.3.1.1 Encounter Collection and Processing

Collecting the encounter information can be performed using one of several network architectures. Some architecture uses a centralized server (or group of servers) to which nodes push their sensed data. The server then processes the data using sanitation, aggregation, prediction, and trace-back algorithms. This architecture was explained in Section 16.3 and represented by block B in Figure 16.1. Semicentralized architecture is another useful architecture when there is limited infrastructure; it uses small throw boxes in parts of the area (e.g., floors of a hospital building). When a node encounters with a throwbox, it pushes its collected data. The throw boxes communicate with each

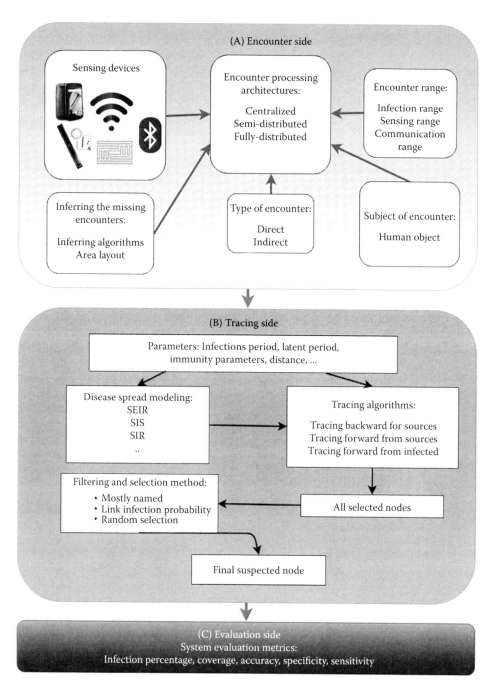

FIGURE 16.2

The infection-tracing framework contains: Block (A) shows the Encounter sensing parts for the collection and processing of encounter data. Block (B) shows the tracing parts, where all suspected nodes are traced and then group of nodes will be selected by using any of the filtering methods. Block (C) shows the evaluation metrics for tracing algorithms.

other during a distributed trace-back process. Another possible architecture is a fully distributed one, in which nodes save their encounters locally and only share it on-demand as requested by other mobile nodes. The mobile nodes can also have different roles or designations based on the heterogeneity of their characteristics (e.g., staff/doctors versus patient/visitor). Clearly, there is a trade-off between the cost, energy, delay, and accuracy of information between the architectures.

16.3.1.2 Encounter Range

There are three different ranges:

1. *Sensing range*: The range in which two devices can sense each other.
2. *Communication range*: The range in which two devices are able to communicate directly with each other.
3. *Infection range*: The distance between an infectious person and others that allows for the infection to be transmitted.

For each range, the reported encounters are different. Hence, when the encounters are ordered on the basis of frequency and duration of encounters, the ranking will differ between ranges. For example, using Bluetooth encounters to identify individuals with maximum patient-contact time may not return the accurate list of people that have encountered the patient the most. This is because the range of agent infection is much smaller than Bluetooth. Thus, such approach may not obtain the correct ranking of our group of interest. This issue needs to be taken into consideration when utilizing encounters in infection trace-back.

16.3.1.3 Direct/Indirect Encounter

Direct encounters are when nodes come in encounter range of each other. These encounters can be recorded directly using mobile devices, unlike indirect encounters with a patient. An example of an indirect encounter is when an individual enters a room after a patient has left, though the disease agent is still present. Indirect encounters cannot be recorded directly as the two nodes are no longer within a sensing range of each other but can be inferred using encounter records, distance, time, and the layout of the hospital.

16.3.1.4 Encounter with Human and Objects

A point—usually neglected but important in our case—is the human–object encounter. Usually, humans are the main carrier of the devices that recognize the contacts and save them, but they are not the only source of infection. Encountering contaminated equipment can be as dangerous as well. Therefore, there is a need for recording encounters with sharing equipment such as wheelchairs, operation tables, X-ray devices, among others.

These devices usually are tracked in smart hospitals. Therefore, the encounter between them and human can be easily found and should be taken into consideration when there is an infection case.

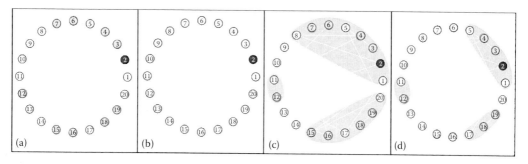

FIGURE 16.3
Illustration for the contact-tracing concept. In (a) the nodes after the disease spread, nodes from 1 to 10 are infected, and 11 to 20 are not, (b) the beginning of the tracing where we only know one infected node (node 1), (c) shows the result of tracing all the suspected nodes, and (d) random selection from the traced nodes.

16.3.2 Tracing Side

The contact tracing process is invoked as soon as an infected case is confirmed (in the hospital). Figure 16.3 illustrates the concept of tracing. The infected nodes after the infection spread is shown in Figure 16.3a in which colored nodes are the infected ones. In Figure 16.3b, the confirmed infected case is represented by node 2. Tracing starts immediately after one case has confirmed. In Figure 16.3c, the tracing has been completed, and all the suspected nodes will be included in the coverage. However, the covered nodes will include some noninfected nodes. Therefore, a selection method might be necessary to identify the population that is more likely to be infected. Figure 16.3d shows the result of a random selection method in which the coverage is reduced; however, there are multiple missing infected cases.

The tracing process consists of three steps: finding sources by tracing back, tracing forward from sources, and tracing forward from the infected population.

16.3.2.1 Tracing Parameters

Several main parameters of the infection tracing process are informed by knowledge of the disease itself. Examples of parameters include the immunities of the population to specific diseases, and the number of confirmed infected cases to trace-back from, latent period, infection period. As a result, there will be various forms of the infection tracing techniques for different circumstances and diseases. Main parameters that shape the tracing that explained in this chapter are as follows:

- *Infectious period* (p_i): The period length in which an infected node is infecting others.
- *Latent period* (p_l): The period between getting the infection and start infecting others.

16.3.2.2 Disease Spread Modeling

The disease spread modeling is an important component in tracing framework. Tracing infection will mimic in reverse to trace the infection spread. There are several models for disease spread in a population. The models depend on the disease itself, for example, if the patients will be immune after recovery or still be susceptible after recovery, and they might be infected again.

The following words are usually used to define the models:

Susceptible (S): The state in which the node might get infected.

Infected (I): The state in which the individuals have the disease and may infect others whom encounter with.

Recovered (R): The state after the node has recovered from the infection and will not infect others while in this state.

Exposed (E): The state in which the node has already infected. However, during this state, the node will not infect others.

Examples of models that have been used to simulate the disease spread in a population are as follows:

Susceptible-Infected (SI) model: Susceptible nodes may get infected, and infected nodes do not recover during the simulation.

Susceptible-Infected-Susceptible (SIS) model: In this model, the infected nodes may recover from the disease and get infected again.

Susceptible-Infected-Recovered (SIR) model: Infected nodes may recover, and recovered nodes will not be infected again.

Susceptible-Infected-Recovered (SEIR) model: When the nodes got an infection, they become exposed in a number of days for the latent period then become infectious for a period of time. After the infectious period, the nodes are recovered and will not be infected again.

More explanation of various mathematical modeling for the disease spread can be found in the mathematical theory of infectious diseases and its applications book (Bailey, 1975).

Also, there are other factors that should be taken into consideration when modeling the disease spread, especially if the information is available. Examples are immunity factors, like if the individual was immunized or not. Another factor when simulates the disease is the ages of persons in which elderly and less immune people may be at higher risk of infection than others.

16.3.2.3 Infection Tracing

The process of tracing begins from an identified case of infection. The tracing backward and forward concepts are illustrated in Figures 16.4 and 16.5. The figures show the encounter of nodes during specified days. Nodes are identified by numbers, and the edge between two nodes represents an encounter between the two nodes during that day. For example, on day 1, there is an encounter between node 20 and node 19, and node 19 encountered with nodes 3 and 12 on the same day.

Node 2 in the figures is the confirmed cases of infection. Therefore, the tracing process will start from this node. We will assume for now that the trace window is only one-day backward or forward. The trace-back process begins on day 3 by tracing node 2 encounters, which are nodes 8 and 12. Then, the encounters of 8 and 12 on day 2 will be traced, which are nodes {6,4,19}. The tracing back will continue from 6, 4, and 19, and tracing back its

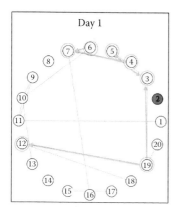
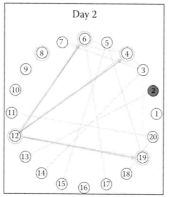
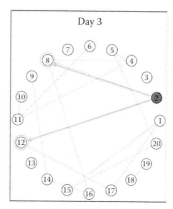

FIGURE 16.4
Back tracing example for three days, starts from day 3.

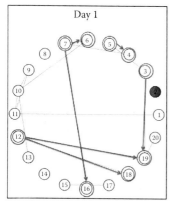
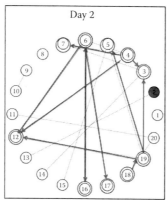
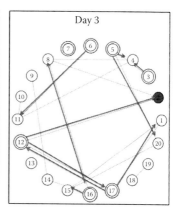

FIGURE 16.5
Forward tracing example for three days starts from day 1.

encounters on day 1. At the end of tracing back process, six nodes will be selected, which are {3,4,5,6,7,8,12,19}. Table 16.1 shows the summary of tracing back each day:

The trace-forward will start from traced nodes that were selected on the trace back process. It starts from day 1 and trace the encounters of {3,5,7,12} as they are the nodes traced on day 1 on the tracing back. The dark gray arrows in Figure 16.5 show the trace-forward process. Table 16.2 summarizes the trace forward process:

TABLE 16.1

Summary of Tracing Back; the Nodes it Starts from and the Traced Nodes on Each Day

Day	Trace from	Traced Nodes
3	2—the confirmed case	{8,12}
2	{8,12}	{6,4,19}
1	{6,4,19}	{7,5,3,12}
	Total:	{3,4,5,6,7,8,12,19}

TABLE 16.2

Summary of Tracing Forward; the Nodes it Starts from and the Traced Nodes on Each Day

Day	Trace from	Traced Nodes
1	{7,5,3,12}	{6,16,4,19,18}
2	{6,16,4,19,18}	{3,5,6,7,12,16,17}
3	{3,5,6,7,12,16,17}	{1,2,4,8,11,15,17,20}
	Total->	{1,2,3,4,5,6,7,8,11,12,15,16,17,18,19,20}

Identifying the Infection sources, which are the nodes that infect other nodes, is achieved by tracing backward. The population at risk is identified by tracing forward from sources. The population at risk might or might not infect others. The difference between finding the sources or population at risk is the process of tracing and the tracing window, as we will explain in the next section. Parameters of disease like (p_i) and (p_l) are used to identify the tracing backward or forward windows.

16.3.2.3.1 Tracing Windows

There are three tracing process types: Tracing backward for a source, trace forward from a source, or tracing forward from an identified infected. Different windows of tracing are estimated on the basis of tracing process type. Figure 16.6 shows the tracing windows.

1. *Trace-Back process for a node*: It starts from an observed infected node. It aims to find the source of infection for this node. The process produces a group of nodes that contains the source. The windows of tracing depend on our knowledge of infection day. If the infection day is known, then the encounters will be traced

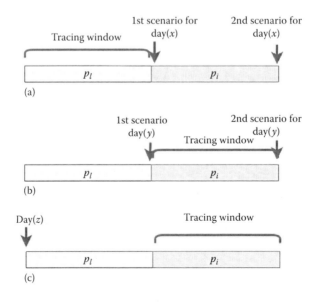

FIGURE 16.6

Time lines that shows the scenarios and the tracing windows. (a) shows the trace back for sources scenarios and tracing windows. (b) shows the first- trace forward process scenarios and tracing windows. (c) shows the second-trace forward process tracing window where the infection day is assumed to be known.

for that day only. However, if it is not, we need to estimate a tracing window. For example, on the day(x), we only know that this node has been infected, but the day of infection is unknown, and we do not have other knowledge of where the nodes are on its infectious period. Consequently, we estimated the knowledge of tracing windows on two scenarios, in which every possible day of infection will be included, and the source of infection will be on the traced list.

a. *First scenario*: The infectious node is on the first day of its infectious period. Consequently, the node was infected on day(x) − p_l.

b. *Second scenario*: The infected node is on the last day of its infectious period. Hence, the nodes' infection day was on the day(x) − (p_l + p_i) + 1.

Figure 16.6a shows the day(x) on both cases.

All other scenarios are in between two assumptions. As a result, all of the nodes encounter during this window will be traced:

$$\{day(x) - (p_i + p_l) + 1, day(x) - p_l\}$$

The nodes that resulted from this window of tracing will be considered as suspected sources of infection for an input node.

2. *Trace forward from a source*: The node that we trace from here is a suspected source of infection. We aim to identify the nodes that might be infected by encountering this node. Also, the window of tracing depends on our knowledge of this nodes infectious period. The worst case is when the only known is the node that was infectious on the day(y) and when the node started the infectious period or ended it are unknown. Consequently, we have two possible scenarios.

a. *First scenario*: The node is on the first day of its infectious period; hence, it will continue infecting its contact until the day(y) + p_i − 1.

b. *Second scenario*: The node is on the last day of its infectious period, so it has infected others since the day(y) − p_i + 1.

Figure 16.6b shows the day(y) on both cases.

All other scenarios are in between two scenarios. As a result, all of the nodes encounter during this window will be traced:

$$\{day(y) - p_i + 1, day(y) + p_i - 1\}$$

The resulted nodes will be considered as population with the risk of being infected by encountering this node.

3. *Trace forward from an infected node*: It applies to a node in which the day of infection for this node is known. The process aims to find a set of nodes that might be infected by this node. Day(z) is assumed to be the day when the node got infected, and the node will start infecting its encounters after its latent period until it finishes its infectious period. Therefore, the contacts in the following window will be traced.

$$\{day(z) + p_l, day(z) + p_l + p_i - 1\}$$

Figure 16.6c shows the estimated scenarios and the tracing window.

16.3.2.3.2 Tracing Approaches

The tracing approaches used the tracing windows that have been described earlier. We will describe one approach that consists of two parts:

- Tracing all suspected nodes S.
- Selecting groups of nodes from the S.

16.3.2.3.2.1 Tracing All Suspected Nodes (S) The result S will include all the infected nodes, but it will also include some noninfected ones, the false positive cases, as shown in Figure 16.3c in which nodes {1,5,8,11,17,20} are included in the trace, even they are not infected.

Tracing S has three phases: tracing sources, tracing the infected nodes from sources, then final tracing of infected nodes.

a. *Tracing sources phase*: The result of this process is a group of nodes that are suspected to be the sources of infection on specific days. The process starts with the confirmed infected case "I" that is known to be infected on day N. Day 1 is the first day where the infection started in the network.

b. *Trace-forward from the sources phase*: This process works on the result of the previous phase. Let us call the result of the previous phase *Sources*. Sources[i] contains the list of nodes that are suspected to be sources of infection on the day(i). As a result, Sources[i] were infectious on the day(i), and they are transmitting the infection to their encounters during their suspected period of infection.

c. *Trace-forward from the infected phase*: The previous phase will list nodes that are suspected to be infected on defined days. If we assume that they are infected, they are in return will spread the infection. In this phase, we trace from nodes that their days of infection are assumed to be known, whereas the previous phase trace from nodes that the days they got their infection are unknown.

The above tracing phases will produce all suspected nodes S. S will include all the infected nodes along with other noninfected nodes. For the purpose of controlling the disease spread, nodes in S are monitored. However, it might be infeasible to contact and monitor all of them due to their large numbers and the limitation of the resources such as time, money, and staff. Consequently, selection methods have been advised to select a group of S. The selection method aims to select the infected nodes in S and ignore the noninfected ones.

16.3.2.3.2.2 Selecting Subset Nodes of S The selection methods attempt to classify nodes in S as infected or not, then select the ones that are classified as infected. This section will introduce three selection methods: random selection, mostly named* (Mn*), and encounter infection probability (EncP). Figure 16.3d is an example of selecting nodes randomly from S.

a. *Random selection*: This method is used mainly for reference purposes. The subgroup of S is selected randomly where each one has the same probability *pr* to be selected.

b. *The mostly named* (Mn*)*: Mostly named as the contact tracing method has been proposed by Armbruster and Brandeau (2007a). The index case is asked to name its contacts. Then, the named contact is assigned a score based on the number of index cases who named it. Each time, an index case has to be identified to continue the contact tracing and update the scores. The named contacts are ranked from highest to lowest, and the top *K* contacts are selected. This policy is modified in i-hospital infection tracing to work on mobile and wireless data, and we called it mostly named* (Mn)*. In Mn*, the patient will not be asked about their contacts. After an infected case is detected, the contact tracing is invoked; the tracing continues automatically through wireless encounter data, and attempt to find all the populations at risk. During tracing, a counter is assigned to each node. The node counter will be incremented by one each time the node is traced. After that, the nodes are ranked on the basis of their counters from highest to lowest, and the top "K" will be selected as the most suspected ones.

c. *Encounter infection probability (EncP):* The probability of infection for a node during its encounter with an infected node is *p* for each second, and $(1 - p)$ is the probability for the node to not be infected in this second. As a result, if the node encounter with an infected node for a duration (*d*), the probability of infection (p_e) will be

$$p_e = 1 - (1-p)^d$$

Consequently, the probability of infection for a node depends on its encounter duration with other infected ones. During the tracing process, the probability is computed and updated each time the node is traced. The probability of infection is computed for multiple encounters with different durations (p_n) as

$$p_n = 1 - \left(\prod_{i=1}^{k} (1-p)^{d(i)} \right)$$

where:
  k is the number of encounters
  d(*i*) is the duration of encounter *i*
  p is the probability of infection per second

Finally, the nodes will be ranked from highest to lowest based on the probability of infection and the top *K* will be selected.

16.3.2.3.3 Assigning Score to Nodes Based on Their Encounter History

Assigning a score to the node may help one to cluster nodes into defined groups. It also will help one to understand how much the epidemic is spread given it starts from a node from a specific cluster. Also, it could be used efficiently in the process of tracing. The score summarizes the node's mobility, and its encounters during its presence in the i-hospital.
  The score could be based on:

- The number of the nodes' distinct encounters
- The nodes' encounter durations
- Frequency of encounters regardless of its duration
- Combination of any of the above
- The number of distinct places the nodes visited during its presence

16.3.3 Evaluation for i-Hospital Infection Tracing System

This part will define several metrics that are used to evaluate the tracing metric. Then, a simulation that used real extensive data is shown. The result of the simulation will be discussed.

16.3.3.1 Evaluation Metrics

Various metrics will be used to evaluate the tracing system. Examples are coverage, specificity, sensitivity, true positive, true negative, and accuracy.

Coverage corresponds to the overhead, and it is defined as the number of nodes that have been traced during the trace-back process to the total number of nodes.

The traced nodes are considered as the infected ones, and the nodes that are not traced are classified as noninfected. Therefore, it defines the true positives (TP) as the number of infected nodes that are classified as infected during a tracing phase. Similarly, the nodes that are not traced (i.e., classified as noninfected in the trace-back phase) and not infected are called the true negatives (TN).

Sensitivity (SV) is computed on the basis of TP and the number of all infected nodes population. Sensitivity reflects the ability of the system to identify all infected cases. In our system, when the tracing all suspected node ends, it will have sensitivity equal to 1. However, it will have high coverage that will affect the specificity.

Specificity (SP) is computed based on TN and the number of nodes that are not infected. It reflects the tracing system ability to filter out noninfected nodes.

Accuracy is the ratio of (TN + TP) to all nodes.

16.3.3.2 Simulation and Evaluation

Real data were used for i-hospital infection tracing evaluation. On campus, Wi-Fi traces are used for simulation purposes. The data were collected by our group from our university campus for years. For this study, we use one week from April 2 to April 8, 2012. Six buildings were chosen where they have different mobility characteristics and different density of users. The traces reflect when mobile devices associate with an access point. Two nodes are considered to be encountered if they associate with the same access point in an overlapping time frame. Although it does not represent the continuous mobility trajectory, the traces are believed to reflect many instances in which two nodes are in proximity of each other and are likely to encounter. Information about the buildings' users and record numbers are shown in Figure 16.7. The information is processed to reflect the contact information. We assume the data encounter include the encounters within infection range. The simulation consists of two phases: disease propagation phase and infection tracing phase.

16.3.3.2.1 Disease Propagation Simulation

The infection is assumed to spread through direct contact from human-to-human. The infection range R_i is the same as the encounter range R_e; that is, $R_i = R_e$. To simulate the disease propagation, SEIR disease propagation model is used. In SEIR model, nodes flow between four states: Susceptible (S), Exposed (E), Infectious (I), and Recovered (R). The node is susceptible until it contacted an infectious node, then it becomes exposed with the rate β. It remains exposed a number of days for the latent period. After the latent period, the node becomes infectious and starts infecting its encounter for a number of days for the infectious period. Then, the node will be recovered. When the node is recovered, it will not be infecting others or getting infectious again. The disease simulation uses the parameters ($\beta = 0.005/s$, p_l = one day, p_i = two days). A random node is chosen to be in an infectious state. Then, it infects the contacts; the disease will propagate between nodes using the above parameters. At each building the percentage of infected nodes (p_f)

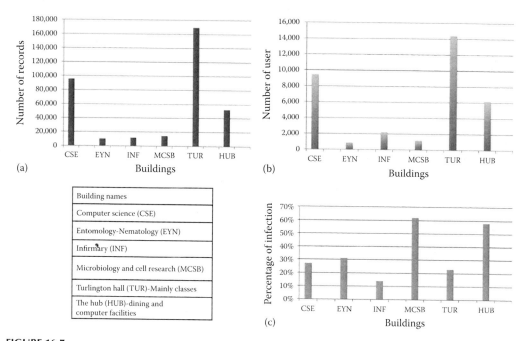

FIGURE 16.7
Buildings information: User and record numbers; and the infection percentage per building. In (a) records numbers, (b) building's users, and (c) infection percentage.

is computed. The percentage of infected nodes is the ratio of the number of infected nodes and the total number of nodes that have been in the building. It varies for each building as it is shown in Figure 16.7.

16.3.3.2.2 Tracing Simulation

The simulation starts with a random node that was infected on the last day. We run the tracing simulation for 10 times, each time the tracing starts from a new random node. First, the group of suspected nodes S is traced, and then the selection methods that were described in Section 16.3 are applied to S to produce the final selected nodes. The following metrics are computed: coverage, true positive, true negative, and accuracy. Figure 16.8 shows the result of the simulation.

i-Hospital infection tracing system will find all infected cases, even if it starts with only one confirmed case of infection. The problem is the overhead in terms of number of the traced nodes. The result of tracing will include a large number of nodes: all of the infected cases and some-noninfected cases as well. Minimize the coverage (the number of selected nodes) and hence reducing the overhead is an important goal of this system. As a result, the selection methods that were explained before are applied to the traced nodes to identify the nodes that are most likely to be the infected nodes. The results of the simulation are shown in Figure 16.8. The figures show the coverage versus accuracy for the studied algorithms in six buildings. Mn* or EncP selection methods achieve higher accuracy than using random selection. The accuracy of Mn* and EncP reaches up to 80% in some cases, especially when the coverage is closer to the infection percentage. The accuracy in some buildings increases with coverage, but it decreases in others, as it is seen in Figure 16.8.

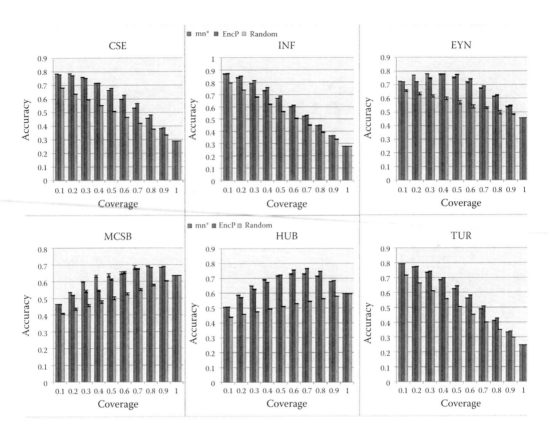

FIGURE 16.8
Result of tracing in six different buildings.

The reason for these various trends of accuracy is related to the node that was chosen to be the source of infection and the encounter patterns of population in each building that will affect the percentage of infections. As the percentage of the infected population gets larger, the accuracy will increase with coverage and vice-versa.

To question is if the choice of starting node (the confirmed case of infection from which the tracing starts) has an effect on the accuracy versus coverage trend or not, we have chosen multiple different random nodes to start tracing from and we compared the results. The figures show that the result has similar trend of accuracy versus coverage with very little variation. On other words, the graph increasing and decreasing trends with coverage seem to remain quite consistent with different choices of nodes where the tracing starts from.

16.4 Conclusion

The first part of this chapter presented i-hospital framework that utilizes different heterogeneous sensing devices in IoT for collecting and processing encounter information. Then, i-hospital infection tracing framework is explained in the next section. In this work, encounter-based information in infrastructure-based server architecture was used for tracing of sources of infection and at-risk population. Our systematic framework of building blocks offers a suite of trace-back algorithms that mimic (in-reverse) the infection spread model, to provide probabilistic forward and backward search techniques. We also provide one of the most extensive studies on evaluation of disease spread using WLAN traces of over 34k users. IoT equipment such as RFID readers and tags, IoT-based wristbands, and IoT-enabled beds encounter data can apply the same tracing algorithms seamlessly. Findings show the potential promise of this method to reach high accuracy while investigating a relatively small population.

References

Armbruster, B., & Brandeau, M. L. (2007a). *Who do you know? A simulation study of infectious disease control through contact tracing.* In proceedings of the 2007 International Conference on Health Sciences Simulation (ICHSS), part of the 2007 Western MultiConference on Modeling & Simulation (WMC) (pp. 79–85). San Diego, CA.

Armbruster, B., & Brandeau, M. L. (2007b). Contact tracing to control infectious disease: When enough is enough. *Health Care Management Science*, 10(4), 341–355.

Bailey, N. T. J. (1975). *The mathematical theory of infectious diseases and its applications.* Charles Griffin & Company Ltd, 5a Crendon Street, High Wycombe, Bucks HP13 6LE.

Brown, C., Efstratiou, C., Leontiadis, I., Quercia, D., Mascolo, C., Scott, J., & Key, P. (2014, September). *The architecture of innovation: Tracking face-to-face interactions with ubicomp technologies.* In proceedings of the 2014 ACM international joint conference on pervasive and ubiquitous computing (pp. 811–822). Seattle, WA: ACM.

CDC (Center for Disease Control and Prevention). (2007). *Healthcare infection control practices advisory committee (HICPAC).* Retrieved from http://www.cdc.gov/hicpac/index.html.

Cisco (2014). *Wi-Fi Location-based services 4.1 design guide.* Retrieved from http://www.cisco.com/c/en/us/td/docs/solutions/Enterprise/Mobility/WiFiLBS-DG.html.

Farrahi, K., Emonet, R., & Cebrian, M. (2015, February). *Predicting a community's flu dynamics with mobile phone data.* In proceedings of the 18th ACM conference on computer supported cooperative work & social computing (pp. 1214–1221). Vancouver, BC: ACM.

Fuhrer, P., & Guinard, D. (2006). *Building a smart hospital using RFID technologies: Use cases and implementation.* Fribourg, Switzerland: Department of Informatics-University of Fribourg.

Kim, Y., & Helmy, A. (2010). CATCH: A protocol framework for cross-layer attacker traceback in mobile multi-hop networks. *Ad Hoc Networks*, 8(2), 193–213.

Kumar, U., & Helmy, A. (2012, November). *Discovering trustworthy social spaces.* In proceedings of the third international workshop on sensing applications on mobile phones (p. 7). Toronto, ON: ACM.

Luo, W., & Tay, W. P. (2013, May). *Finding an infection source under the SIS model.* In 2013 IEEE international conference on acoustics, speech and signal processing (pp. 2930–2934). Vancouver, BC: IEEE.

Magill, S. S., Edwards, J. R., Bamberg, W., Beldavs, Z. G., Dumyati, G., Kainer, M. A., ... & Ray, S. M. (2014). Multistate point-prevalence survey of health care–Associated infections. *New England Journal of Medicine*, 370(13), 1198–1208.

Peng, T., Leckie, C., & Ramamohanarao, K. (2007). Survey of network-based defense mechanisms countering the DoS and DDoS problems. *ACM Computing Surveys* (CSUR), 39(1), 3.

Reddy, E., Kumar, S., Rollings, N., & Chandra, R. (2015). Mobile application for dengue fever monitoring and tracking via GPS: Case study for Fiji. *arXiv preprint arXiv*:1503.00814.

RFID journal (2016), *Frequently asked questions.* Retrieved from http://www.rfidjournal.com/site/faqs#Anchor-Is-53555.

Salathé, M., Kazandjieva, M., Lee, J. W., Levis, P., Feldman, M. W., & Jones, J. H. (2010). A high-resolution human contact network for infectious disease transmission. *Proceedings of the National Academy of Sciences*, 107(51), 22020–22025.

Shah, D., & Zaman, T. (2011). Rumors in a network: Who's the culprit? *IEEE Transactions on Information Theory, 57*(8), 5163–5181.

Stehlé, J., Voirin, N., Barrat, A., Cattuto, C., Colizza, V., Isella, L., … Vanhems, P. (2011). Simulation of an SEIR infectious disease model on the dynamic contact network of conference attendees. *BMC Medicine, 9*(1), 1.

Subramony, H., Lai, F. Y., Ang, L. W., Cutter, J. L., Lim, P. L., & James, L. (2010). An epidemiological study of 1348 cases of pandemic H1N1 influenza admitted to Singapore Hospitals from July to September 2009. *Annals Academy of Medicine Singapore, 39*(4), 283.

Team, W. E. R. (2014). Ebola virus disease in West Africa—The first 9 months of the epidemic and forward projections. *The New England Journal of Medicine, 2014*(371), 1481–1495.

Tina, H. S. (2016). *How one patient spread MERS to 82 people.* Retrieved from https://www.sciencenews.org/blog/science-ticker/how-one-patient-spread-mers-82-people.

Weiser, A. A., Gross, S., Schielke, A., Wigger, J. F., Ernert, A., Adolphs, J., … Appel, B. (2013). Trace-back and trace-forward tools developed ad hoc and used during the STEC O104: H4 outbreak 2011 in Germany and generic concepts for future outbreak situations. *Foodborne Pathogens and Disease, 10*(3), 263–269.

Yoneki, E., & Crowcroft, J. (2014). EpiMap: Towards quantifying contact networks for understanding epidemiology in developing countries. *Ad Hoc Networks, 13*, 83–93.

17

Emergency Department, Sustainability, and eHealth: A Proposal to Merge These Elements Improving the Sanitary System

Andrea Malizia, Laura Morciano, Jacopo Maria Legramante, Pasqualino Gaudio, Sandro Mancinelli, Francesco Gilardi, Carlo Bellecci, and Leonardo Palombi

CONTENTS

17.1 Introduction

With a population of nearly 60 million, Italy is the sixth most populous country in Europe and the 23rd in the world, representing 0.81% of the global population. The territory covers 301,316 km^2, with a population density of 193 inhabitants per km^2. A range of indicators shows that the health of the population has improved over the last decades. Average life expectancy reached 79.4 years for men and 84.5 years for women in 2011, the second highest in Europe (compared with 77.4 years for men and 83.1 years for women for the European Union as a whole). Italy has one of the lowest total fertility rates in the world: in 2011, it was 1.4 births per woman, far below the replacement level of 2.1. The population growth rate is, therefore, very low (0.3% in 2012), one of the lowest in the European Union, and immigration is the source of most of this growth. Italy is divided into 20 regions (Figure 17.1), each region has a certain number of provinces (that to date are 93), and the provinces are divided into municipalities (comuni) that are around 8100 in all the country. Regions differ in terms of demography, economic development, health care infrastructure, and health expenditure. There are marked regional differences for both men and women in most health indicators, reflecting the economic and social imbalance between the north and south of the country.

FIGURE 17.1
Regions of Italy.

The World Health Organization has ranked the Italian health care system second best in the world with only French system ranked higher (WHO 2000). Italy has a National Health Service established in 1978 with the declared goal of providing uniform and comprehensive care, financed by general taxation.

The system is regionally based and organized at three levels: national, regional, and local. At national level, the Ministry of Health (supported by several specialized agencies) sets the fundamental principles and goals of the health system, determines the core benefit package of health services guaranteed across the country, and allocates national funds to the regions. The regions are responsible for organizing and delivering health care: They have virtually exclusive responsibility for the organization and administration of publicly financed health care (Ferré et al., 2014). At local level, health services are delivered through a network of population-based health local unit deliver public health, community health services, and primary care directly, and secondary and specialist care directly or through public hospitals or accredited private providers. The main source of financing is national and regional taxes, supplemented by copayments for pharmaceuticals and outpatient care. In 2012, total health expenditure accounted for 9.2% of

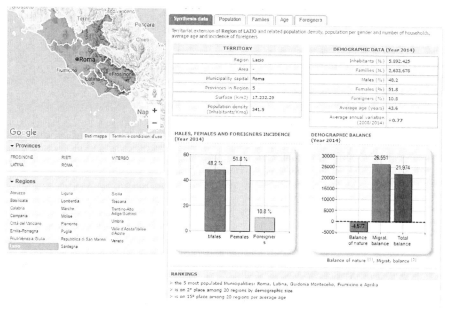

FIGURE 17.2
Lazio demographic situation.

gross domestic product (GDP) (slightly below the European Union average of 9.6%). Public sources made up 78.2% of total health care spending.

Lazio is a region placed in the center of Italy; it is characterized by a number of inhabitants that is lower than 6 million, with an average age of 43.6 years and a quasi-equally division between the percentage of men (48.2%) and women (51.8%) (Figure 17.2). The young population of the region (0–24 years) is the one with the lowest number of people, whereas around the 75% of the Lazio's population is in the range of 25 to over 75, as is shown in Figure 17.3 (year 2014). The total number of foreign persons is around the 10% (Figure 17.2), and there are several foreign countries represented in the Lazio population (Figure 17.4, referred to 2010). For this reason, for each initiative/implementation proposed, it is essential to consider cultural differences. Lazio region has 43 medical structures with emergency department (ED). A website developed by the health service of the Lazio Region monitors, continuously, the number of access per single structures.

In this chapter, the authors analyzed the problem of the EDs. As previously mentioned, self-funding is an important option for the regions, and the reduction of costs for the ED is a proper solution for increasing the economic capability of the regional health systems. To get this result, the authors proposed the following series of implementations based on the main concept of the sustainability:

- Interventions with a social impact
- Interventions with an environmental impact
- Interventions with an economical impact

To achieve this, each intervention should considering cultural differences (Figure 17.5).

Sustainable development is development that meets the needs of the present without compromising the ability of future generations to meet their own needs (Brundtland definition, 1987).

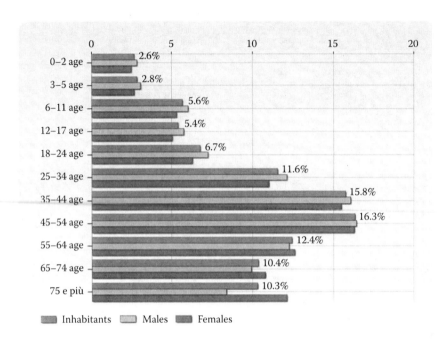

FIGURE 17.3
Lazio inhabitants age distribution (2014).

Nationality (Year 2010)				
Nationality	(n.)		% on foreigners	% on population
Romania	196,410		36.19	3.43
Filippine	32,126		5.92	0.56
Polonia	24,392		4.49	0.43
Albania	23,337		4.30	0.41
Ucraina	18,922		3.49	0.33
Bangladesh	16,161		2.98	0.28
Perù	14,895		2.74	0.26
Cina Rep. Popolare	14,890		2.74	0.26
India	14,586		2.69	0.25
Moldova	12,413		2.29	0.22
Marocco	11,606		2.14	0.20
Ecuador	9,676		1.78	0.17
Egitto	9,209		1.70	0.16
Sri Lanka	8,150		1.50	0.14
Bulgaria	7,722		1.42	0.13

Rankings
> is on 5° place among 20 regions per % foreigners on total inhabitants
> is on 11° place among 20 regions per % underage foreigners
> Foreigners growth rate[1]: 31.6% (2° place among 20 regions)

FIGURE 17.4
Lazio foreign inhabitant's distribution.

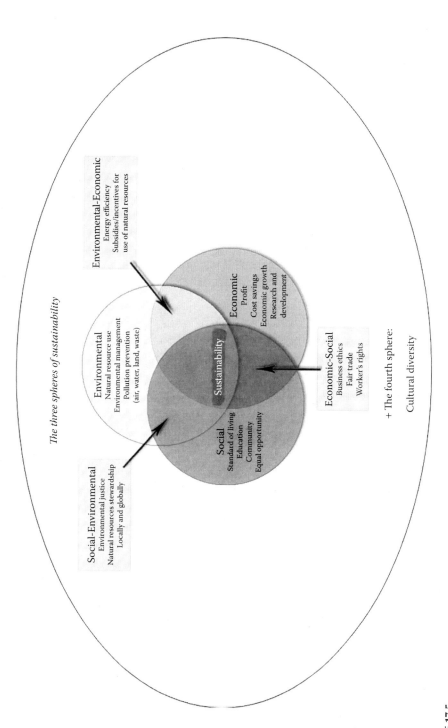

FIGURE 17.5
Sustainability scheme.

The case study considered in this work will be the ED of the "Policlinico Tor Vergata" (PTV), the University of Rome Tor Vergata's hospital. The PTV is in the south area of Rome, in the Local Health Unit Roma2, one of the biggest of the region and that with the lowest socioeconomic level.

The authors analyzed a 1-year access to the ED, which comprises an area of about 1.5 million people and almost 45,000 annual accesses.

A skilled-nurse staff based on Italian Triage guidelines assesses the first patient evaluation:

- *White tag*: Nonurgent condition
- *Green tag*: Less urgent condition/low priority
- *Yellow tag*: Urgent, potentially life-threatening emergency conditions
- *Red tag*: Critical, immediately life-threatening emergency condition

A total number of 46,820 patients' visit occurred during the study period. The ED was accessed by 38,016 users, with a mean age of 49.6 ± 21.6 years. Patients with an age ≥65 years make more than 25% of those accesses (about 11,000 accesses/year). The average distribution of the Triage code in the overall ED population was the following:

1.	Green tag	59%
2.	Yellow tag	26%
3.	White tag	10%
4.	Red tag	4%
5.	Not executed	1%

The hospitalization percentage of the patients entering in the ED is 17%, the percentage of patients that are sent home (home discharge) is 56% and the percentage that leave the ED without a visit (left without being seen [LWBS]) is 15%. Among those who are LWBS in the ED, the percentage rises three-fold among patients defined as *frequent users*. Frequent users are those patients with four or more accesses in ED per year. Despite they are poorly represented in the examined ED, comprising 6% of the overall accesses, they greatly contribute to the ED overcrowding. Recent data suggest a strong correlation between older age and frequent use of the ED, due to the multiple chronicle conditions of these patients and a lack of alternative care choices.

One of the main indicators of the ED overcrowding is the percentage of LWBS, which suggests too long waiting time. This percentage in the ED of the PTV is the higher of the Lazio region (15% vs 6%).

Older patients have a strong influence on the numbers mentioned above. Elderly patients require significantly more emergency care resources than their younger counterpart. This means both higher number of visits and the use of more complex resources, due to the higher number of existing comorbidities. The greater complexity of this kind of patients is also clear from rate at which they arrive in the ED with grey or black tag (OR 3.1: LC 2.9–3.2), as shown in Figure 17.6.

Another important aspect is the number of hospitalization occurred in this kind of patients. Due to their complexity, the elderly were more often hospitalized than younger adults. In our case study, the risk of hospitalization increased 5-fold in patients aged ≥ 65 (OR 5.2; LC 4.9–5.3). This is one of the highest rates of hospitalization ever reported in the literature (Table 17.1 and Figure 17.7).

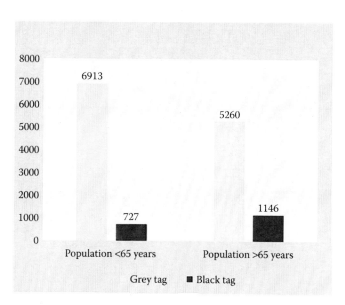

FIGURE 17.6

Yearly number of black and grey code tags at PTV divided per age.

TABLE 17.1

Comparison of Odds Over >65 Years versus Odds <65 Years between PTV and Other Foreign Hospital Structures

Reference	Year	Type of Study	% of Over >65 Years	Odds >65 Years versus Odds <65 Years
Aminzadeh	2002	Systematic revision	32–47	2,5–4,5
Yim VWT	2009	Retrospective analysis	45	3
Sona A	2012	Prospect	39, 8	–
Lowthian JA	2012	Retrospective analysis	39	3,9
Albert M	2013	Data collection CDC	36, 5	–
Latham LP	2014	Retrospective transversal	21	–
Keyes DC	2014	Case-control	51	2
Legramante	2016	Retrospective analysis	44	5,2

As highlighting in the previous data, older individuals and frequent users had a significantly higher number of ED accesses, contributing to the overcrowding of EDs. Although numbers of interventions have been introduced to limit this problem, PTV is the second structure per number of access in short-time period (in Figure 17.8 is reported, for example, a bar graph taken in 2 min of observation of the Institutional website).

Each access to the ED has been estimated costing 360 € per patient, whereas the hospitalization is calculated in 700 € per day (with an average length of stay of 5 days per patient). The accesses cost in the PTV ED settle around 20–25 on millions € per year. Considering the workdays lost due to each access, these numbers seem to further increase.

The authors estimate, for PTV, a catchment area of patients divided in the following:

- 25% from Rome
- 20% from Frosinone
- 20% from the other provinces of Lazio

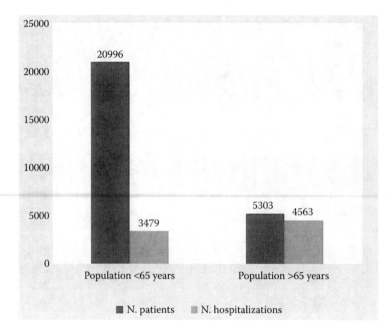

FIGURE 17.7
Yearly number of hospitalizations per patients at PTV per age.

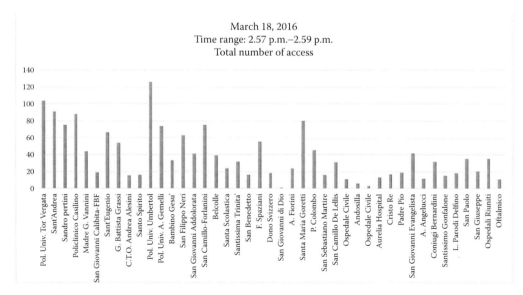

FIGURE 17.8
Number of access in the Lazio ED & First Aid in 2 min, March 18, 2016.

The total estimation is around 1.5 million of potential patients. Considering that the estimated number of accesses to the first aid of PTV is around 540,000, the estimated percentage of accesses is around the 33% of the potential catchment area of the patients. It is plausible considering a percentage of potential accesses in a range from 18% to 33%.

FIGURE 17.9
Contamination produced by one travel of 38 min of car (33 of train).

The impact of these huge numbers of accesses has consequences even on the environment, as people usually use public or private transportation to reach the PTV. The authors have interviewed (from April 2016 to July 2016) the patients to investigate how they arrive at PTV, and how far they live from the hospital. The results are that the 85% of the interviewed patients reach the PTV using a private transportation, and the average time of travel is around 35–42 min.

The authors used these data as boundary conditions to run a calculation with the free-license software ecopassenger, considering a travel of 38 min (from the city of Colleferro near to Rome). The results obtained are shown in Figure 17.9.

It is clear that multiplying this value for the number of the yearly accesses, the contamination values produced to reach the PTV (considering the estimation previously done) are very high (Table 17.2).

TABLE 17.2

Estimated Contaminations due to the Travels of the Patients to Reach the PTV ED

Data	Values
Carbon dioxide (kg/year)	1.320300–2.430000
Energy resource consumption (l/year)	586800–1.08000
Particulate matter (kg/year)	130–238
Nitrogen oxides (kg/year)	5428–9990
Nonmethane hydrocarbons (kg/year)	440–810

Data from the state-of-the-art suggest that reducing the accesses in the ED is the main factor to reduce costs and improve quality of the services through the following:

- Reduction of both length of stay and waiting time
- Reduction of diagnostic and therapeutic errors
- Improvement of psychological states of both patients and operators
- Reduction of air pollution

17.2 Design of Interventions to Optimize the ED

The interventions to optimize the PTV ED need to have the following triple impact:

- *Social impact*: Accesses reduction, improves the appropriateness of access points, improves the capability to do right diagnosis, and assigns right therapies, waiting times reduction.
- *Environmental impact*: Reduction of pollutants into the atmosphere, reduction of fossil fuels.
- *Economic impact*: Reduction of the costs, both for the hospital and the patients.

To do that the main interventions proposed are based on the following:

- Hi triage
- H24 medical assistance
- eHealth

17.2.1 Hi Triage

To determine the quality of patient care, there are four main actions of the emergency process that have to work correctly: *triage, testing and evaluation, handoffs, and admitting* (Eisenberg et al., 2005).

Triage derived from the French word *trier*, and it is defined as the sorting of patients according to urgency and need for care (Conrand Stoppler, 2014). To date, triage is used to assess priority for health care in many different conditions, which are as follows:

- In mass causalities, it is used to identify priorities and decide who need to be transported in medical structure
- In the emergency rooms to decide who treat urgently
- To optimize the use of space in crowded medical structures

At the hospital, a skilled-nurse staff assesses triage, and it could be influenced by overcrowding. O'Connor, Gatien, Weir, and Calder (2014) hypothesized that patients would be more often undertriaged if the ED was crowded. This has striking impacts on the health of patients and the quality of care. In fact, undertriaged ones could lead the patients to leave the hospital without being visited increasing the length of stay. It can provokes adverse outcomes and adverse events such as an increase in mortality or delayed antibiotic administration in patients with infectious diseases (Apker, Mallak, & Gibson, 2007; Edwards, & Sines, 2008; Cheung et al., 2010; Maughan, Lei, & Cydulka, 2011).

> *In a simulated emergency situation, quantitative information requiring frequent updates and information generated early were most likely to be inaccurate* (Bogenstatter et al., 2009). *Metaanalyses* (Bost, Crilly, Wallis, Patterson, & Chaboyer, 2010; Calleja, Aitken, & Cooke, 2011) *reveal that structured handovers that involved both written and oral communication improved information exchange.* (Lori, Eric, & Forde, 2016)

From this analysis, emerges the main role of the nurse, which cannot be substituted. He provides to

- Evaluate the conditions of the patient according to the main vital functions, symptoms, and vital parameters.
- Assign to the patient the personal data, the reason of the access in the emergency room, the name of the family doctor, and the pharmacologic history.
- Assign the priority code to the patients.
- Check periodically the conditions of the patient to evaluate variation that can determine a different priority.

This work can be improved by a more efficient organization and the development of new technologies as follows:

- The introduction of innovative diagnostic tools with automatized measures of the following:
  - Body temperature, during the check in, through a proper collocation of thermocameras and customized software
  - Heart rate
  - Blood pressure
  - Hematologic and biochemic fast-check to improve the evaluation of vital functions
  - Saturometry
  - Stress index
- Adoption of hardware and software to monitor patients with green tag codes to constantly check the body parameters and the psychological conditions of the patients.
- Hardware and software (similar to a pc station) allow the patients in nonurgent conditions to partially compile their medical records, realized in different languages: These tools could improve communication between patients and medical staff during the assessment of health status.
- Touch screen for nurse and operator that can process a first/rapid medical register with a direct access to the previous data of the patients and the information get during his passage through the diagnostic paths.

In the United States of America, a lot of hospitals have started an important use of technologies in ED (Geisler et al., 2010). The patient that is registered through an electronic medical record is easily controllable, but it has to pay attention to the incapability to register important details that can be excluded (or hidden) by the software. Another critical issue could be represented by poor flexibility by which the electronic systems log information (Atak et al., 2005; Bogenstatter et al., 2009; Eisenberg et al., 2006; Schubert et al., 2013).

Hi triage ones hypothesized by the authors combined the essential role of the skilled nurses in the ED and the improved use of new technologic tools to simplify, speed up, and optimize the triage process, while reducing errors caused by overcrowding.

17.2.2 H24 Medical Assistance

The last report from the Italian Society of Emergency Medicine depicted that older adults need hospitalization as the main cause of ED overcrowding. As mentioned above, older adults represent about 25% of all ED users, and they are hospitalized in almost 50% of the cases. One critical point of this elderly patients flow is the availability of continuing care after hospital discharge. Indeed, many of them require, once discharged, other forms of assistance that would not allow them to return to their homes. To avoid this issue, the authors hypothesized two relevant measures. First, an early taking charge of frailty patients, with a discharge service directly managed by the hospital to reduce patients' visits to the ED and further reducing the number of hospitalizations and the average length of stay. Second, introducing the concept of health smart home, a home equipped with several medical devices, that can be remotely controlled, to monitor the patients 24 h. (Bailenson et al., 2008; Banerjee et al., 2015; Kientz et al., 2008; Lehmann et al., 2004; Rialle et al., 2002; Riva, 2003).

There are many solutions studied; a (main) resume is reported in Table 17.1 from Mano et al. (2016).

17.2.3 eHealth

eHealth refers to the use of tools based on information and communication technologies to support and promote the prevention, diagnosis, treatment, and monitoring of disease and the management of health and lifestyle (Ministry of Health). To improve ED performances, the authors suggest the introduction of technologies tools to the PTV hospital, which are as follows:

- *Electronic health record*: An electronic health record (EHR), or electronic medical record, refers to the systematized collection of patient and population electronically stored health information in a digital format. These records can be shared across different health care settings. Records are shared through network-connected, enterprise-wide information systems, or other information networks and exchanges. EHRs may include a range of data, including demographics, medical history, medication and allergies, immunization status, lab results, diagnostic images, vital signs, detection of anthropometric data, and billing information.

  In Italy, the use of EHR is regulated by the Ministry of Health (Italian Ministry of Health, 2011). The use of EHRs has been widely demonstrated, and this system is applied in a lot of medical structure worldwide. (Asan, 2017; Blumenthal, 2010; Bates & Bitton, 2010; Embi et al., 2013; Collins et al., 2011; Carayon et al., 2014; Carayon, 2012; Häyrinen et al., 2008; Weir et al., 2011).

- *Telemedicine*: Telemedicine is the use of telecommunication and information technology to provide remote medical care. It helps us to eliminate distance barriers and can improve access to medical services that would often not be consistently available in distant rural communities. It is also used to save lives in critical care and emergency situations. Although there were distant precursors to telemedicine, it is essentially a product of the twentieth century telecommunication and information technologies. These technologies permit communications between patient and medical staff with both convenience and fidelity, as well as the transmission of medical, imaging, and health informatics data from one site to another. Early forms of telemedicine achieved with telephone and radio have been supplemented with video telephony, advanced diagnostic methods supported by distributed client/server applications, and additionally with telemedical devices to support in-home care. (Audebert et al., 2005; Bairagi & Sapkal, 2013; Castiglione et al., 2015; Hsu, 2017; Legris et al., 2016; Tao et al., 2016; Zhu et al., 2015).

Actually, in Italy, Minister of Health has prepared telemedicine guidelines. The authors want to propose this particular branch of ehealth for PTV because it can be used for the following services:

- Providing a consultation with a patient or a specialist assisting the primary care physician in rendering a diagnosis. This may involve the use of live interactive video or the use of store and forward transmission of diagnostic images, vital signs, and/or video clips along with patient data for later review.

- Remote patient monitoring, including home telehealth, uses devices to remotely collect and send data to a home-health agency or a remote diagnostic testing facility for interpretation. Such applications might include a specific vital sign, such as blood glucose or heart ECG or a variety of indicators for homebound patients. The following services can be used to supplement the use of visiting nurses.

- Consumer medical and health information includes the use of the Internet and wireless devices for consumers to obtain specialized health information and on-line discussion groups to provide peer-to-peer support.

- Medical education provides continuing medical education credits for health professionals and special medical education seminars for targeted groups in remote locations.

- *Consumer Health Informatics*: Consumer Health Informatics (CHI) is a subbranch of health informatics that helps us to bridge the gap between patients and health resources. The American Medical Informatics Association defines it as "the field devoted to informatics from multiple consumer or patient views." The Consumer Health Informatics Working Group of the International Medical Informatics Association defines it as "the use of modern computers and telecommunications to support consumers in obtaining information, analysing unique health care needs and helping them make decisions about their own health." CHI includes patient-focused informatics, health literacy, and consumer education. The focus of this field is to allow consumers to manage their own health, using Internet-based strategies and resources with consumer-friendly language. Currently, CHI stands at a crossroads between various health care-related fields such as nursing, public health, health promotion, and health education.

- *Virtual Health Care teams*: Virtual Health care teams (VHT) are groups of health experts cooperating and digitally sharing information about patients. The creation of VHTs would be fundamental in big medical structures such as PTV to create good communications between colleagues and between operator of PTV and the family doctors.

The acquired data coming from the eHealth application on PTV ED can allow the development of software, based on self-learning codes, to be able to give a preliminary evaluation of clinical risk, needs of hospitalization, and social assistance to support the following:

1. The regional health record of Lazio (project in progress in the Department of Biomedicine and Prevention)
2. The health system
3. The research and teaching activities of University of Rome Tor Vergata

17.3 Feasibility Study and Creation of Indicators

Following two phases are essential to get the proposals effective:

1. *A feasibility study*: Necessary to support the evaluation that is necessary to adopt the choices and increase the operability spectra. The first phase is a technical verification of the real capacity to realize all the proposals from an economic-managerial and organizational point of view. Each proposal has to be deeply analyzed to evaluate the real possibility to be realized to calculate the time for its realization and schedule the so called phases of *investment recovery*.
2. *The creation of sustainability indicators*: As instrument to support all the sustainability procedure and policies adopted to realize the proposals. The authors have identified a list of indicators divided per themes and types (Table 17.3).

TABLE 17.3

List of Indicators to Reach a Sustainable ED for PTV

Theme	Subtheme	Indicator	Type of Indicator
Social indicators	PTV ED	People influx reduction	Descriptive
		Quality improve of operator's work	Performance
		Access improvements	Descriptive
	Patients manage	Understanding of patients problems	Descriptive
		Correct sorting wards	Performance
		Reduction of waiting times	Performance
Environmental indicators	Energy	Reduction of energy consumption inside PTV	Composite
		Reduction of energy consumption of the people	Composite
	Pollutants and contamination	Reduction of pollution emission in atmosphere	Composite
		Reduction of consumption of fossil combustibles	Composite
Economic indicators	PTV costs	Reduction of cost for operators	Performance
		Reduction of accesses	Descriptive
		Reduction of FU	Composite
		Reduction of managerial costs	Composite
	People costs	Reduction of work days lost	Performance
		Reduction of patient costs	Descriptive

17.4 IoT and Health Care

The Internet of Things (IoT) concept entails the use of devices to collect data and connect those data in private or public cloud, enabling them to automatically trigger certain events and be available to properly manage emergency situations.

Internet-connected devices have been introduced to patients in various forms. Whether data comes from fetal monitors, electrocardiograms, temperature monitors, or blood-glucose levels, tracking health information is vital for some patients. Many of these measures require follow-up interaction with a health care professional. This creates an opening for smarter devices to deliver more valuable data, lessening the need for direct patient–physician interaction. By embedding IoT-enabled devices in medical equipment, health care professionals will be able to monitor patients more effectively—and use the data gleaned from the devices to figure out who needs the most hands-on attention. In other words, by making the most of this network of devices, health care professionals could use data to create a system of proactive management—as they say, prevention is better than the cure. A number of technologies can reduce overall costs for the prevention or management of chronic illnesses. These include devices that constantly monitor health indicators, devices that autoadminister therapies, or devices that track real-time health data when a patient self-administers a therapy. Many patients have started to use mobile applications (apps) to manage various health issues because they have an increased access to high-speed Internet and they have tablets or smartphones to do that. These devices and mobile apps are now increasingly used and integrated with telemedicine and telehealth via the medical IoT. A new category of *personalized preventative health coaches* (Digital Health Advisors) will emerge. These workers will possess the skills and the ability to interpret and understand health and well-being data. They will help their clients to avoid chronic and diet-related illness, improve cognitive function, achieve improved mental health, and achieve improved lifestyles overall. As the global population ages, such roles will become increasingly important (Dimitrov, 2016).

17.5 Conclusion

The analysis of the available data has been necessary to properly evaluate a number of interventions that can improve the performance of the whole ED of PTV in a sustainable direction. These interventions are useful to optimize the functionalities of PTV ED not only in condition of normal operability but also in extraordinary condition such as earthquakes, tsunami, and chemical, biological, radiological, and nuclear (CBRNe) events (Di Giovanni et al., 2014; Malizia, 2016).

The attended results are as follows:

- Risks reduction.
- Attention to social and cultural diversities through software and hardware for the patients available in different languages but able to immediately translate each communication between operators and patients.
- Usage of the indicators in Table 17.3 to evaluate the ED of PTV to propose it as a pilot model in Italy.

- University courses creation to form specialized personnel with a multidisciplinary preparation to also launch multidisciplinary and innovative research projects.
- Reduction, around 15%, of patients accesses that means (according to the number previously mentioned)
  - Reduction of access cost (and of the connected hospitalizations) around 2.2–3 million €/year that can be reinvested to recover the expenses.
  - Pollution reduction that, considering the number in Figure 17.9, could be of 3,500 kg/year of carbon dioxide, 3.5 kg/year of particulate, 150 kg/year of nitrogen oxides, and 12 kg/year of nonmethane hydrocarbons.
  - Reduction of fuel consumption around 16,000 liters/year that, together with the working days, has a direct impact on the patient's costs.
  - Environmental study to evaluate the impact on humans, environment, animals, materials, and cultural resources.

The conclusion of this paper is that the abovementioned proposals and the results attended provoke an improvement in term of the following:

- Appropriateness of access in ED
- Reduction of inflows in ED
- Education of waiting times in ED
- Appropriate selection of patients' problems
- Best operator performances
- Best and most appropriate care for the patient
- Availability of supplementary educational tools with the development of remote or on-site presentations of cases, their parameterization, the results of clinical and surgical interventions, and so on
- Improvement of the economic environment through the reduction of personnel costs, reduction of overall costs (visits, hospitalizations, and management structure), and reduction of the costs incurred by the patient
- Reducing the impact on the environment thanks to the reduction of patients' travel to reach the ED has a direct impact on reducing fuel consumption, reducing pollutants emissions into the atmosphere and reducing the energy consumption in general

The future development for/of this work is the punctual analysis of each proposal from a sustainability point of view through a collection of experimental evidences worldwide collected.

References

Albert M. (2013). National Hospital Ambulatory Medical Care Survey: Emergency Department Summary Tables. Available from: https://www.cdc.gov/nchs/data/ahcd/nhamcs_emergency/2013_ed_web_tables.pdf

Aminzadeh, F., & Dalziel, W.B., (2002). Older adults in the emergency department: a systematic review of patterns of use, adverse outcomes, and effectiveness of interventions. *Annals of Emergency Medicine, 39*(3), 238–247.

Apker, J., Mallak, L. A., & Gibson, S. C. (2007). Communicating in the "gray zone": Perceptions about emergency physician hospitalist handoffs and patient safety. *Academic Emergency Medicine, 14*, 884–894.

Asan, O. (2017). Providers' perceived facilitators and barriers to EHR screen sharing in outpatient settings. *Applied Ergonomics, 58*, 301–307.

Atak, L., Rankin, J. A., & Then, K. L. (2005). Effectiveness of a 6-week online course in the Canadian triage and acuity scale for emergency nurses. *Journal of Emergency Nursing, 31*, 436–441.

Audebert, H. J., Kukla, C., Clarmann von Claranau, S., Kühn, J., Vatankhah, B., Schenkel, J., ... TEMPiS Group. (2005). Telemedicine for safe and extended use of thrombolysis in stroke: The Telemedic Pilot Project for Integrative Stroke Care (TEMPiS) in Bavaria. *Stroke, 36*(2), 287–291.

Bailenson, J. N., Pontikakis, E. D., Mauss, I. B., Grossd, J. J., Jabone, M. E., Hutchersond, C. A. C., ... John, O. (2008). Real-time classification of evoked emotions using facial feature tracking and physiological responses. *International Journal of Human-Computer Studies, 66*(5), 303–317.

Bairagi, V. K., & Sapkal, A. M. (2013). ROI-based DICOM image compression for telemedicine. Sadhana. *Academy Proceedings in Engineering Sciences, 38*(1), 123–131.

Banerjee, T., Keller, J. M., Popescu, M., & Skubic, M. (2015). Recognizing complex instrumental activities of daily living using scene information and fuzzy logic. *Computer Vision Image Understand, 140*, 68–82.

Bates, D. W., & Bitton, A. (2010). The future of health information technology in the patient-centered medical home. *Health Affairs, 29*(4), 614–621.

Blumenthal, D. (2010). Launching HITECH. *New England Journal of Medicine, 362*(5), 382–385.

Bogenstatter, Y., Tschan, F., Semmer, N. K., Spychiger, M., Breuer, M., & Marsch, S. (2009). How accurate is information transmitted to medical professionals joining a medical emergency? A simulator study. *Human Factors, 51*, 115–125.

Bost, N., Crilly, J., Wallis, M., Patterson, E., & Chaboyer, W. (2010). Clinical handover of patients arriving by ambulance to the emergency department—A literature review. *International Emergency Nursing, 18*, 210–220.

Calleja, P., Aitken, L. M., & Cooke, M. L. (2011). Information transfer for multi-trauma patients on discharge from the emergency department: Mixed-method narrative review. *Journal of Advanced Nursing, 67*, 4–18.

Carayon, P. (2012). Sociotechnical systems approach to healthcare quality and patient safety. *Work, 41*, 3850–3854.

Carayon, P., Li, Y., Kelly, M. M., DuBenske, L. L., Xie, A., McCabe, B., Orne, J., ... Cox, E. D. (2014). Stimulated recall methodology for assessing work system barriers and facilitators in family-centered rounds in a pediatric hospital. *Applied Ergonomics, 45*(6), 1540–1546.

Castiglione, A., Pizzolante, R., De Santis, A., Carpentieri, B., Castiglione, A., & Palmieri, F. (2015). Cloud-based adaptive compression and secure management services for 3D healthcare data. *Future Generation Computer Systems, 43–44*, 120–134.

Cheung, D. S., Kelly, J. J., Beach, C., Berkeley, R. P., Bitterman, R. A., Broida, R. I., & White, M. L. (2010). Improving handoffs in the emergency department. *Annals of Emergency Medicine, 55*, 171–180.

Collins, S. A., Bakken, S., Vawdrey, D. K., Coiera, E., & Currie, L. (2011). Model development for EHR interdisciplinary information exchange of ICU common goals. *International Journal of Medical Informatics, 80*(8), 141–149.

Conrand Stoppler, M. (2014). Medical triage: Code tags and triage terminology. *MedicineNet.com*. Available from: http://www.medicinenet.com/script/main/art.asp?articlekey=79529

Di Giovanni, D., Luttazzi, E., Marchi, F., Latini, G., Carestia, M., Malizia, A., ... Gaudio, P. (2014). Two realistic scenarios of intentional release of radionuclides (Cs-137, Sr-90)—The use of the HotSpot code to forecast contamination extent. *WSEAS Transactions on Environment and Development, 10*, 106–122.

Dimitrov, D. (2016). Medical internet of things and big data in healthcare. *Healthcare Information Research, 22*(3), 156–163.

Edwards, B., & Sines, D. (2008). Passing the audition—The appraisal of client credibility and assessment by nurses at triage. *Journal of Clinical Nursing, 17*, 2444–2451.

Eisenberg, E. M., Baglia, J., & Pynes, J. (2006). Transforming emergency medicine through narrative: Qualitative action research at a community hospital. *Health Communication*, 19, 197–208.

Eisenberg, E. M., Murphy, A. G., Sutcliffe, K., Wears, R., Schenkel, S., Perry, S., & Vanderhoef, M. (2005). Communication in emergency medicine: Implications for patient safety. *Communication Monographs*, 72, 390–413.

Embi, P. J., Weir, C., Efthimiadis, E. N., Thielke, S. M., Hedeen, A. N., & Hammond, K. W. (2013). Computerized provider documentation: Findings and implications of a multisite study of clinicians and administrators. *Journal of the American Medical Informatics Association*, 20(4), 718–726.

Ferré, F., de Belvis A. G, Valerio, L., Longhi, S., Lazzari, A., Fattore, G., Ricciardi, W., & Maresso, A. (2014). Italy: Health system review. *Health Systems in Transition*, 16(4), 1–168.

Geisler, B. P., Schuur, J. D., & Pallin, D. J. (2010). Estimates of electronic medical records in U.S. Emergency departments. *PLoS One*, 5(2), e9274.

Häyrinen, K., Saranto, K., & Nykänen, P. (2008). Definition, structure, content, use and impacts of electronic health records: A review of the research literature. *International Journal of Medical Informatics*, 77(5), 291–304.

Hsu, W. Y. (2017). Clustering-based compression connected to cloud databases in telemedicine and long-term care applications. *Telematics and Informatics*, 34(1), 299–310.

Italian Ministry of Health. (2011). The electronic Health Record National Guidelines. *Official website of Italian Minister of Health*. Available from: http://www.salute.gov.it/portale/documentazione/p6_2_2_1.jsp?lingua=italiano&id=1654

Keyes, D. C., Singal, B., Kropf, C. W., & Fisk, A. (2014). Impact of a new senior emergency department on emergency department recidivism, rate of hospital admission, and hospital length of stay. *Annals of emergency medicine*, 63(5), 517–524.

Kientz, J. A., Patel, S. N., Jones, B., Price, E., Mynatt, E. D., & Abowd, G. D. (2008). *The Georgia tech aware home*. CHI'08 extended abstracts on human factors in computing systems, *ACM*, Florence, Italy, pp. 3675–3680.

Latham, L. P., & Acroyd-Stolarz, S. (2014). Emergency department utilization by older adults: a descriptive study. *Canadian Geriatrics Journal*, 17(4), 118–125.

Legramante, J. M., Morciano, L., Lucaroni, F., Gilardi, F., Caredda, E., Pesaresi, A., Coscia, et al. (2016). Frequent use of emergency departments by the elderly population when continuing care is not well established. *Plos One*, https://doi.org/10.1371/journal.pone.0165939.

Legris, N., Hervieu-Bègue, M., Daubail, B., Daumas, A., Delpont, B., Osseby, G.V., ... Béjot, Y. (2016). Telemedicine for the acute management of stroke in Burgundy, France: An evaluation of effectiveness and safety. *European Journal of Neurology*, 23(9), 1433–1440.

Lehmann, O., Bauer, M., Becker, C., & Nicklas, D. (2004). *From home to world-support-ing context-aware applications through world models. In* proceedings of the second IEEE annual conference on pervasive computing and communications, 2004. PerCom 2004, *IEEE*, Orlando, FL, pp. 297–306.

Lori, A. R., Eric, M. E. & Forde, C. (2016). The role of patients' stories in emergency medicine triage. *Health Communication*, 31(9), 1155–1164.

Lowthian, J. A., Curtis, A. J., Jolley, D. J., Stoelwinder, J. U., McNeil, J. J., & Cameron, P. A. (2012). Demand at the emergency department front door: 10-year trends in presentation. *Medical Journal of Australia*, 196(2), 128–132.

Malizia, A. (2016). Disaster management in case of CBRNe events: An innovative methodology to improve the safety knowledge of advisors and first responders. *Defense and Security Analysis*, 32(1), 79–90.

Mano, L. Y., Faiçal, B. S., Nakamura, L. H. V., Gomes, P. H., Libralon, G. L., Meneguete, R. I., ... Ueyama, J. (2016). Exploiting IoT technologies for enhancing health smart homes through patient identification and emotion recognition. *Computer Communications*, 89–90, 178–190.

Maughan, B. C., Lei, L., & Cydulka, R. K. (2011). ED handoffs: Observed practices and communication errors. *The American Journal of Emergency Medicine*, 29, 502–511.

O'Connor, E., Gatien, M., Weir, C., & Lisa Calder, L. (2014). Evaluating the effect of emergency department crowding on triage destination. *International Journal of Emergency Medicine*, 7, 16.

Rialle, V., Duchene, F., Noury, N., Bajolle, L., & Demongeot, J. (2002). Health "Smart" home: information technology for patients at home. *Telemedicine Journal and e-Health, 8*(4), 395–409.

Riva, G. (2003). Ambient intelligence in health care. *Cyber Psychology Behaviour, 6*(3), 295–300.

Romero, E., Araujo, A. M., Moya, J. M., Goyeneche, J. M., Vallejo, J. C., Malag, J., … Fraga, D. (2009). Image processing based services for ambient assistant scenarios. *Distributed Computing, Artificial Intelligence, Bioinformatics, Soft Computing, and Ambient Assisted Living. Lecture Notes in Computer Science, 5518,* 800–807.

Schubert, C. C., Denmark, T. K., Crandall, B., Gnome, A., & Pappas, J. (2013). Characterizing novice-expert differences in macrocognition: An exploratory study of cognitive work in the emergency department. *Annals of Emergency Medicine, 61,* 96–109.

Sona, A., Maggiani, G., Astengo, M., Comba, M., Chiusano, V., Isaia, G., Merlo, C., Pricop, L., Quagliotti, E., Moiraghi, C., Fonte, G., Bo, M. (2012). Determinants of recourse to hospital treatment in the elderly. *European Journal of Public Health, 22*(1), 76–80.

Tao, L., Paiement, A., Damen, D., Mirmehdi, M., Hannuna, S., Camplani, M., Burghardt, T., & Craddock, I. (2016). A comparative study of pose representation and dynamics modelling for online motion quality assessment. *Computer Vision and Image Understanding, 148,* 136–152.

Weir, C. R., Hammond, K. W., Embi, P. J., Efthimiadis, E. N., Thielke, S. M., & Hedeen, A. N. (2011). An exploration of the impact of computerized patient documentation on clinical collaboration. *International Journal of Medical Informatics, 80*(8), 62–71.

Yim, V. W. T., Graham, C. A., & Rainer, T. H. (2009). A comparison of emergency department utilization by elderly and younger adult patients presenting to three hospitals in Hong Kong. *International Journal of Emergency Medicine, 2*(1), 19–24.

Zhu, N., Diethe, T., Camplani, M., Tao, L., Burrows, A., Twomey, N., … Craddock, I. (2015). Bridging e-Health and the internet of things: The SPHERE project. View at Publisher. *IEEE Intelligent Systems, 30*(4), 39–46.

18

The Contiki Operating System: A Tool for Design and Development of IoT—Case Study Analysis

B. Venkatalakshmi, A. Pravin Renold, and S. Vijayakumar

CONTENTS

18.1 Introduction

The Contiki operating system (OS) is an open-source OS specially designed for Internet of Things (IoT). It supports Internet connectivity to monitor and control the low cost and resource-constrained embedded devices. The Contiki OS is based on an event-driven kernel that supports multithreading. It supports a micro-transmission control protocol/internet protocol (TCP/IP) stack for IoT and Rime communication stack for wireless sensor networks. The Contiki OS is designed using the C language. The advantage of uIP (micro IP) is to provide TCP/IP protocol suite even for resource constrained 8-bit microcontrollers. As the embedded device able to run TCP/IP makes it suitable to connect to Internet. The uIP has minimal features needed for a full TCP/IP stack. The main control function of uIP does the following activities repeatedly:

1. Check for the arrival of packet from the network
2. Check for the periodic timeout

The memory is effectively managed in uIP. The explicit dynamic memory allocation is not supported. When a packet reaches a node, the device driver places it in the global buffer (long enough to hold one packet of maximum size). To avoid overwriting, the application will process the data immediately or save it in the secondary buffer for later processing. To further optimize the memory usage, during collision, the application will be able to generate sent data rather than retransmitting the data from buffer as in TCP/IP. The application program interface specifies the mechanism of application program that interacts with the uIP TCP/IP stack.

Contiki OS has an inbuilt simulator called Cooja from version 2.0 of Contiki. It is a Java-based simulator. The simulator aims to minimize the step between code development and executing it on different platforms. Cooja is a cross-level simulator. It supports simulation and emulation; Cooja executes the program on the host CPU or in an instruction-level TI MSP430 emulator. Cooja performs simulations at three different levels such as application level, OS level, and machine-code instruction level (Osterlind et al., 2006).

1. *Application level*: It deals with the design and implementation of routing protocols, radio medium, radio devices, and duty cycles of sensor nodes.

2. *Operating system level*: It includes the entire Contiki OS, the process, and its associated libraries.

3. *Machine-code instruction level*: It includes the emulation of Tmote sky board's execution using the assembly code pertaining to the architecture of the board.

The two main components of the Cooja simulator are interface and plug-in. The interface denotes the property of a node such as radio, position of nodes, and hardware peripherals. The plug-in takes care of the interactions with the simulation and the nodes in the simulator.

The various features of Contiki OS are as follows:

1. Support for recently approved standards by IETF (Internet engineering task force) in the domain of 6LoWPAN such as RPL (routing protocol for low-power lossy networks), CoAP (constrained application protocol).

2. Support for emulation by means of the simulator Cooja.

3. Usage of lightweight, stackless thread called protothread. Protothreads avoid conditional blocking and reduce explicit state machines in event-driven programs. A protothread is stackless, that is, all protothreads in a system run on the same stack.

4. Mechanism to operate nodes in low power state and to support tool for monitoring energy consumption by the nodes.

5. Support for a light weight flash file system called as coffee. Coffee file system is used as storage device in resource-constrained system. It uses application program interface to provide the functionality such as storage, access, and handle data on storage.

6. Usage of shell-based debugging environment. The shell commands allow text-based interaction with sensor nodes based on Unix-like terminal.

18.2 Architecture of Contiki OS

As mentioned in above section, Contiki OS is based on C language. So the input file supports the C language which contains the information like operating mode of the node such as time driven or event driven. The application level configuration such as processor details, sensor details, sensing duration, network, and medium access control (MAC) type are specified in input file. The core components of Contiki OS shown in Figure 18.1 includes kernel, program loader, supporting libraries such as shell, flash file system, energy monitoring module, and the communication stack. The core of Contiki OS is implemented in C language. The input file is mapped with the core files component such as communication stack and other associated files. The compiled program could be loaded to the Contiki OS supported hardwares such as MSP430, Atmel, Tmote sky boards, and so on.

The communication stack refers to a group of protocols that implements network protocol suite. The communication stack of Contiki OS supports sensor networks communication using Rime and uIP with 6LoWPAN network communication framework as shown in Figure 18.2.

The device drivers act as an interface between the application and the hardware. Moreover, they are responsible for reading the packet and forwarding them to the higher layers. The communication stack processes the packets and forward them to the application for which the packet is intended for. In the case of any response to be done, the application program responds via the communication stack.

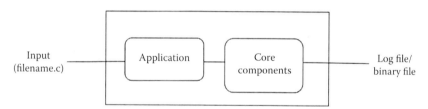

FIGURE 18.1
Overview of Contiki OS.

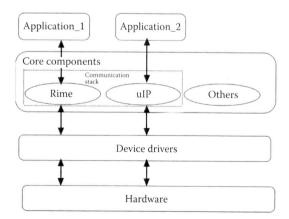

FIGURE 18.2
Communication stack interaction with other modules.

18.3 Communication Stack in Contiki OS

a. *RIME*: The protocols in Rime communication stack shown in Figure 18.3 support networking paradigm in wireless sensor networks. The wireless sensor networks are considered as the basic building blocks of IoT. The Rime communication stack provides a set of lightweight communication primitives. The Rime communication stack supports single-hop, multihop, unicast, mesh, tree routing, and broadcast-based communication primitives. The packets forwarded by a node to neighboring nodes are maintained in packet queues, whereas each packet queue has a lifetime. The packet is removed from the queue once the lifetime expires. The packet queue is defined with the name of the queue and the maximum size of the queue. The nodes are addressed on the basis of an unsigned integer data type with a width of 8 bits (uint8_t). The sink node is always provided with node id as "1" by default. The routing protocols determine the path based on the occurrence of event or based on the expiration of time toward the sink node. Once the path is determined, the packet is transmitted in unicast fashion on the chosen path.

b. *Micro-IP (uIP)*: uIP is the world's smallest full TCP/IP stack. Intended for tiny microcontroller systems in which code size and RAM are severely constrained, uIP only requires 4–5 kB of code space and a few hundred bytes of RAM. uIP has been ported to a wide range of systems and has found its way into many commercial products.

uIP has the network frameworks as 6LoWPAN. 6LoWPAN open standard was defined by IETF to address the need of wide address space required by IP-driven connected devices that can only be provided by IPv6. **6LoWPAN** is an acronym of IPv6 over Low-Power Wireless Personal Area Networks. IETF developed this standard to make IPv6 to be used in low power and lossy networks that are based on IEEE 802.15.4. 6LoWPAN contains an adaptation layer that allows IPv6 packets can be carried out through IEEE 802.15.4 link layer frames.

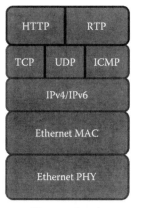

IP protocol stack

6LoWPAN protocol stack

FIGURE 18.3
6LoWPAN protocol stack.

The 6LoWPAN protocol stack shown in Figure 18.3 is equivalent to the traditional IP protocol stack with some differences; 6LoWPAN supports IPv6 only, so a LoWPAN adaptation layer is introduced to optimize the IPv6 over IEEE 802.15.4. Next, the commonly used transport layer protocol is user datagram protocol (UDP); TCP is not used because of performance reasons.

Following are the main features of 6LoWPAN stack:

- Physical layer (IEEE 802.15.4 PHY) is used to convert data bits into signals that are transmitted and received through radio frequency (RF) communication.

- The data link layer provides reliable connection between two nodes. The data link layer also contains LoWPAN adaptation layer for adapting the IPv6 packets over IEEE 802.15.4.

- Network layer initiates the assigning of IP address to individual nodes and routing the packets either through mesh under (link layer) or route over (network layer) routing protocols. The packets are forwarded hop-by-hop from source to destination. Packets are fragmented and sent to next hop whereas these fragments are collected in the receiving hop and sent to either upper layer or to the next hop according to routing table information.

- Transport layer is responsible for the applications to have their own communication sessions. Usually, two types of transport layer protocols are used—TCP and UDP. But due to some performance reasons of TCP-like power consumption, packet overhead, UDP is preferred over TCP as a transport layer protocol. Other protocols such as transport layer security and the Internet control message protocol are also used.

- Application layer uses sockets for communication between applications. Each 6LoWPAN application uses sockets for sending and receiving the packets that are associated with protocols such as UDP and ports (source/destination). Application layer is also used for data formatting. Widely used application protocol is HTML. But due to its text-based approach, it has a large overhead that is not suitable for 6LoWPAN systems. For this, an alternative application protocol such as CoAP and Message Queue Telemetry Transport (MQTT) is used. The sicslowpan module supports the use of IPV6 on IEEE802.15.4 in Contiki OS. The functionalities such as addressing, fragmentation, and header compression are implemented.

18.4 Case Studies

For a beginner, to experience coding in the Cooja simulator, a small sample program exercise has been provided as Case Study 18.4.1.

18.4.1 A Case Study on the Cooja Simulator with Its Program

```
#include "contiki.h" // Core of contiki operating system
#include "stdio.h"
PROCESS(first,"HELLO_WORLD_PROCESS"); // Process control block
AUTOSTART_PROCESSES(&first); // Autostart of process during system boot
PROCESS_THREAD(first, ev, data) // Protothread with name of the process,
event identifier and pointer to data
```

```
{
PROCESS_BEGIN(); // Beginning of protothread
printf("First Program \n")
PROCESS_END(); // End of process protothread
}
```

The Contiki-built system is composed of a number of makefiles. These are as follows:

- *Makefile*: The project's makefile, located in the project directory.
- *Makefile.include*: The system-wide Contiki makefile, located in the root of the Contiki source tree.
- *Makefile.$(TARGET) (where $(TARGET) is the name of the platform that is currently being built)*: Rules for the specific platform, located in the platform's subdirectory in the platform/directory.
- *Makefile.$(CPU) (where $(CPU) is the name of the CPU or microcontroller architecture used on the platform for which Contiki is built)*: Rules for the CPU architecture, located in the CPU architecture's subdirectory in the cpu/directory.
- *Makefile.$(APP) (where $(APP) is the name of an application in the apps/directory)*: Rules for applications in the apps/directories. Each application has its own makefile.

The Makefile of the above code is as follows:

```
CONTIKI =../..
all: hello
include $(CONTIKI)/Makefile.include
```

The first line specifies the location of Contiki followed by the name of the application and system wide makefile location that contains definitions of the core of Contiki system. The latest version of Contiki OS can be downloaded at www.contiki-os.org. This website provides the Contiki OS and its simulator Cooja along with VMware. The website provides step-by-step tutorial for the execution of program.

The rest of the case studies of the chapter demonstrate an approach for exploring the communication stacks of Contiki OS. For example, Case Study 18.4.2 helps in understanding that RPL protocol in uIP stack that can be suitably programed for overall network performance enhancement using network coding. Case Study 18.4.3 explores feasibility of RPL protocol for body area networks. CoAP, an application layer protocol of uIP stack, has been customized for an application through Case Study 18.4.4. A secure data aggregation based on privacy homomorphism (PH) technique has been demonstrated in Rime stack of Contiki through Case Study 18.4.1.

18.4.2 A Case Study on the Routing Protocol for Low Power Lossy Networks with Network Coding

IoT demands next-generation access networking with cost effective, high reliability, and secure access networking trends for industrial applications (Rojas, 2015). This can be achieved through a novel technique called network coding that was briefed into work done by Weiner et al. (2014). It insisted the need for PHY/MAC layer enhancement to address low-latency and high-reliability requirements of wireless nodes. Swamy et al. (2016) ensured that network-coding technique could be a solution for this. The case study

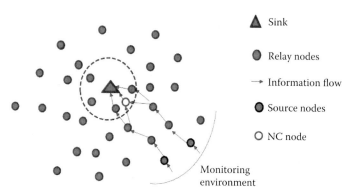

FIGURE 18.4
Network model for network coding.

given below also demonstrates that the network-coding technique can be used at IoT nodes with reduced retransmissions and increased throughput at the sink node.

Network coding is a technique which can be used to improve a network's throughput, efficiency, and scalability, as well as resilience to attacks and eavesdropping. Instead of simply relaying the packets of information, the nodes of a network can combine the packets together for transmission. This can be used to increase the maximum possible information flow in a network. Moreover, it may help for resilient transmission of packets. Network coding is a method of optimizing the flow of digital data in a network by transmitting digital evidence about messages. The case study considered a network scenario, as shown in Figure 18.4.

Based on the position and the status of incoming contents, the nodes are configured for modes like simple forward or encode and forward. Simple Ex-Or based encoding has been used. In the Cooja simulator, the RPL at the network layer has been modified for achieving this status variation of nodes. A function called forwarding function in IPV6 has been updated in synchronization with timers to diagnose the contents intended for same destination. If the condition is true and if the node is an eligible encoding node, then the packets are exported and then forwarded. The procedure of this case study includes three modules of programming, namely source node, forwarding node, and receiving node coding. Using this procedure, we can obtain higher throughput at receivers, packet backup under node failures, effective distributed storage, and so on. The result statistics can be obtained by integrating the Cooja simulator with wireshark analyzer. More details on this case study on network coding are available in the work done by Venkatalakshmi and Vasanth Kumar (2015).

18.4.3 A Case Study on the Routing Protocol for Low Power Lossy Networks for Body Area Network

The wireless body area network is used for continuous monitoring of vital signs of human body. The network consists of implanted sensor nodes and a sink for data collection as shown in Figure 18.5. The sensor nodes are responsible for sensing and transmitting the data to the sink. In implanted sensor nodes of wireless body area network, the heat generated by the node's circuitry causes damage to the tissues. As communication consumes more power and heat, designing an effective routing protocol for wireless body area networks by incorporating temperature in the routing metrics becomes essential. The thermal

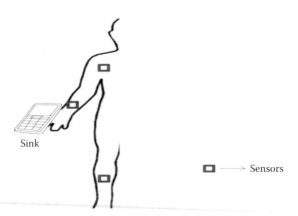

FIGURE 18.5
Architecture of body area network (BAN).

aware protocols maintain the node in active state if the node's temperature is below the safety level.

Thermal aware RPL leaves the hotspot node while forwarding packets. The neighbor nodes are determined on the basis of the metrics such as distance and temperature. The nodes with minimal distance and having temperature value below the threshold are chosen for forwarding the packets. The objective function of RPL is modified to include the new metrics temperature along with distance to build the destination-oriented directed acyclic graph. If the temperature of the node is greater than the threshold, an alternate path is chosen to forward the packet. The node with high temperature is allowed to go to sleep state for certain time duration. The performance of thermal aware RPL gives better packet delivery ratio (Renold, 2014).

18.4.4 A Case Study on the Constrained Application Protocol for Smart Parking

6LoWPAN usage for parking lot monitoring reduces the time taken by users to determine the free parking lot. The advantage of this approach is that the drivers can decide their duration time of parking their vehicles via Internet. The parking lot is allocated on the basis of requirement. The system comprises of parking lot, sky motes, and mobile devices with Internet connectivity. The free parking slots are intimated to the users via web address as per request (Aarthi & Renold, 2014). The data transmission from Internet to that of user uses a protocol termed CoAP. The motes used for the parking application can be classified into three categories such as one mote acts as server, one as border router, and clients may be in any number as shown in Figure 18.6.

- *Server*: A Restful server shows how to use the REST layer to develop server-side applications.
- *Border router*: Border router keeps radio turned on. Enabling of it helps in connection between that of client as well as server to that of CoAP web address. Border router has the same stack and fits into mote memory.
- *Client*: A CoAP client that polls the/actuators/toggle resource for certain-time duration.

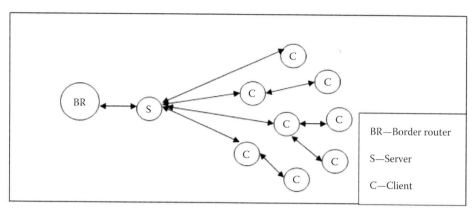

FIGURE 18.6
Categories of motes.

Parking lot is considered to be a service in random order queue, and Markov chain is used to explain memory-less property. The following equation shows the characteristic of nondependence of the past:

$$p_{ij}(t) := P(X_{\tau+t} = j \mid X_\tau = i) = P(X_t = j \mid X_0 = i)$$

where $P(X_{\tau+t} = j \mid X_\tau = i)$ represents the future occurrence

$P(X_t = j \mid X_0 = i)$ represents the present occurrence. If the user wants to know the free parking lot, he enters the vehicle number and the duration of time to park the vehicle. CoAP uniform resource identifier (URI) is made for this work was coap: //vec:6000. The communication between server and clients are of multihop fashion. The data from the motes are sent to that of Copper Web Browser that can be viewed by the different users (Kovatsch, 2011). Based on the duration of parking, the slots are allocated for the vehicle. The methodology achieves less time in determination of free parking lot.

18.4.5 A Case Study on the Privacy Homomorphism-Based Secured Data Aggregation

Data-aggregation schemes are vulnerable to various security attacks. In such cases, attackers maliciously change the aggregated data collected from the sensor nodes and send them to cluster heads (CHs). Data-aggregation schemes combined with PH give better security than traditional methods. In PH-based encryption, the incoming data from sensor nodes are encrypted and fed to the CHs that in turn directly aggregate the ciphered texts without decryption (Steffi Diana and Vijayakumar, 2014).

In data-aggregation scheme, the sensor nodes collect information from the end nodes. Whenever the base station requires data, at data aggregator "A" that collects the information from the sensor nodes, computes the addition, subtraction, and exclusive-or of the incoming data and sends the aggregated data to the base station "B" over a multihop path, and node "E" is leaf node in this path as shown in Figure 18.7. It provides better bandwidth and energy consumption in the network. In the base station, these aggregated data are decrypted. Without proper encryption, the adversary nodes can inject malicious data into the network. For this reason, we are using data aggregation combined with PH as a standard mechanism for encryption. PH enables us to do direct computation on encrypted data. Here, a specific type of computation is performed on the cipher text and obtained an encrypted result in the form of cipher text that is a result of the input plaintext. It can do addition and multiplication of the encrypted data and obtain the decrypted result without knowing the actual value

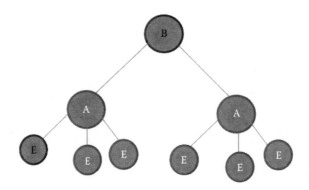

FIGURE 18.7
Sensor nodes controlled by base station.

of the data. Moreover, the sensor nodes collect differential data rather than raw data from the sensors for energy efficiency and minimize the burden of CHs in processing the data.

The case study follows a network model described as follows. The protocol executes in two sessions—reference data transfer session and subsequent data transfer session. In reference data session, the reference data are transferred from sensor nodes to base station in a secure and energy-efficient way and perform verification of it in the base station. Base station can keep the reference data as a correspondence to the node id for recovery of original sensed data in differential data transmission session. The data-transfer session transfers the differential data from sensor nodes to base station in a secure and energy-efficient way. Hence, the base station can recover the individual sensed data and can check the integrity of all sensing data. To demonstrate the proposed protocol, a cluster 1 may be considered. Here, wth sensor node (SN_w) is chosen as CH of cluster 1, and the remaining sensor nodes ($SN_1 \ldots SN_{w-1}$) are considered as cluster members. Both CH and cluster members perform sensing and the CH's performs data aggregation. The entire scenario is well depicted in Figure 18.8.

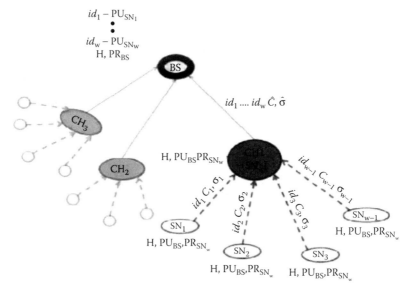

FIGURE 18.8
Cluster formation in sensor networks.

In the setup phase, the base station uses this phase to generate and install all necessary secrets for each sensor node and for them too.

$$PU_{SNi} = PR_{SNi} \times g2$$

where:

PR- is the private key
PU- is the public key
g2 is the generator

In the initial phase, the sensor node executes this phase for transferring cipher text appended to the node id with digital signature as a pair to CH.

$$\text{cipher text of sensed data} = (C_i) = (r_i, s_i) = (k_i \times G, M_i + k_r \times \gamma)$$

$$\text{digital signature } \sigma_i = PR_{SN_i} \times h_i$$

In aggregate phase, the CH_i uses this phase to aggregate the cipher texts, signatures coming from its cluster members, and to generate aggregated cipher text and aggregated signature without performing any decryptions and encryption at the CH_i.

$$\text{Aggregated Cipher text } (\hat{C}_1) = (\hat{R}_1, \hat{S}_1) \Big| = \sum_{i=1}^{w} r_i, \sum_{i=1}^{w} S_i$$

$$\text{Aggregated signature } (\check{\sigma}) = \sum_{i=1}^{w} \sigma_i = \sigma_1 + \cdots \sigma_w$$

In the recovery phase, the base station goes through this phase to recover each individual sensing data from the aggregated cipher text reached at the base station.

Decryption: Base station generates M' by decrypting the ciphertext with Private key of base station (PR_{BS}).

$$\hat{C}_1 = (\hat{R}_1, \hat{S}_1) \quad \text{and} \quad M' = -PR_{BS} \times \hat{R}_1 + \hat{S}_1$$

Remapping: Remapping using rmap() function to remap elliptic curve points into plain text.

$$\text{Remapped data } m' = rmap(M^+) = M^+/G$$

Decoding: Base station recovers the sensing data.

$$d_i = m'[l. (i-1), (l,i)-1]$$

In the verify phase, the base station checks the integrity and authenticity of all sensing data in this phase. So that base station can validate the correctness and the source of the sensing data. The entire scenario is well depicted in Figure 18.9.

The ID_i of each recovered data at the base station is checked with the IDs stored in base station. If it matches, base station confirms that the data are from a valid node (ID).

Integrity of each recovered data is checked by base station.

$$e_n(\hat{\sigma}, g_2) = \prod_{i=1}^{w} e_n(h_i, PU_{SN_i})$$

$$h_i = H(d_i)$$

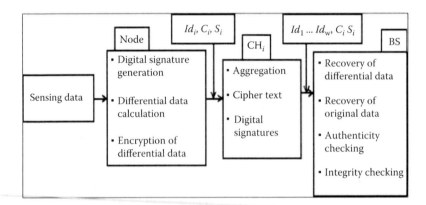

FIGURE 18.9
Nodes secured communication.

This network scenario can be simulated in the Cooja simulator of Contiki OS as shown below.

1. The sensor node collects the differential temp data from sensors and encrypts it

```
difdata = llabs(data-refvalue);
printf("Sensor data value = %llu \n",difdata);
difdata = difdata << betavalue;
difdata = binary(difdata);
difdata = difdata * G;
return difdata;
```

2. And then it generates digital signature and sends ciphertext and digital signature with $ appended to the aggregator.

```
unsigned long long r, encryptdata;
difdata = encode();
r = k * G;
encryptdata = difdata + (k * PUBS);
return encryptdata;
```

3. The aggregator collects the differential data from sensor nodes and sends the aggregated info to base station.

```
if(((UIP_IP_BUF->srcipaddr.u8[sizeof(UIP_IP_BUF->srcipaddr.u8)
- 1])==nodeid[i])&&(flag[i]==0))
{
  s1=s1+aggdata;
  s2=s2+aggsig;
  flag[i]=1;
  aflag=0;
}

uip_udp_packet_sendto(client_conn1, buf, strlen(buf),
&server_ipaddr1,UIP_HTONS(UDP_BASE_PORT));
```

4. The original data is recovered in the base station.

```
data[i]=mapdata%1000;
mapdata=mapdata/1000;
num=binary(data[i]);
printf("\noriginal data recovered at Base station = %llu",num);
```

5. Finally, it is decrypted in the base station.

```
data1=(long long*)malloc(sizeof(long long));
Mapdata1=(-PRBS*r1)+s1;
Mapdata1=Mapdata1/G;
data1 = remapping(Mapdata1);
return data1;
```

Thus, the above technique recovers sensing data from the aggregated result and verifies data integrity and authenticity. It improves the network lifetime by achieving energy and bandwidth efficiency while preserving security requirements such as confidentiality, data authentication, and integrity. It reduces the energy consumption by transferring differential data rather than raw data from nodes so that the CH also wants to process small amounts of bits.

18.5 Conclusion

Contiki OS is more user friendly and users can configure for different core operations of a wireless networks. All the layers can be comfortably configured for the application-specific requirements. Its compatibility with Skymotes, Zolertia (Z1) motes, Open motes, and so on provides real-time application development using the Contiki OS and Cooja simulator as an effective tool for IoT.

References

Aarthi, R., and Renold, A. P. (2014). COAP based acute parking lot monitoring system using sensor networks. *ICTACT Journal on Communication Technology*, 5(2): 923–928.

Kovatsch, M. (2011). Demo abstract: Human-coap interaction with copper. In *Distributed Computing in Sensor Systems and Workshops* (DCOSS), 2011 International Conference on (pp. 1–2). IEEE.

Osterlind, F., Dunkels, A., Eriksson, J., Finne, N. and Voigt, T. (2006). Cross-level sensor network simulation with Cooja. In *Proceedings of 31st IEEE Conference on Local Computer Networks* (pp. 641–648). IEEE.

Renold, P. (2014). Routing protocol for low power lossy networks. In *Advanced Communication Control and Computing Technologies* (ICACCCT), 2014 International Conference on (pp. 1457–1461). IEEE.

Rojas, B. (2015). Toward next-generation access networking technologies in industrial/enterprise internet of things. Retrieved from: http://download.peplink.com/resources/IDC-iot-whitepaper.pdf.

Steffi Diana, B. and Vijayakumar S. (2014). A concealed data aggregation using privacy homomorphism in wireless sensor networks. *International Journal of Innovative Research in Science, Engineering and Technology (IJIRSET)*, 3(3): 2404–2410.

Swamy, V. N., Rigge, P., Ranade, G., Sahai, A., and Nikolic, B. (2016). Network coding for high-reliability low-latency wireless control. In *Wireless Communications and Networking Conference (WCNC), 2016 IEEE* (pp. 1–7). IEEE.

The Contiki OS. Retrieved from http://dunkels.com/adam/software.html (accessed January 11, 2017).

The Contiki OS. Retrieved from http://www.contiki-os.org/ (accessed January 11, 2017).

Venkatalakshmi B. and Vasanth Kumar C. (2015). Design and development of network coding algorithm for IoT devices. *Australian Journal of Basic and Applied Sciences*, 9(16): 367–372.

Weiner, M., Jorgovanovic, M., Sahai, A., and Nikolie, B. (2014). Design of a low-latency, high-reliability wireless communication system for control applications. In *Communications (ICC), 2014 IEEE International Conference on* (pp. 3829–3835). IEEE.

19

Novel Trends and Advances in Health Care Using IoT: A Revolution

Awanish Kumar and Archana Vimal

CONTENTS

19.1 Introduction

The health care sector is a major considerate area worldwide. The government of all over the world invests a huge amount of GDP in the health care sector for public health and welfare. According to World Bank Report, 2014, the total percent of GDP spent on health care throughout the world was 9.9%. However, high-income countries are paying more attention toward the public health compared with low-income countries. In Figure 19.1, a comparative GDP expenditure graph is shown (The World Bank, 2014).

In India, there are only 40 doctors per 100,000 people (Charania et al., 2016). However, the health care sector is growing at the rate of 15% annually and is expected to cover market value United States Dollar (USD) 280 billion by the year 2020 (Nishith Desai Associates, 2016). It is changing by adopting emergent technology-based devices and services. In such situation, the intervention of Internet of Things (IoT) in health care would be useful to improve people's life. It is expected to earn USD 14tn revenue globally (Srinivasa, 2016). The main aim of IoT in the health care sector is to enhance the quality of people's life and that too in an economical way. At the same time, it should provide equity and ease of access. The revolution in the mode of information and communication technologies (ICTs) has a great impact in addressing global health problems. Such modes include computers, the Internet, and smartphones that transform the way of communication between individuals, pursue and exchange of information, and enrich their lives (WHO, 2009). With the intervention of advancing technology, it gets easier to connect physical objects to the Internet, opening a new world of opportunities. IoT-based health care systems make use of a set of interconnected devices that produces a network system. The function of this system is health care assessment such as monitoring patients and automatically detecting situations of medical emergency. The general architecture of IoT begins with different sensing devices that collect patient data by using different devices and sensors. Then there is a middleware that provides interoperability and security. The next level is monitoring devices that are connected

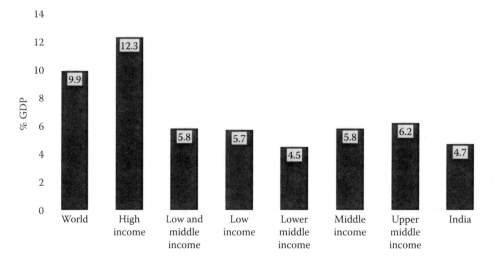

FIGURE 19.1
Percent of GDP expenditure on the health care sector for the year 2014.

via wireless-networking technology. It functions in a similar fashion as of a network manager that is responsible for storing, aggregating, consolidating, and security alarming (Fan et al., 2014). IoT interconnects all the objects and makes them smart, which could be termed the next technological revolution (Fan et al., 2014). These technologies empower public-health care system through various advanced applications but altogether it brings some challenges. The major concerns are ethical, legal, and social issues such as patient's privacy and safety (Omogbadegun & Ayo, 2007).

19.2 Health Care Implications

19.2.1 Wireless Patient Monitoring

The introduction of wireless technology and the mobile system plays a major role in the transformation of the health care industry. With the introduction of wireless networking, the limitation imposed by a traditional wired solution like Ethernet and USB are overcome. The present era is the third generation of wireless technology starting from pagers to basic mobile phones and now to smartphones. The health care professionals substitute personal digital assistant with smartphones for their routine work within the clinical environment. They are taking benefits of smartphones with cellular integration and built-in Wi-Fi, and Bluetooth wireless functionality. It helps them to maintain patient's records and to provide other health care services in a better and efficient way.

It was appraised by Ericsson Mobility Report and Allied Business Intelligence (ABI) Research that there will be 30 billion gadgets connected through a wireless network by the year 2019 (Charania et al., 2016). As the development in wireless technology advances, it upsurges the quality of services provided to the patients. It serves with patients with more individual care, enhances the efficiency of caretaker, and decreases the chances of medical negligence. In spite of the well establishment of wireless technology in various other domains, few problems arise in the health care sector. The reason is that the requirement of the medical domain is completely different. It demands high quality and assurance in terms of operation and implementation (Ngoc, 2008). Freescale provides microcontrollers that support wireless connectivity for devices. The system relies on the use of wireless standard connections. Personal area networks used for personal gadgets employs Bluetooth and Bluetooth low energy. For local area network used in health care center, Wi-Fi and Bluetooth-based connections are used (Niewolny, 2013).

19.2.2 Mobile System Access

In recent years, a drift from PC to mobile devices is observed. The mobile devices may be smartphones, notebook, or tablet computers. Such devices are equipped with sensors and actuators. This makes the gadget smart by empowering them to compute, act, sense, and interchange information with the associated network. Apart from this, mobile devices are provided with radio-frequency identification, quick response codes, global positioning system, and/or radio-frequency identification scanners or quick response codes readers. All these technology enhancers act as a connector between the physical world and cyberspace (Coetzee & Eksteen, 2011).

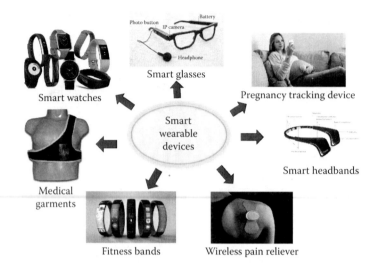

FIGURE 19.2
Some smart health/medicine-related wearable devices and gadgets available in the market.

19.2.3 Medical Devices and Sensors

IoT-based medical devices and sensors have drastically changed the health care monitoring and treatment in a positive way. It supports a better living standard by giving personalized and optimized services. It is also helpful in providing timely and cost-effective services, and many times prove as a life saver in critical medical conditions. The recent development of technologies such as sensor, the Internet, cloud, mobility, and big data technologies add on the benefits of the existing medical devices. In addition, it supports better connectivity to various health programs across the world (Khanna & Misra, 2013). The developing countries such as India emerges a huge market for wearable devices that includes smartwatches, fitness bands, health and fitness trackers, smart glasses and others. Some of the smart wearable devices are shown in Figure 19.2.

International Data Corporation revealed that about 400,000 units of wearable devices were sold during the first quarter of 2016 in India (Wearable Tech News, 2016). The smart wearable market share captured by the fitness band was 87.7%, whereas the devices running on third-party applications hold 12.3%. An increase of 67.2% delivering wearable devices is observed from the year 2015. The International Data Corporation also suggests that smartwatches will singly occupy 70% market share of wearable devices by the year 2019 (Singh, 2016).

19.2.4 Telemedications

The term *telemedicine* was introduced in the 1970s that mean *healing at a distance* (Strehle & Shabde, 2006). The World Health Organization has accepted the following wide depiction of telemedicine and explained it as "The delivery of health care services, where distance is a critical factor, by all health care professionals using information and communication technologies for the exchange of valid information for diagnosis, treatment and prevention of disease and injuries, research and evaluation, and for the continuing education of health care providers, all in the interests of advancing the health of individuals and their communities" (World Health Organization, 1998). It can be explained as an approach of delivering distant health care with the help of telecommunication and information technology. It is a real-time, two-way communication between patient and clinician at a distant site. The communication takes place by means of advanced telecommunication devices such as smartphones or real-time video conferencing. Health care practitioners evaluate, diagnose, and treat patients by sharing medical data with peers and specialist from worldwide. This increases the reach of the clinician to remote areas. This proves as a boon for the communities that are deprived of health care benefits due to the geographical barrier (Med, 2016). Other than this, it is also a mean to modernize the health care itself by the involvement of the user in the management. Moreover, it offers a new insight to develop a healthy standard of living (Puskin, Johnston, & Speedie, 2006). To provide the benefits of eHealth care or telemedicine facilities in rural areas, many health care centers have implemented the public–private partnership mode. This sector is growing very rapidly in India with a compound annual growth rate (CAGR) of 20% and is expected to grow USD 18.7 million by 2017 (eHealth, 2015). Indian Government has also come up with the scheme such as National eHealth Authority to encourage the concept of telemedicine in India (Nishith Desai Associates, 2016). The various advantages that telemedicine offers are summarized in Figure 19.3.

There are broadly three categories of telemedicine (Smith, 2015) that are as follows:

- Store and forward
- Remote monitoring
- Real-time interactive

The various fields of medication in which telemedicine transformed the health care sector are summarized in Figure 19.4.

FIGURE 19.3
The various advantages offered by telemedicine in improving the health of people.

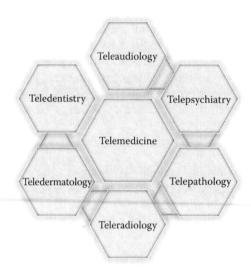

FIGURE 19.4
The various applications of telemedicine in the different medical fields.

One of the major applications is in telecardiology. It revolutionizes the cardiac care system and has the potential to save time money and lives. It provides the means for the remote specialist interpretation of electrocardiographic recordings through telephone transmission. It brings the efficient way of diagnosis and management of acute and chronic cardiac disease along with identifying patient at risk. It is beneficial for both physicians and patients as it serves benefits in terms of ease of access, the speed of diagnosis, the efficiency of management, and the freeing up of resources. It reduces the number of secondary referrals, misdiagnosed cases, and reducing door-to-balloon time. It empowers clinicians to improve clinical assessment of the patient and enables them to safely manage more of their own patients (Backman et al., 2010).

19.3 Enable Technologies Making IoT a Reality

The execution of IoT for practical use is possible through the use of several enabling technologies. For the well-organized functioning of smart sensors and microcontroller, components with certain capabilities are needed. That device should have low power operation, integrated precision-analog capabilities, and graphical user interfaces (GUIs). Other than this the device should be small with extended battery life to increase its usability and should offer high accuracy at the relatively cheaper rate (Ullah et al., 2016).

19.3.1 Low Power Operations

For making IoT a reality, there are some features that must be satisfied. Low power demand is one of them. This is particularly useful when devices have no or limited access to power sources. The devices should only be active when there is a need for its operation.

Its communication with large network devices demands huge power; therefore, it must be occasional. Moreover, human and system intervention should be less frequent. The practical implementation of this is possible when it is supported by control over idle and sleep mode, signaling, uplink power adaptation to link and others. Collaboration with other devices is also an alternative to decrease power consumption (Wu et al., 2011). The example is IP (Internet protocol) stack, a light protocol that links a large number of communicating devices all over the world. Its operation required only small and battery-operated embedded devices (Atzori et al., 2010).

19.3.2 Integrated Precision-Analog Capabilities

For making IoT implementation practically possible, it is a mandatory to have high accuracy technology achievable at low cost. Integrated precision-analog capabilities are one of the ways by which sensors can achieve this. There are technologies enabling this by the use of microcontrollers. Such microcontrollers have analog components such as low-power op-amps and analog-to-digital converters with high-resolution capacity (Niewolny, 2013). This can be clarified with the assistance of a case of temperature control. Temperature control (a specific temperature) is required for a large number of applications whether it is storing milk at the cold, a geyser maintaining optimum temperature, or glucose meter having a temperature within specified limit for correct result prediction. The temperature control in all such devices is possible with the help of microcontroller-based temperature sensors (Smith, 2013).

19.3.3 Graphical User Interface

Along with above mentioned technologies, user-friendly and easily accessible GUIs are compulsory for people to take benefit of smart technologies. A good GUI is the one that enables display devices to deliver more information with vivid detail and makes it easy to access that information (Niewolny, 2013). Presently, smartphone app provides a GUI for accessing and controlling the devices from home through server real IP. For that users need any Wi-Fi- or 3G/4G-enabled smartphone that supports Java (Piyare, 2013).

19.3.4 Connectivity Standards

The term IoT can be interpreted in a simple way as a world-wide network of interconnected objects that exchange information based on standard communication protocols. It is predicted that three-time growth in the Internet connected IoT devices will be observed by the year 2020. It is expected to increase from 10 to 34 billion (BI Intelligence, 2016). To deal with increasing burden, the ideal connectivity should be in such a way that it is accessible anytime, anyplace, by anyone, and for anything. The intelligence of the network should rely on ensuring trust, privacy, and security in addition to the lower requirement of resources. It should provide better computation capability at low-energy consumption (Atzori et al., 2010).

19.3.5 Health Care Solutions Using Smartphones—Medical IoT Apps

The health care solution provided by smartphones is termed mobile health or m-Health. It is described as the use of mobile and wireless technologies in the practice of medicine and the monitoring of public health. The most recent example is a smartphone with health sensors as accessories like wrist gear (Sermakani, 2014). To serve Indian people, Indian Medical Associations (IMA) come forward with a digital partnership scheme with Lybrate (private mobile health care provider). Around 0.25 million doctors are associated with it, get trained to serve the people better through advanced technology based services. It will be a bridging step toward connecting rural and urban heath care system (WI Bureau, 2015). The Government of India launched a mobile app called National Health Portal India targeting rural residents. The Gram Panchayats (a local self-government organization) are connected to Government through broadband and assist semiliterate or illiterate rural in accessing health-related information (National Health Portal of India, 2015). NGOs such as Gramvaani brought mobile and interactive voice response-based technology to facilitate health benefits to 700 billion rural population of India (Gramvaani, 2013).

19.4 Challenges to Overcome

19.4.1 Scalability

In spite of the fast growth of IoT in the health care sector, there are only 6% people are aware or taking benefit out of it (Madhuri & Sowjanya, 2016). There are few challenges need to be addressed for its universal application. One of such obstacles is scalability. In simple terms, scalability is the ability of a system to maintain or increase its efficiency with increased workload. The increased workload can be in the form of a large volume of data storage, its processing and analysis. With the enhancement of technology-based facility in medical and health care sector, a large volume of data are collected. The raw data generated demands big data analytics and cloud storage for meaningful interpretation of data (Sermakani, 2014). To meet the dynamically growing requirements/demands of people, society, and health organizations, IoT networks need to be scalable and adaptable. (Islam et al., 2015). Scalability in IoT can be attained by bearing in mind few important recommendations. Whenever deploying an IoT system, long-term support for devices based on technology should be incorporated. The potential longevity of the network technology should be considered and must be compared with expected lifespans of devices. The system must be created in such a way that its expansion is possible in an easy way, thinking of future prospects. The next considerate parameter is device durability. Ensure quality of device to cut down the operational cost for the device along with the network for long-term operation (John Horn, 2016).

19.4.2 Interoperability

Health care information and management systems society (HIMSS) explained the term interoperability in the context of health care as the potentiality of different information technology systems and software applications to communicate, exchange data, and use the information that has been exchanged. The data exchanged are shared between clinicians, lab, hospital, pharmacy, and patient. The aim is to improve the effective delivery of health care for individuals and communities (HIMMS, 2013). As wireless technology is emerging and changing rapidly, it causes interoperability in IoT more difficult. This raises intercommunication problems leading to integration issues. The devices used in IoT ranges from low power to high power, and capabilities are a major problem that arises during integration. Apart from the heterogeneity of devices and technology, few more interoperability associated problems are security and privacy-related issues (Elkhodr et al., 2016). The solution of this can be achieved in the form of cloud computing that aids in IoT data collection and data processing, supports rapid setup, and integration of new things at low costs (Hodkari & Aghrebi, 2016).

19.4.3 Safety of Patients

One of the major concerns of IoT is patient safety. Some legal issues arise from time to time with respect to patient's safety. There are apps and wearables available in the market that is not subjected to food and drug administration (FDA) regulation. There are reliability issues like whether an app or device provides the quality service that the developer claims. Such mobile apps and medical devices pose a risk to a patient's safety (Terry, 2016). There must be strict Government regulation and policy to have a check on the launching of such devices and apps to secure patient's safety.

19.4.4 Security and Personal Privacy

Privacy is another issue that should be dealt with while expanding IoT. Knowingly or unknowingly, everyone has to counter face IoT at some point. Sometimes, huge data generated by the means of IoT application remain invisible and unnoticeable causing loss of control over it. Some other problems also arise like shifting of control toward devices and algorithm due to automatic decisions causes security issues (Gong, 2016). There are works going in the direction of providing security and personal privacy to the user. One of such examples is the patient-centered access to secure systems online project. It was intended to apply cutting edge security to the correspondence of clinical data throughout the Internet (Masys et al., 2002). The use of encryption algorithm and hashing techniques will also ensure data privacy and integrity, respectively (Devi & Muthuselvi, 2016).

19.4.5 Lack of Government Support

IoT in the health care sector is still going through developing phase. Industry hesitates to invest in the lack of assurance of profitable returns. The government should come up with financial support for social welfare and prosperity. Taking inspiration and assurance from Government projects, more private firms will participate in IoT-based schemes for development of people and society (Kim & Kim, 2016). Many countries like the United Kingdom, the United States, China, and Japan came forward with an investment plan in IoT sector. UK government has funded a £5 million project, whereas China invests USD 800 million in the IoT industry. Such type of motivation is needed for growth and development in IoT industry (Xu et al., 2014).

Another demand from Government is to set regulatory bodies such as FDA for people's safety and security. Such authorities must actively participate in setting up the standards committee, bringing up the regulations, and to make sure strict persuasion of laws by people and industry (Sermakani, 2014).

19.5 Accomplishments through IoT

19.5.1 Glucose Level Sensing

Diabetes is a chronic disorder affecting a large number of the population worldwide. To address this global problem, it is needed to have non-invasive glucose-level sensing technique. Its effective management could be possible with the help of a combination of IoT and m-health (mobile health). It is called m-IoT (m-healthy things). Researchers have developed a new technique that connects IPv6-based communication technologies such as 6LoWPAN with emerging 4G networks. It comprises a temperature sensor to quantify the body temperature of the patient. Along with this, it also has a noninvasive optophysiological sensor attached for glucose-level monitoring in the body of the patient. By using an IPV6 access point that is connected to an IP-based health care center makes clinicians access patient's data. This helps in real-time monitoring and management of glucose level in patients (Istepanian et al., 2011).

19.5.2 Electrocardiogram Monitoring

According to a WHO report, cardiovascular diseases are the top most cause of death throughout the world. With the expansion of Internet facility, secure sharing of patient's information and clinical data are possible even for remote patient's monitoring applications. A web-based ECG system helps in real monitoring of cardiac patient from home. This is an improved and cheaper way of cardiac-disease management from home and can be applied to other diseases such as diabetes and asthma (Magrabi et al., 1999). Continuous ECG monitoring is required to tackle and manage an emergency situation. In such a case, wireless body sensor networks emerge as an economical, continuous, and remote health monitoring for next generation of ambulatory personal telecardiology or e-cardiology systems. With the help of such mini and wearable sensors, doctors are able to record and monitor real-time cardiogram of chronic patient. This wireless technology is providing personalized care and is a boon for such patient (Kanoun et al., 2011). Another such tool is Bluetooth low-energy technology that helps in wireless monitoring of ECG. This method lessens power utilization as well as gives long monitoring strategy with the assistance of cell phones (Yu et al., 2012).

19.5.3 Rehabilitation System

IoT-based rehabilitation system comes with the solution of better supervision of patients and old-age people. IoT is a proficient way to interconnect all the accessible resources and provide the immediate action required. The benefit it offers over conventional hospital rehabilitation is ease-effective treatment, adequate interaction, and quick reconfiguration. It is a better way to fulfill patient's specific requirements by maximum utilization of medical resources (Fan et al., 2014). One of such examples includes smart rehabilitation system for poststroke patients. Around 80% of poststroke patients face difficulty in body movement. It is difficult for them to perform day-to-day activity due to muscle weakness, paralysis, stiffness, or changes in sensation. The intervention of technology helped patients with impaired motor activity to regain their health. The introduction of smart devices helps patient to do exercise on their own, tracking of their performance became easy, and reduces their dependency on occupational therapists (Harris & Sthapit, 2016).

A progress toward smart rehabilitation system is exemplified through brain–computer interface controlled functional electrical stimulation that has been proposed for spinal cord injury and stroke patients. It comes with a durable and easily accessible system that activates the motor pathways by means of motor imagery and sensory pathways (Al-Taleb & Vuckovic, 2016). Another example includes the use of ankle–foot rehabilitation system using the robotic system to assist elderly people suffering from joint spasticity (Chen et al., 2016).

19.5.4 Medication Management

Smart medication system can empower elderly people with independent and quality life. Elderly patients suffering from one or more chronic diseases need timely medication, specific therapies, periodic/routine monitoring of health parameters, specific diet plan, personal care, and attention. Information and communication technology and robotic technologies provide a new care process (Fiorini et al., 2016). An automatic drug dispensers or talking pillboxes constantly remind about timely intake of medicine and help to avoid any ambiguity (Powe, 1999). A step-ahead solution is cloud robotics that enables lucrative relationships among patients, family members, caregivers, pharmacists, and physicians. It served with a combined service of cloud computing along with a residential robot, an android robot application, and an online interface (Fiorini et al., 2016). Another good example of IoT-based medication system is in the case of Parkinson's disease and multiple sclerosis management in which wearable sensors are used for quality improvement of patient's life (Uem et al., 2016). Apart from this, the personalized care could be provided by interconnecting existing medical device products such as inhalers and insulin pens with sensors and connectivity technologies (Dimitrov, 2016).

19.5.5 Wheelchair Management

Another application of IoT is supporting patients with declining motor and sensory capabilities by using wheelchair-management techniques. The traditional assistive devices such as canes, crutches, walkers, and wheelchairs embedded with modern technology ease patient's life. Such devices with the help of sensor gain knowledge about the patient's physiological and physical state and functions accordingly (Ko et al., 2010). Use of such systems not only decreases patient's dependency on others but also lessens family member's worries about his/her safety. This can be clarified with the assistance of an illustration in which scientists designed wheelchair with a locator and ultrasonic radar. The locator made it easy to detect the patient's activity and inform the responsible authority, such as if patients fall from a wheelchair in the toilet, the emergency signal reaches the nurse to take immediate action to help the patient, whereas the radar system uses ultrasonic pulses to monitor patient's activity in their beds (Alemdar & Ersoy, 2010). Tracking of wheelchairs at airports has also become easier by the use of IoT (Srinivasa, 2016).

19.5.6 Monitor an Aging Family Member

The data collected by a poll of 12,000 experts from health care sector unveil that aging population would be the greatest challenge of this domain by the year 2025. The respondents are majorly from Asia Pacific (20%), Europe, and Middle East Africa (30%) (Staff, 2015). Aged family members required continuous monitoring to tackle any emergency medical situation. Their activity can be traced by the deployment of emergent technology such as ultrasound-based technology. It is very helpful in a critical situation such as detecting person fall. There is an emergency-call facility available based on battery-operated system. Apart from this, there are waterproof sensor systems that can be worn like a wristwatch. The sensor is programed to locate user at every 15–20 s and send relevant data to the homecare gateway by means of inbuilt-wireless-wide area network connection. Authority is warned if any critical incidence (such as cardiac arrest) occurs to provide medical assistance without delay (Rghioui et al., 2014).

19.5.7 Other IoT Technologies

Other than above mentioned assistances, IoT technologies impart other benefits in the health care sector. The advantage it offers includes collecting data and its automatic processing, identification, and verification of individuals, tracing patients' activities, wearable gadgets, and others (Atzori et al., 2010). One of such example is Zio Patch. It is a cardiac rate observer/recorder for transient ischemic attack (TIA) or stroke patient. During its implication, a waterproof patch is attached to the patient chest for 14 days. The data gathered during this period and the patch are supposed to be mailed to a central data processing center. It is then analyzed by the cardiologists to take required remedial steps (Tung et al., 2014).

19.6 Conclusion and Future Perspective

Although many accomplishments are achieved in the health care sector using IoT, still there are many shortcomings that need to be remedied for the well-being of people and society. There are many supporting technologies and devices that make IoT, a reality. However, few challenges are there as a barrier which need to be overcome in future. One of such challenges is electrical power requirement by IoT-based devices, especially in remote areas or inaccessible areas. Providing benefits of IoT-based services that too at an affordable price is a major challenge. Solar powering in the form of photovoltaics (PV) to power devices seems to be a potent futuristic solution for this problem. Other sustainable energy sources could also be used as energy-scavenging technique such as kinetic, thermal, or wind energy. These sources could be a striking alternative of photovoltaics (Haight & Haensch, 2016). Other than this, the future of IoT also lies on Internet of nano-things. The development of nanomachines with communication capabilities and their interconnection with micro- and macrodevices will enable the Internet of nano-things. This internet of nano-things will be effective in terms of existing technology by faster mode of communication, less power consumption, and economically beneficial (Akyildiz & Jornet, 2010).

Introducing wearable gadgets linked to mobile apps meet current medical demand and will be advantageous for public health. It is estimated that by the year 2020, smart wearable devices will achieve business returns of USD 22.9 billion (Jegan, 2016). However, some problems persist and are imperceptible. For example, a large number of Indian population (around 63 million people) is suffering from hearing impairment (Mohan Kumar, 2015).

To improve their life quality, mobile phone linked with the wearable device can be used as an audiometry device. Another example of wearable technology is *smart lens* that is in the phase of development by the joint collaboration effort of Google and Novartis. It could be useful for 1.7 billon people suffering from presbyopia (vision defect). It will also detect glucose level in a diabetic patient through tear (Senior, 2014). Another milestone in the medical and health care science that is yet to be achieved is the introduction of surgical robots. To make this true, renowned name from medical and IT field (Johnson & Johnson and Google) has come ahead with a tie up *Verb Surgical*. They are developing high-end technology smart surgical robots of relatively smaller size (20% less) and that too at a relatively low cost (Pierson, 2015).

References

Akyildiz, I. F., & Jornet, J. M. (2010). The Internet of nano-things. *IEEE Wireless Communications, 17*(6), 58–63.

Alemdar, H., & Ersoy, C. (2010). Wireless sensor networks for healthcare: A survey. *Computer Networks, 54*(15), 2688–2710. doi:10.1016/j.comnet.2010.05.003

Al-Taleb, M. K., & Vuckovic, A. (2016, May). *Home-based rehabilitation system using portable brain-computer interface and functional electrical stimulation.* In International brain-computer interference meeting (p. 100). doi:10.3217/978-3-85125-467-9-100

Atzori, L., Iera, A., & Morabito, G. (2010). The Internet of things?: A survey. *Computer Networks, 54*(15), 2787–2805. doi:10.1016/j.comnet.2010.05.010

Backman, W., Bendel, D., & Rakhit, R. (2010). The telecardiology revolution: Improving the management of cardiac disease in primary care. *Journal of the Royal Society of Medicine, 103*(11), 442–446. doi:10.1258/jrsm.2010.100301

BI Intelligence. (2016). *There are two major challenges to implementing IoT solutions in healthcare.* Retrieved from http://www.businessinsider.com/there-are-two-major-challenges-to-implementing-iot-solutions-in-healthcare-2016–7?IR=T

Charania, C., Nair, G., Rajadhyaksha, S., & Shinde, A. (2016). Healthcare using Internet of things. *International Journal of Technical Research and Applications, 41*(March), 42–45.

Chen, G., Zhou, Z., Vanderborght, B., Wang, N., & Wang, Q. (2016). Proxy-based sliding mode control of a robotic ankle-foot system for post-stroke rehabilitation. *Advanced Robotics, 30*(15), 992–1003. doi:10.1080/01691864.2016.1176601

Coetzee, L., & Eksteen, J. (2011, May). *The Internet of things-Promise for the future? An Introduction.* IST Africa, Gaborone, Botswana (pp. 1–9).

Devi, K. N., & Muthuselvi, R. (2016). Secret sharing of IoT healthcare data using cryptographic algorithm. *International Journal of Engineering Research, 5*(4), 790–991.

Dimitrov, D. V. (2016). Medical Internet of things and big data in healthcare. *Healthcare Informatics Research, 22*(3), 156–163.

eHealth. (2015). *Telemedicine: Metamorphosing Indian healthcare.* Retrieved from http://ehealth.eletsonline.com/2015/11/telemedicine-metamorphosing-indian-healthcare/

Elkhodr, M., Shahrestani, S., & Cheung, H. (2016). The Internet of things: New interoperability, management and security challenges. *International Journal of Network Security and Its Application, 8*(2), 85–102.

Fan, Y. J., Yin, Y. H., Xu, L. D., Zeng, Y., & Wu, F. (2014). IoT-based smart rehabilitation system. *IEEE Transactions on Industrial Informatics, 10*(2), 1568–1577. doi:10.1109/TII.2014.2302583

Fiorini, L., Esposito, R., Bonaccorsi, M., Petrazzuolo, C., Saponara, F., Giannantonio, R., ... Cavallo, F. (2016). Enabling personalised medical support for chronic disease management through a hybrid robot-cloud approach. *Autonomous Robot, 40*(7), 1–14. doi:10.3837/tiis.0000.00.000

Gong, W. (2016, July). *The Internet of things (IoT): What is the potential of the Internet of things (IoT) as a marketing tool?* 7th IBA Bachelor Thesis Conference, Enschede, The Netherlands (pp. 1–13).

Gramvaani. (2013). *Rural health care: Towards a healthy rural India.* Retrieved from http://www.gramvaani.org/?p=1629

Haight, B. R., & Haensch, W. (2016). Solar-powering the Internet of things. *Science, 353*(6295), 124–125.

Harris, N. R., & Sthapit, D. (2016). *Towards a personalised rehabilitation system for post stroke treatment.* Sensors application symposium (pp. 1–5).

HIMMS. (2013). *Definition of interoperability.* Retrieved from http://www.himss.org/library/interoperability-standards/what-is

Hodkari, H. K., & Aghrebi, S. G. M. (2016). Necessity of the integration Internet of things and cloud services with quality of service assurance approach. *Bulletin de La Société Royale Des Sciences de Liège, 85,* 434–445.

Islam, S. M. R., Kwak, D., & Kabir, H. (2015). The Internet of things for health care: A comprehensive survey. *Access, IEEE, 3,* 678–708.

Istepanian, R. S. H., Hu, S., Philip, N. Y., & Sungoor, A. (2011). *The potential of Internet of m-health Things m-IoT for non-invasive glucose level sensing.* Proceedings of the annual international conference of the IEEE Engineering in Medicine and Biology Society, EMBS (pp. 5264–5266). doi:10.1109/IEMBS.2011.6091302

Jegan. (2016). *Transforming healthcare through IoT-TVS NEXT.* Retrieved from http://tvsnext.io/blog/transforming-healthcare-through-iot/

John Horn. (2016). *3 keys to ensuring scalability in the IoT.* Retrieved from http://www.itproportal.com/2016/02/02/3-keys-to-ensuring-scalability-in-the-iot/

Kanoun, K., Mamaghanian, H., Khaled, N., & Atienza, D. (2011). *A real-time compressed sensing-based personal electrocardiogram monitoring system.* Design, automation & test in Europe conference and exhibition, Grenoble, France (pp. 1–6).

Khanna, A., & Misra, P. (2013). *The Internet of Things for Medical Devices-Prospects, Challenges and the Way Forward (A white paper).* Retrieved from http://234w.tc.tracom.net/resources/white_papers/Pages/Internet-of-Things-Medical-Devices.aspx

Kim, S., & Kim, S. (2016). A multi-criteria approach toward discovering killer IoT application in Korea. *Technological Forecasting and Social Change, 102,* 143–155. doi:10.1016/j.techfore.2015.05.007

Ko, B. J., Lu, C., Srivastava, M. B., Stankovic, J. A., Ieee, F., Terzis, A., & Welsh, M. (2010). Wireless sensor networks for healthcare. *Proceedings of the IEEE, 98*(11), 1947–1960.

Madhuri, T., & Sowjanya, P. (2016). A study on public health surveillance using Internet of things and its applications. *International Journal of Innovative Research in Computer and Communication Engineering, 4*(4), 5143–5147. doi:10.15680/IJIRCCE.2016

Magrabi, F., Lovell, N. H., & Celler, B. G. (1999). A web-based approach for electrocardiogram monitoring in the home. *International Journal of Medical Informatics, 54*(2), 145–153. doi:10.1016/S1386-5056(98)00177-4

Masys, D., Baker, D., Butros, A., & Cowles, K. E. (2002). Giving patients access to their medical records via the Internet: The PCASSO experience. *Journal of the American Medical Informatics Association, 9*(2), 181–191. doi:10.1197/jamia.M1005

Med, G. (2016). *What is telemedicine?* Retrieved from https://www.globalmed.com/additional-resources/medical-technology

Mohan Kumar. (2015). *Wearable medical devices and end-to-end healthcare.* Retrieved from https://www.mdtmag.com/blog/2015/06/wearable-medical-devices-and-end-end-healthcare

National Health Portal of India. (2015). *m-Health.* Retrieved from http://www.nhp.gov.in/miscellaneous/m-health

Ngoc, T. V. (2008). *Medical applications of wireless networks abstract?: Table of contents:* Retrieved from http://www.cse.wustl.edu/~jain/cse574-08/index.html

Niewolny, D. (2013). *How the Internet of things is revolutionizing healthcare. Freescale Semiconductor, Inc.* (Vol. October). Retrieved from http://cache.freescale.com/files/corporate/doc/white_paper/IOTREVHEALCARWP.pdf

Nishith Desai Associates. (2016). *Investment in healthcare sector in India*. Retrieved from http://www.nishithdesai.com/fileadmin/user_upload/pdfs/Research%20Papers/Investment_in_Healthcare_Sector_in_India.pdf

Omogbadegun, Z. O., & Ayo, C. K. (2007). *Impact of mobile and wireless technology on healthcare*. In *3GSM & Mobile Computing: An Emerging Growth Engine for National Development* (pp. 164–171). Covenant University, Ota, Nigeria.

Pierson, R. (2015, December 10). *J&J, Alphabet aim for smarter, smaller, cheaper surgical robot*. Retrieved from http://www.reuters.com/article/us-alphabet-johnson-johnson-robots-idUSKBN0TT1SB20151210#xG8lkORQElOQ6M7U.97

Piyare, R. (2013). Internet of things: Ubiquitous home control and monitoring system using android based smart phone. *International Journal of Internet of Things*, 2(1), 5–11. doi:10.5923/j.ijit.20130201.02

Powe, P. S. (1999). Date of Patent: US Patent No. 5,954,225. Washington, DC: U.S. Patent and Trademark Office. Retrieved from https://www.google.es/patents/US5954225.

Puskin, D., Johnston, B., & Speedie, S. (2006). *Telemedicine, Telehealth, and Health Information Technology*. Washington, DC. Retrieved from http://www.who.int/goe/policies/countries/usa_support_tele.pdf?ua=1

Rghioui, A., L'aarje, A., Elouaai, F., & Bouhorma, M. (2014). The Internet of things for healthcare monitoring: Security review and proposed solution. *2014 Third IEEE International Colloquium in Information Science and Technology (CIST)*, 5(5), 384–389.

Senior, M. (2014). Novartis signs up for Google smart lens. *Nature Biotechnology*, 32(9), 856.

Sermakani, V. (2014, April). *Transforming healthcare through Internet of things*. Project management practitioners' conference (pp. 1–26). Banglore, India. Retrieved from http://pmibangalorechapter.in/pmpc/2014/tech_papers/healthcare.pdf

Singh, A. (2016). *India is becoming an important market for wearable devices*. Retrieved from http://www.sundayguardianlive.com/opinion/7513-india-becoming-important-market-wearable-devices

Smith, D. (2013). *An introduction to MSP430™ microcontroller-based temperature-sensing solutions. Texas Instruments Inc.* Retrieved from http://goo.gl/UKaURK

Smith, Y. (2015). *Types of telemedicine*. Retrieved from http://www.news-medical.net/health/Types-of-Telemedicine.aspx

Srinivasa K. (2016). View from India: Educate today for tomorrow's internet of medical things. Engineering and technology. Retrieved from https://eandt.theiet.org/content/articles/2016/11/view-from-india-educate-today-for-tomorrow-s-internet-of-medical-things/

Staff, F. (2015). *IoT, Big data key to overcoming healthcare bottlenecks by 2025: Survey-Firstpost*. Retrieved from http://www.firstpost.com/business/iot-big-data-key-overcoming-healthcare-bottlenecks-2025-survey-2373440.html

Strehle, E., & Shabde, N. (2006). One hundred years of telemedicine: Does this new technology have a place in paediatrics? *Archives of Disease in Childhood*, 91(12), 956–959. doi:10.1136/adc.2006.099671

Terry, N. P. (2016, April). *Will the Internet of things disrupt healthcare?* Indiana University Robert H. McKinney School of Law Research Paper No. 2016–21. doi:10.2139/ssrn.2760447

The World Bank. (2014). *Health expenditure, total (% of GDP)*. Retrieved from http://data.worldbank.org/indicator/SH.XPD.TOTL.ZS?locations=IN

Tung, C. E., Su, D., Turakhia, M. P., & Lansberg, M. G. (2014). Diagnostic yield of extended cardiac patch monitoring in patients with stroke or TIA. *Frontiers in Neurology*, 5, 266. doi:10.3389/fneur.2014.00266

Uem, J. M., Maier, K. S., Hucker, S., & Scheck, O. (2016). Twelve-week sensor assessment in parkinson's disease: Impact on quality of life. *Movement Disorder*, 1–2. doi:10.1002/mds.26676

Ullah, K., Shah, M. A., & Zhang, S. (2016). *Effective ways to use Internet of things in the field of medical and smart health care*. International conference on intelligent system engineering, Islamabad, Pakistan (pp. 1–8).

Wearable Tech News. (2016). *India wearable market hits 400,000 units in Q116 with fitness bands the key.* Retrieved from http://www.wearabletechnology-news.com/news/2016/jun/09/india-wearable-market-hits-400000-units-q116-fitness-bands-key/

WHO. (2009). *Telemedicine-Opportunities and developments in Member States* (Vol. 2).

WI Bureau. (2015). *Mobile healthcare platform lybrate ties-up with IMA; aiming to turn doctors technically trained.* Retrieved from http://www.franchiseindia.com/wellness/Mobile-healthcare-platform-Lybrate-ties-up-with-IMA;-aiming-to-turn-doctors-technically-trained.5717

World Health Organization. (1998). *A health telematics policy. Report of the WHO group consultation on health telematics.* Retrieved from http://scholar.google.com/scholar?hl=en&btnG=Search&q=in title:A+health+telematics+policy#5

Wu, G., Talwar, S., Johnsson, K., Himayat, N., & Johnson, K. D. (2011). M2M: From mobile to embedded Internet. *IEEE Communication Magazine, 49*(4), 36–43.

Xu, L. Da., He, W., & Li, S. (2014). Internet of things in industries: A survey. *IEEE Transactions on Industrial Informatics, 10*(4), 2233–2243.

Yu, B., Xu, L., & Li, Y. (2012). *Bluetooth Low Energy (BLE) based mobile electrocardiogram monitoring system.* IEEE International Conference on Information and Automation (pp. 763–767). doi:10.1109/ICInfA.2012.6246921

20

Mining Ubiquitous Data Streams for IoT

Nitesh Funde, Meera Dhabu, Ayesha Khan, and Rahul Jichkar

CONTENTS

20.1 Introduction

People send and receive information anytime, anywhere all across the globe with the help of Internet. It is not only just people that are using the Internet; nowadays, physical objects also connect with each other in the internetwork with the help of Internet. Internet of Things (IoT) basically connects all the physical objects in the world to the Internet (Gubbi et al., 2013). Machine-to-machine communication is useful in the manufacturing industry to monitor machinery operations, detect fault, and raise alerts. It will technologically revolutionize the next generation having a similar impact brought by the revolution of the computer and the Internet.

With the advancement in the networking industry, more number of people access to the network that generates a huge amount of data, making it difficult to extract useful knowledge. The data generated by IoT are massive; it contains potentially infinite high-speed data streams from various heterogeneous sources. Therefore, the traditional data-mining models are not sufficient to mine this new type of data. Big data analytics is one of dynamic areas in IoT, which satisfies the characteristics of this new type of data. The main motivation of this chapter is the real-time analytics of the data generated by the sensors in the various IoT-based systems. The real-time data are nothing but the data streams that are sequential, online, and multidimensional in nature. In the real-world application of data streams, underlying distribution of data changes over the time. These changes will make a greater impact on the prediction accuracy because of the default static predictive model. The state is called concept drift. Thus, the predictive model will have to be updated as the data distribution changes over the time. This area of data mining has acquired the attention of the researchers to develop new data-stream-mining techniques since the last decade. Various excellent techniques have been proposed to address these issues of analyzing fast data streams (Nguyen et al., 2015; Pinage et al., 2016).

With ever-increasing computational power and wireless-networking capability of small ubiquitous devices such as mobile, sensor devices, the area of ubiquitous computing provides the best technological infrastructure for the IoT. It provides an excellent opportunity to perform intelligent data analysis in the ubiquitous computing environments. Ubiquitous data mining is the process of performing data analysis on embedded, mobile, and ubiquitous devices to support applications such as mobile-activity recognition, smart homes, and health care (Gama & May, 2011). Thus, ubiquitous data-stream-mining (UDM) algorithms must be computationally efficient, resource-aware, context-aware, and energy-efficient (Gaber et al., 2014). This chapter aims to present the detailed survey of recent research developments of data stream mining in the ubiquitous computing environment and discuss the challenges of ubiquitous data stream mining in IoT.

This chapter is organized as follows. It begins with an overview of big data fundamentals in IoT. Then, a comprehensive survey of ubiquitous data-stream-mining algorithms is presented. Finally, the ubiquitous applications and challenges in IoT are discussed, followed by the conclusion.

20.2 IoT and Big Data Fundamentals

The next era will be of IoT computing differently from the traditional desktop computing. In IoT, many of physical things that surround us will be in the network of each other. Business and individuals will be able to take control of their environment with the help of IoT. Modern day home, vehicles, and office environment can be controlled remotely with just a smartphone. This new challenge of high degree of connectivity is addressed by radio-frequency identification (RFID) and sensor networks technologies. Thus, it is possible to collect huge amount of data about these devices, specifically the information provided by the sensors embedded in the environment. IoT refers to the world of Internet-connected devices by which big data are gathered, stored, processed, and managed. Big data are having main 3 V's—volume, variety, and velocity. The data generated by devices in IoT are massive and contain potentially infinite high-speed data streams from heterogeneous, distributed sensor sources. Thus, IoT needs the big data methodology for IoT analytics.

TABLE 20.1

Difference between Traditional Data and Data in IoT

Characteristics	Traditional Data	Data in IoT
Data schema	Static	Dynamic
Available resources	Flexible	Limited
Computation results	Accurate	Approximate findings
Data modeling	Persistent	Modeled as transient data stream
Algorithms	Offline processing in which time is not constraint	Online processing in which time is important as data may skip
Dimension	One dimensional	Multidimensional

20.2.1 Comparisons between Traditional Data and Dynamic Data in IoT

With the continuous advancement in the hardware and software technologies, nature of the data is always changing in the today's world. Infinite volume of dynamic data are generated in various domains such as sensors, Twitter, Facebook, network monitoring and traffic engineering, financial market and web logs, web page click streams, and so on. These data are nothing but the data streams that are different from the conventional static database. The difference between conventional and new type of data is given in Table 20.1. Thus, it is interesting how to process and manage the data streams in a computationally efficient way. This becomes a hot topic of data mining in IoT environments.

20.2.2 Concept Drift

In the data-mining community, the concept is defined to represent the overall data distribution that is used to perform classification, clustering, and other problems at any time. Generally, the underlying data distribution is expected to be stable, that is, concept does evolve over time. However, in many real-world applications of data stream, the statistical properties of the target environments change over time. This problem is known as concept drift.

Thus, concept drift occurs when underlying data distribution changes over time in an unpredictable ways. For example, the behavior of customer in an e-commerce website may change over time. This shopping behavior of customer changes seasonally. The concept drift of the unknown probability function can be defined as follows.

$$P_i(x, \omega) \neq P_j(x, \omega) \tag{20.1}$$

where:
x is a data instance
ω represents the class, and the change occurs from time t_i to time t_j, where $t_i < t_j$.

This mean that optimal predictive model for $P_i(x, \omega)$ is no longer suitable (or optimal) for $P_j(x, \omega)$ in a changing environment (Ang et al., 2013). Concept drift also compromises the prediction accuracy in the above settings.

There are four types of concept drifts such as gradual, incremental, abrupt, and recurring concepts. In gradual concept drift, concept is gradually replaced by the new concept.

Here, both the concepts coexist in a time period in which changes are detected by using the methods based on time windows that can scan the data. In incremental concept drift, the concept evolves slowly over time. Here, the new concept replaced the old concept completely, and both the concept cannot exist in the same time period. In abrupt concept drift, old concept is suddenly replaced by the new concept. In recurring concept, old concept disappears but may arrive in the future.

20.2.3 Ubiquitous Data Stream Mining

The effort taken by the researchers in the creation of human-to-human interaction in the late 1980s resulted in the creation of ubiquitous computing in which the objective is to bring the technology in the everyday life. Thus, it opens a new opportunity to make intelligent data analysis in the ubiquitous computing environment.

UDM is the process of analyzing real-time sensory data coming from heterogeneous sources onboard ubiquitous devices (Gaber et al., 2014). UDM is the part of knowledge discovery in ubiquitous computing environment. It consists of different technologies such as data mining, machine learning, sensor networks, and activity recognition.

The ubiquitous knowledge discovery is having following characteristics:

- *Limited resources*: Battery power, bandwidth, computational power, and memory.
- *Real-time*: The data and other constructs are real time in nature, and learning models should to be updated in accordance with the evolving environment.
- *Locality*: The ubiquitous devices know only their local spatiotemporal environment.
- *Information processing power*: The devices are capable of information processing.
- *Mobility*: These devices can change their location.
- *Distributed*: These devices are distributed in nature because of their exchange of information with other devices.

It is necessary to use data-mining techniques to extract meaningful information from data in ubiquitous devices with the above mentioned characteristics. The learning takes place locally within the ubiquitous devices. However, the learning algorithm is quite different than the traditional data-mining algorithms. The algorithm should be online, incremental, anytime, and anywhere rather than offline batch scenario in traditional data-mining algorithm. In the UDM, learning takes place under unreliable, highly resource-constraint environment such as battery power and bandwidth. Thus, the algorithm should be lightweight in terms of resource utilization. The next section describes the existing ubiquitous data-stream processing and mining algorithm onboard ubiquitous devices.

20.3 Ubiquitous Data Stream Processing and Mining Algorithms

The real world consisting of interconnected physical ubiquitous objects, exchanging the information constantly, the volume and velocity of the data generated, and the procedure involved in handling of those data become critical. There is a range of IoT applications that requires mining of data streams in ubiquitous environments such as eHealth and

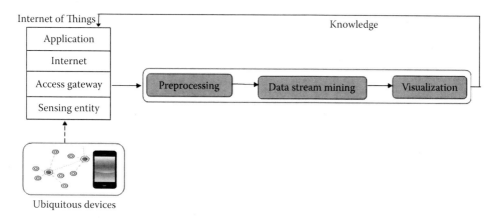

FIGURE 20.1
Conceptual architecture of ubiquitous data stream mining in the Internet of Things (IoT).

intelligent transportation systems (ITSs). The data generated by the ubiquitous devices need systematic approach to extract knowledge for successful IoT applications.

Nowadays, efforts have been taken by researchers on big data analytics in the ubiquitous devices and IoT. The conceptual architecture of UDM process in IoT is shown in Figure 20.1. The ubiquitous devices such as sensor network and smartphones consist of various sensors; inbuilt sensors (accelerometer in case of smartphones) collect the raw data that need to be preprocessed. The data-stream-mining techniques consist of various classifications, clustering, and frequent pattern mining (FPM) techniques; extract the useful information that can be visualized; and appropriate decision is made from this. This knowledge can be shared with other ubiquitous devices in the network.

The knowledge-discovery process in the ubiquitous computing environment is explained in detail in the next upcoming subsections.

20.3.1 Preprocessing

Preprocessing always plays an important role in the process of data mining. Here, in IoT of ubiquitous computing environment, raw data come from heterogeneous, sensor sources that contain noisy, missing, incomplete values, and the data may be structured, semistructured, and unstructured. Preprocessing is one of the basic, important components of big data mining system. The raw data should be well prepared before it is feeding to data-stream-mining process.

The raw data can be related to the context data (temperature, GPS-data), data coming from different applications or data of different behaviors of users. There are three basic traditional preprocessing approaches that need to apply to the data stream in ubiquitous environments. They are as follows: outlier removal, normalization, and handling missing values.

Outliers are referred to as the data points that are far from the mean of the corresponding normal variables. The behavior of outlier is different from the normal data point. This type of data points are erroneous values and have to be removed from further process of analysis. The basic way to remove outliers is to use hypothesis testing method when the data are normally distributed. However, researchers found this method not to be robust.

Normalization is an important step of preprocessing that scales down or scales up the data range before it feeds to the further stage. It is given by computing the mean $\bar{x}$ and the variance σ^2 of the data points x_i using following equations:

$$\bar{x} = \frac{1}{N}\sum_{i=1}^{N} x_i \quad \text{and} \quad \sigma^2 = \frac{1}{N-1}\sum_{i=1}^{N}(x_i - \bar{x}_i)^2 \tag{20.2}$$

where N is the number of data points. Then, the original values are transformed using the following equation.

$$z_{ik} = \frac{x_{ik} - \bar{x}_k}{\sigma_k} \tag{20.3}$$

The variable z_{ik} is normally distributed with a mean of 0 and a variance of 1. Hence, all the data points will have identical values of mean and variance. If the mean is subtracted from the data points, then that process is called centering.

The scaling technique is a special case of normalization that reduces the data pints to a predefined interval [0, 1] or [−1, 1]. Sigmoid scaling is a renowned scaling technique that reduces all data points to interval [0, 1] using nonlinear transformation on the original data points.

$$z_{ik} = \frac{1}{1 + \exp(-x_{ik})} \tag{20.4}$$

Moreover, some simple functions f (·) can be applied on the data values instead of converting it into the interval of 0 to 1. The common function is the logarithmic function that transforms original data points to the logarithmic scale.

The third approach of preprocessing is the handling of missing values. The most popular technique is the missing data imputation which replaces the missing value with the random or average value. For example, the average of data points can be used in the place of missing value.

Figo et al. (2010) described various preprocessing techniques for the context recognition of smartphone user from the accelerometer data. The key focus of context-aware services is the capability of mobile devices to collect, process, and manage the raw sensor data for extracting useful information. From this, device must be able to determine the features of sensor signal. Sensors in the mobile devices generate a large volume of raw-data points contaminated with noise in environment that require to be filtered out. The built-in accelerometer sensor gives the signals instead of a value (given by other sensor) and requires complex preprocessing techniques to characterize the physical activity of the user, that is, running and walking (context) within a certain time frame. The authors have classified sensor signals processing techniques in three broad domains for feature extraction, namely, the time domain, the frequency domain, and discrete representation domains.

In the time domain, mathematical and statistical functions are used to extract information from raw sensor data. It includes several mathematical and statistical functions such as mean, median, standard deviation, min, max, range, root mean square metric, correlation, cross correlation, and integration. They are mostly used to select the main features in the raw sensor data.

In the frequency domain techniques, the repetition of the sensor signal is captured. It often correlates with the periodic nature of a particular activity such as running or walking. Fourier transform is the commonly used technique of signal transformation that allows the important features of time-based signal such as its average, coefficient sum, and information entropy to represent it in the frequency domain.

In the discrete representation domains, the transformation of accelerometer signal and other sensor signals into strings of discrete representation symbols. Although there is an information loss after discretization process, the sensor signal is compressed well into string. For example, consider a case of a 3-axis accelerometer signal, in which a series of n input samples split into windows of w consecutive samples. The first step is to compute the average value for each of window followed by the discretization process over an alphabet of fixed size. The domain value function is used by the discretization process that defines the values of intervals corresponding to a given symbol. The symbolic aggregate approximation is developed recently, which uses piecewise aggregate approximation for mapping range of values to string symbols (Bondu et al., 2013). Edit distance techniques (such as Euclidean distance) are used to match the string with known patterns of user activity. Plötz et al. (2011) used principal component analysis and deep learning-based feature learning method for activity recognition in ubiquitous computing.

After preprocessing, the stream data are now well prepared for analysis. In the next section, the data-stream-mining techniques in the ubiquitous computing environments are reviewed.

20.3.2 Ubiquitous Data-Stream-Mining Algorithms

As the UDM is the process of analyzing real-time sensory data coming from heterogeneous sources onboard resource-constraint ubiquitous devices, there is a requirement of algorithms that are cost efficient in terms of computations and adaptive to their varying and dynamic operational context. Here in UDM, the primary and important is not only the limitation of resources but also these resources and factors such as data rates are subjected to continuous variability. Thus, algorithm needs to be adaptive to the changing concept. There are three approaches for adaptation in UDM algorithms, namely resource-aware approach, situation-aware approach (or context-aware), and the combination of resource and situation-aware approach, that is, hybrid approach.

In the ubiquitous computing environments, various devices are connected in the network. Analyzing the real-time data in this environment is a challenging task because of the limited resource constraints such as battery power, memory, and network bandwidth in the ubiquitous devices. Thus, the adaptation in UDM implies that there is continuous processing or mining of data streams onboard a ubiquitous device and continuous variability in resource availability. For example, the adaptation of various parameters in data stream processing algorithms of real time environmental monitoring such as resource levels, data rates etc. The common approach is of temporal window size of data streams. The size of the temporal window can be increased or decreased on the basis of the current status. Thus, the larger window states that the processing of the stream data occurs less frequently, whereas the smaller window performs processing more frequently. Hence, higher frequency of processing of small temporal window leads to higher consumption of resources, and accuracy of the stream processing will be higher during the window. This happen vice-versa to the larger temporal window. Researchers have proposed the granularity-based adaptation algorithms that adapt according to availability of resource and data stream rate (Gaber et al., 2014).

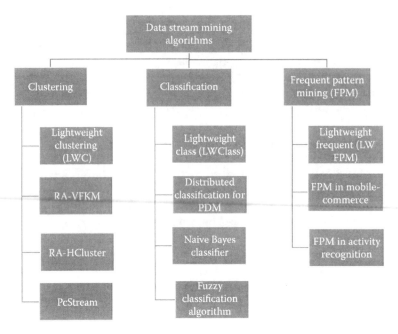

FIGURE 20.2
Ubiquitous data stream mining algorithms.

This approach is classified into three different forms of variation such as algorithm input granularity (AIG), algorithm processing granularity, and algorithm output granularity (AOG). AIG is the process in which data rates feeding into the mining algorithm are adapted according to battery charge. AOG is the process in which the adaptivity is achieved by adjusting the algorithm output rate (number of clusters). This process is discussed in more detail in the next paragraph. Algorithm processing granularity does the adaptation of processing of algorithm according to CPU usage.

As discussed, the data-stream-mining algorithm requires performing granularity-based processes that adjust its parameters operations to deal with reduced or increased resource availability and speed of data rates. This adjustment will have the impact on the accuracy of mining algorithms. Therefore, resource availability and the accuracy of mining algorithms have a strong corelationship.

All the UDM algorithms are reviewed, which are lightweight and cost-efficient in terms of resource-utilization. The algorithms follows the granularity-based approaches specifically the AOG form and resource-aware adaptation, situation-aware approach, and hybrid of two. The lightweight clustering (LWC), lightweight class, and lightweight FPM algorithm follow the AOG approach particularly. The structure of UDM algorithms review is as shown in Figure 20.2.

20.3.2.1 Clustering

Clustering is the process of grouping the high similar data points in one group, whereas these data points are dissimilar to other data points that exist in the different group or cluster. It is based on the principle of maximizing the intraclass similarity and minimizing the interclass similarity. The clustering is the unsupervised learning method. Various clustering algorithms are proposed till now. The algorithms in the ubiquitous environments are reviewed as follows.

1. *Lightweight Clustering* (*LWC*): The LWC follows the AOG-based approach in which algorithm rate, data rate, time needed to fill the available memory, and accuracy are interrelated (Gaber et al., 2005). In AOG, algorithm threshold is the parameter for controlling the creation of the new outputs based on the three factors that vary over the time. They are available memory, time remaining to fill the available memory, and the rate of data stream. The threshold value is the maximum acceptable distance element of the data stream and the mean of the group (centroid). The higher the threshold, the lower will be the output size produced. The threshold of algorithm used the Euclidean or Manhattan distance functions. This threshold of algorithm is also used in the lightweight classification and FPM.

   LWC is a one-pass LWC algorithm. The basic idea of LWC proposed by Gaber et al. (2005) is to add the new data points to existing clusters incrementally in accordance with adaptive threshold value that is the smallest distance from the data point to the centroid of cluster. If the distance between the new data element of the data stream and the centroid of existing cluster is greater than the threshold values, then create the new cluster. After creation of cluster, if the memory is full then perform the merge operation to accommodate the new cluster in memory. If the measured distance is not greater than the threshold, then increase the weight of existing centroid. The flowchart of LWC is shown in Figure 20.3.

2. *RA-VFKM* (*Resource-Aware Very Fast K-Means*): Shah et al. (2005) proposed RA-VFKM which is the extended version of Very Fast K-Means (VFKM) algorithm for execution on resource-constrained ubiquitous devices. VFKM (Domingos & Hulten, 2001) is a data stream-clustering algorithm developed as faster extended version of K-means. VFKM uses only particular number of data item in each step for faster execution instead of using all the available data items used by the K-means. Thus, VFKM observes an error ε_i with the probability δ_i in the results obtained after mining the particular number of data items in each step i as compared with the results that would have been obtained by mining all the data items in K-means. The goal of VFKM is to perform a clustering in minimum time with atmost ε^* and atleast $1-\delta^*$ probability with the given parameters.

   In a RA-VFKM, resource-awareness and adaptation concepts are applied using AOG approach. The two new error parameters are added such as ε^{**} the maximum allowable error limit and δ^{**} maximum allowable probability. These two parameters are controlled and managed in each processing cycle based on the availability of resource (when the available memory decreases or in critical stage). The algorithm will consume less computational resources, if these parameter values are increased by some calculated factor to satisfy the goal of fast K-means (FKM) as mentioned above. Overall, if greater accuracy is needed (decrease in values of error parameter), then algorithm consumes computational resources quickly which can result into an application failure. Therefore, RA-VFKM improved the longevity and continuity in resource-utilization onboard a ubiquitous device compared with nonresource-aware version of algorithm.

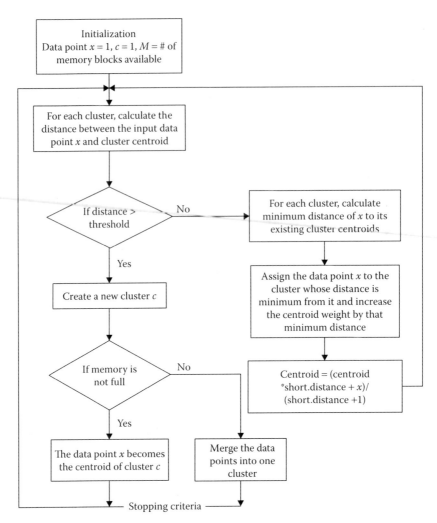

FIGURE 20.3
Lightweight clustering (LWC) flowchart.

3. *RAH-Cluster (Resource-Aware High Quality Clustering)*: It is a clustering algorithm that adapts according to the available resources, so that the ubiquitous device can continue with the process of clustering at acceptable accuracy under limited memory resources (Chao & Chao, 2011). The process is divided into the following two parts: online maintenance and offline clustering. In the online maintenance, the statistics of summary of data stream are computed and used by the process-offline clustering. In the first part, the sliding window model is used for processing the data stream, and summary statistics are computed in the form of microclusters that updates incrementally. The hierarchical summary frame is used for storing the microcluster, and the level of it can be adjusted on the basis of resource availability. The hierarchical summary frame can be adjusted to a higher level by reducing the amount of data to be processed in a limited resources scenario to decrease the resource consumption.

In the second part of offline clustering, the summary statistics are used for the clustering process, and adaptation is carried out on the basis of the available memory resource. Here, first the resource observing module decides whether the memory is sufficient or not by computing its usage and remaining memory. When the memory is critically low, then sliding window size and hierarchical summary frame are adjusted using AIG approach, and threshold distance value is decreased to reduce the amount of data to be processed. The distance threshold can be increased to improve the accuracy of RAH cluster. Compared with RA-VFKM, RAH cluster perform well in terms of accuracy and maintains low and stable usage of memory resource.

4. *PcStream*: PcStream is a stream-clustering algorithm for finding temporal context of an entity in an unsupervised manner (Mirsky et al., 2015). It mines the sensor stream data for contexts. It dynamically detects the present context as well as predicts the coming context. For example, it used to detect the different behavior of data point, that is, anomaly in sensor networks.

The name *pcStream* is used for the principal components of the data distributions in the stream data that are used dynamically to detect and compare contexts. The basic idea of the core pcStream algorithm is to capture the data stream's distribution and determine when it change to known or unknown (new) distribution. There are various parameters of pcStream—threshold distance, the minimum context drift size, available memory size of model, and the percent of variance to retain in the context models. The general process of the algorithm is started with calculation of statistical similarity between the new data point x and the known context. If the result is greater than the defined threshold, then it will be the potential point for outlier or new cluster. If such points are continuous and are able to fill the minimum context drift size, then the new cluster (context) is discovered. If it is not greater than the user defined threshold, then it will belong to known context. Here, the memory resource is maintained stable by merging the context models. The implementation of pcStream algorithm is ongoing on android mobile platform.

20.3.2.2 Classification

The classification is the supervised learning problem, important for the decision making. Give a data point, assigning it to one of predefined target class label or category called classification. The classification aims to predict the class labels for each of the data instances accurately. There are many challenges of ubiquitous stream data learning system for the problem of classification such as limited resource constraints, adaptation, handling recurring concepts, use knowledge from other ubiquitous devices to improve learning local concepts, and others. A large number of classification algorithms are available, but out of them, only a few algorithms are suitable for classifiying data streams in ubiquitous environments. The algorithms of classification are surveyed, which have been customized to operate in UDM based on the principle of anytime and anywhere data stream mining.

1. *Lightweight classification algorithm*: The lightweight classification algorithm starts with setting the number of data points according to available memory (Gaber et al., 2005). The algorithm follows the AOG approach that has been discussed in the above sections. Whenever a new data point comes, the algorithm searches for the

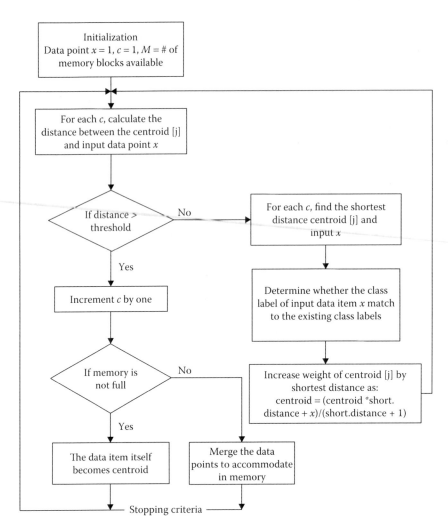

FIGURE 20.4
Lightweight classification (LWClass) flowchart.

instance nearer to it based on predefined distance threshold. The distance threshold represents the acceptable similarity measure to consider the new data point according to existing element's attribute values. If the algorithm finds this data element, then it checks the class label of stored data items and new data element. If it is similar with the accepted distance, then the weight of stored data item is increased by one; otherwise, the weight is decreased by one, and new data item is ignored. If the weight becomes zero, then stored data elements will be released from the memory. The flowchart of the lightweight classification algorithm is shown in Figure 20.4.

2. *Distributed classification for pocket data mining*: Pocket Data Mining (PDM) is the process of analyzing data streams for distributed environments (Stahl et al., 2012). Advancements in wireless mobile computing devices such as laptops, tablets, and smartphones have made it possible to run a wide-range of ubiquitous applications

in this environment. Stahl et al. (2012) proposed the adoption of data-stream-classification techniques such as hoeffding trees and naïve Bayes for PDM. PDM perform collaborative and distributed data mining using stream mining technologies. The PDM architecture consists of different agent miners, mobile resource discoverers, and mobile agent decision makers. Here, agent miner is the mobile, distributed in the environment, which implements the basic stream-mining algorithms. The naïve Bayes classifier is originally developed for batch learning; but its incremental nature made it applicable to data streams. Here, the naïve Bayes classifier uses its implementation from massive online analysis (MOA) tool (Bifet et al., 2010). Naïve Bayes classifier is driven from the Bayes Theorem that states that if E is an event of interest, and event E occurs with $P(E)$ probability, and $P(E|X)$ is the conditional probability that event E occurs under that X occurs then:

$$P(E|X) = P(X|E)P(E)$$

$$P(X)$$

The naïve Bayes classifier is to assign a class label to a data instance to which it belongs with the highest probability. Osmani et al. (2014) used adaptive naïve Bayes classifier with kernel density estimation on smartphone to detect the proximity of two or more individuals (devices) which infers the interpersonal distance (Osmani et al., 2014).

3. *Fuzzy classification algorithm*: The unsupervised data-stream-mining techniques need human intervention for further analysis and understanding of the clustering results. It becomes challenging in UDM applications when real-time data analysis or decision making is required. Horovitz et al. (2005) presented approach to annotate the results obtained from ubiquitous data stream clustering to facilitate interpretation to enable real-time decision making.

The approach is divided into three stages. First stage is the online LWC of the data stream. In the second stage, principles of fuzzy logic in fuzzy labeling clustering algorithm is used to label the clusters in real time based on the domain or expert knowledge. The third stage is the fuzzy classification algorithm that uses the labeled clusters onboard the ubiquitous device. In this stage, new incoming stream data will be assigned to labeled clusters using fuzzy degree of membership. The following factors are considered when classifying an incoming data stream to one of the available classes: distance of new data item from centroid of class, the weight of the class (number of data elements in the class), and the distance of data item from other class centroid. The classification results include the degree of probability that the data item belonging to the particular class.

This fuzzy approach of clustering and classification algorithm is used and verified in the road-safety application to detect the drunk-driving behavior efficiently and accurately.

20.3.2.3 Frequent Pattern Mining

A frequent pattern is a pattern such as a set of items, sequences that occurs frequently in a dataset. For example, a set of items, such as milk and bread that are appearing frequently together in transaction database, is a frequent itemset. The other example is of a subsequence such as buying first a smartphone, and then a memory card and a screen guard, if it occurs frequently in an online shopping database history, then it is called a

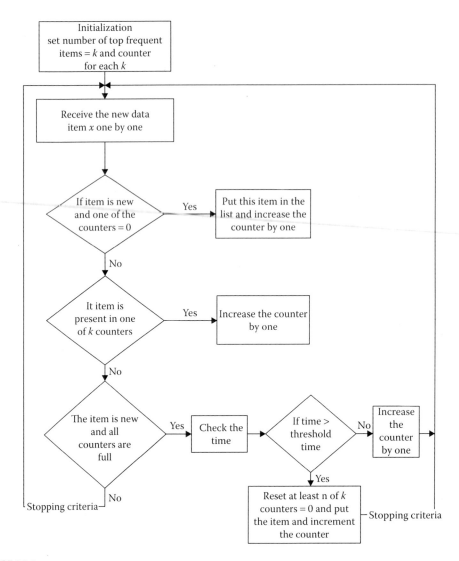

FIGURE 20.5
Lightweight frequent pattern mining (LWFPM) flowchart.

frequent sequential pattern. FPM is a key player in association mining, correlations, and many other relationships among the data. It also helps in clustering, classification and other data-mining tasks. Thus, it has become a vital task of data-mining research. The frequent pattern-mining algorithms were surveyed in the mobile or ubiquitous environments of limited resources for the various applications in mobile commerce and human activity recognition (Figure 20.5).

1. *Lightweight frequent pattern-mining (LWFPM) algorithm*: The lightweight frequent pattern-mining algorithm (LWFPM) sets the number of frequent items that will be estimated according memory available (Gaber et al., 2005). It follows the AOG approach. This number of frequent items changes over the time to cope with high stream data rate. On receiving the new data points one by one, algorithm tries

to find counter and increases the item for the existing registered items. The new item will be ignored if all the counters are occupied. Counters will also have to decrease by one till the LWF algorithm reaches some threshold time. The number of least frequent items will be ignored also, and their counters are reset to zero. If the new data item is similar to one of existing items in memory then counters will be increased by one. The major parameters that may affect the accuracy of algorithm are threshold time, number of calculated frequent items, and the number of items ignored. The flowchart of the LWF algorithm is shown in Figure 20.5.

2. *Frequent pattern mining (FPM) in personal sensing device*: The research of frequent pattern-mining algorithm for personal sensing device based data-mining systems is still ongoing and is at the initial stage. There are only two studies that adapt FPM algorithms in mobile commerce and human activity recognition.

Lu et al. (2012) proposed the Personal Mobile Commerce Pattern (PMCP-Mine) as the part of Mobile Commerce Explorer framework to find the personal shopping pattern online of mobile users. The approach of PMCP-Mine first mines the frequent transactions of a mobile user from a local purchase dataset and removes the infrequent transactions; updates the local transactions database. Finally, PMCP-Mine will predict the new pattern of transactions based on the local updated patterns of transactions.

Wang et al. (2012) proposed hierarchical approach based data mining techniques of emerging patterns (EP) for activity recognition using body sensor networks (BSN). It works at two layers. In the first layer, data are processed at BSN node and transmitted it to a mobile device for further analysis. At BSN node, lightweight algorithms are used for real-time recognitions of gesture. EP is a set of frequent items in one class and a set of infrequent items in other classes. The main assumption of this EP-based technique is that data instances of EP items are most likely belong to the corresponding EP class. The proposed algorithm of EP-based technique performs better than the single layer based and HMM-based algorithms.

The use of learning methods with FPM-based algorithms is gaining popularity in adaptive systems. For example, incremental learning methods are used at first stage to adapt continuously with concept drift. Second stage includes the discovery of context correlations using adaptive a priori algorithms (Kishore Ramakrishnan et al., 2014). It is important to note that the combination of FPM algorithms with different learning methods is investigated rigorously in resourceful environment, but their usefulness in resource-constrained environments (RCE) is still an unexplored research area. Therefore, there is need of detailed study of performance of FPM-based algorithms to find frequent patterns in RCEs.

20.4 Ubiquitous Data-Stream-Mining Applications for IoT

There are several UDM applications that are important for the IoT.

20.4.1 Smart Vehicles

For developing a smart traffic management system, it is necessary for our vehicles to be smart, by incorporating multiple sensors connecting them to the Internet and making it a part of system of systems. Such advanced vehicles will have the capability to sense, speak, and hear. It will have the capacity to collect data about the distance between surrounding vehicles, temperature of the engine. Thus, it will contribute to the ITS or smart traffic management system application of IoT.

Li et al. (2015) designed road vehicle monitoring system based on intelligent visual IoT. The system is able to identify and extract the visual tags of vehicles on the urban roads. Visual tags consist of license plate number, vehicle type, and color. The visual sensor nodes are installed on the urban roads and mobile sensing vehicles for collecting basic information of it. The design of wireless sensor network consists of different intelligent visual sensor nodes that extract the visual tags of vehicles on the roads and transmit video streaming. All the nodes are distributed on the urban roads, and vehicles together construct a large scale intelligent visual IoT. It can effectively help the traffic law enforcement officers to discover the blacklisted, stolen, and illegal vehicles. Each sensor node contains a high-resolution camera and an embedded processor, and a wireless link between these nodes and the central server is established. The central server can receive and analyze the visual tags transmitted by all the nodes. The estimated route of the target vehicle can be chronologically linked with the characteristics of its visual tag that can be associated and mined from the central database. The data mining related tasks can also be done on the sensing vehicles to collect localize important information. The concept of a Smart vehicle can play a major role in road safety by performing real-time data analysis of sensory data in moving vehicles for accident prevention. One may also think of drunk-driving detection through onboard analysis of data streams in moving vehicle.

Horovitz et al. (2005) integrate fuzzy approach in the LWC and classification algorithms for detecting the drunk-driving detection. They have simulated the overall scenario by analyzing the data of blood alcohol concentration and collecting the different characteristics of the drivers such as reaction time to peripheral signals (sec), correct responses to peripheral signals (number), lane position deviation (ft), speed deviation (mph), and times over speed limit (number), collisions (number). Some rules are defined to detect the unusual events. If there are *higher number of correct responses to peripheral signals*, then driver will be the less drunk and if there are higher number of *times over speed limit* then the driver will be more.

20.4.2 Remote Health Monitoring

There are many uses of the systems and products that connect to the IoT, which are changing business in health care industry. Both patients and health care providers benefit from IoT making a bigger presence in health care. Health care IoT comes handy by means of mobile medical applications or wearable devices that can capture health data of patients. Hospitals can use IoT to keep tabs on the location of medical devices, personnel, and patients. Internet-connected devices have been introduced to patients in various forms. Data might come from electrocardiograms, fetal monitors, blood glucose levels, or temperature monitors; tracking this health information is important for patients. Follow-up interaction with a health care specialist is required for many of such measures. This creates a need for smarter devices to deliver more valuable data, reducing the need for direct patient-physician interaction.

Remote health monitoring is an application UDM system can be used to provide 24 × 7 monitoring of patients wearing biosensors. Typical remote health-monitoring applications use the smartphone as a communications device to transfer the data collected from the patients wearing the biosensors. Sensors collected the real-time data about the health condition of a patient. Nowadays, earphones of smartphones can read a patient's blood pressure; accelerometer could detect the patient's physical activity and thus smartphone is used to detect the consciousness of the patient. The sensory data can be mined to find behavioral patterns of the patient. The medical staff could also be equipped with smartphones so, if any critical conditions of patient will send alerts to the nurse. Thus, it is possible to use UDM to analyze the data immediately onboard mobile device and send immediate alerts to the nurse when required. The algorithms need to be scalable and cost-efficient in terms of energy consumption and other computational resources. The LWC algorithm that is discussed above can be used for remote health-monitoring application.

Hassanalieragh et al. (2015) discussed the opportunities and challenges for IoT in realizing this vision of the future of health care monitoring. They also reviewed the different approaches for integrating the remote health-monitoring technologies into the clinical practice of medicine. Here, the LWC type algorithm is used for onboard mining of data stream with fuzzy logic rules.

20.4.3 Public Safety

IoT technologies are transforming our living area into smart environment typical examples such as smart city, smart building. Variety of sensors is deployed in smart environments for capturing and analyzing the data in a smart way, adapting to the behavior of people. Public safety is one of the challenges of smart city. As the sensors are deployed in smart cities to monitor the pulse and report situations all around the cities in an 24 x 7 basis, the safety problems can be identified early and services can be localized quickly that will improve the safety of the citizens. Many studies show that safety issues can be addressed even with large cheap sensors, for example in crowd detection, fire breakout, blocked street, or a burst pipe (Soliman et al., 2010).

There are many challenges in that have been taken into account for public safety such as establishing dynamic communication channels, dealing with big data in real time, and preserving privacy issues (Vermesan & Friess, 2015). The real time data analysis is the first important concern in public safety because critical situation needs to be detected early so that authorities will have enough time to take action to provide safety solutions. Here, various operations of data mining need to be done onboard ubiquitous device in order to transfer the results or data for further analysis.

20.5 Challenges in the Ubiquitous Data Stream Mining toward the Next Generation of IoT

With the rapid development in IoT, big data technologies, the fundamental challenge is to explore the large amount of stream data and extract the meaningful information for further actions. In the ubiquitous data stream systems, following challenges need to be addressed.

1. Large amount of stream data need to collect, mine, and analyze at real-time onboard ubiquitous device thus need the mining data stream techniques in ubiquitous environment.
2. Data come from heterogeneous distributed sensor sources, and communication is done with different devices, the data format may be different or unstructured thus need preprocessing techniques.

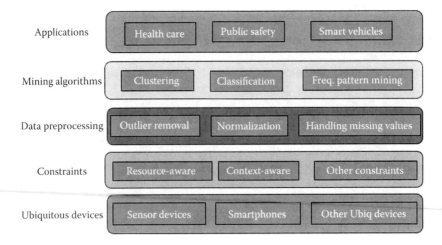

FIGURE 20.6
Big data mining system for ubiquitous devices in IoT.

3. The knowledge is deeply hidden in the data and knowledge extraction is not straight forward so need to analyze different properties of data and find the associations in the data.

4. The computational resources in the ubiquitous environments such as battery power, memory are limited. Thus big data techniques have to be performed well in the RCE of the ubiquitous computing. The UDM methods should also be managed well. The accuracy of method should be compromised according to the situations depending on the specific types of applications. There is need of context-aware systems in today's IoT applications.

A big data mining system for ubiquitous devices in IoT is suggested as shown in Figure 20.6.

20.6 Conclusion

This chapter presents current progresses and challenges associated with UDM in running application; considering time, memory, and resource constraints such as battery power in the ubiquitous devices. Nowadays, ubiquitous computing environment is popular because of its ever-increasing computational power and capacity of wireless networking of the small devices such as sensor devices, smartphones. These ubiquitous devices play an important role in the various IoT applications such as health care, smart homes, ITSs. This chapter aims to present the detailed survey of recent research developments of data stream mining in the ubiquitous computing environment and discuss the challenges of UDM in IoT.

The different UDM methods such as classification, clustering and FPM in RCEs are reviewed. Also the challenges of UDM are presented, which needs to be addressed in the future research of IoT. Big data mining system for ubiquitous data mining in IoT is given. There is a need for further study of this ubiquitous data mining for the next generation of IoT.

References

Ang, H. H., Gopalkrishnan, V., Zliobaite, I., Pechenizkiy, M., & Hoi, S. C. (2013). Predictive handling of asynchronous concept drifts in distributed environments. *IEEE Transactions on Knowledge and Data Engineering*, 25(10), 2343–2355.

Bifet, A., Holmes, G., Kirkby, R., & Pfahringer, B. (2010). Moa: Massive online analysis. *Journal of Machine Learning Research*, 11(May), 1601–1604.

Bondu, A., Boullé, M., & Grossin, B. (2013, August). *SAXO: An optimized data-driven symbolic representation of time series.* Neural networks (IJCNN), The 2013 international joint conference on (pp. 1–9). IEEE.

Chao, C. M., & Chao, G. L. (2011). *Resource-aware high quality clustering in ubiquitous data streams.* Proceedings of the 13th International Conference on Enterprise Information Systems, Beijing, China (Vol. 1, pp. 64–73).

Domingos, P., & Hulten, G. (2001, June). A general method for scaling up machine learning algorithms and its application to clustering. In *ICML* (Vol. 1, pp. 106–113).

Figo, D., Diniz, P. C., Ferreira, D. R., & Cardoso, J. M. (2010). Preprocessing techniques for context recognition from accelerometer data. *Personal and Ubiquitous Computing*, 14(7), 645–662.

Gaber, M. M., Gama, J., Krishnaswamy, S., Gomes, J. B., & Stahl, F. (2014). Data stream mining in ubiquitous environments: state-of-the-art and current directions. *Wiley Interdisciplinary Reviews: Data Mining and Knowledge Discovery*, 4(2), 116–138.

Gaber, M. M., Krishnaswamy, S., & Zaslavsky, A. (2005). On-board mining of data streams in sensor networks. In Advanced methods for knowledge discovery from complex data (pp. 307–335). London: Springer.

Gama, J., & May, M. (2011) Ubiquitous knowledge discovery. *Intelligent Data Analysis*, 15, 1.

Gubbi, J., Buyya, R., Marusic, S., & Palaniswami, M. (2013). Internet of Things (IoT): A vision, architectural elements, and future directions. *Future Generation Computer Systems*, 29(7), 1645–1660.

Hassanalieragh, M., Page, A., Soyata, T., Sharma, G., Aktas, M., Mateos, G., & Andreescu, S. (2015, June). Health monitoring and management using Internet-of-things (IoT) sensing with cloud-based processing: Opportunities and challenges. In *Services Computing (SCC), 2015 IEEE International Conference on* (pp. 285–292). IEEE.

Horovitz, O., Krishnaswamy, S., & Gaber, M. (2005, October). A fuzzy approach for interpretation and application of ubiquitous data stream clustering. In *Proceedings of Second International Workshop on Knowledge Discovery in Data Streams, in conjunction with the 16th European Conference on Machine Learning and the 9th European Conference on the Principals and Practice of Knowledge Discovery ECML/PKDD.*

Kishore Ramakrishnan, A., Preuveneers, D., & Berbers, Y. (2014). Enabling self-learning in dynamic and open IoT environments. *Procedia Computer Science*, 32, 207–214.

Li, Q., Cheng, H., Zhou, Y., & Huo, G. (2015). Road vehicle monitoring system based on intelligent visual Internet of things. *Journal of Sensors*.

Lu, E. H. C., Lee, W. C., & Tseng, V. S. M. (2012). A framework for personal mobile commerce pattern mining and prediction. *IEEE transactions on Knowledge and Data engineering*, 24(5), 769–782.

Mirsky, Y., Shapira, B., Rokach, L., & Elovici, Y. (2015, May). pcStream: A stream clustering algorithm for dynamically detecting and managing temporal contexts. In *Pacific-Asia Conference on Knowledge Discovery and Data Mining* (pp. 119–133). Springer International Publishing.

Nguyen, H. L., Woon, Y. K., & Ng, W. K. (2015). A survey on data stream clustering and classification. *Knowledge and information systems*, 45(3), 535–569.

Osmani, V., Carreras, I., Matic, A., & Saar, P. (2014). An analysis of distance estimation to detect proximity in social interactions. *Journal of Ambient Intelligence and Humanized Computing*, 5(3), 297–306.

Pinage, F. A., dos Santos, E. M., & da Gama, J. M. P. (2016). Classification systems in dynamic environments: An overview. *Wiley Interdisciplinary Reviews: Data Mining and Knowledge Discovery*, 6(5), 156–166.

Plötz, T., Hammerla, N. Y., & Olivier, P. (2011, July). Feature learning for activity recognition in ubiquitous computing. In *IJCAI Proceedings-International Joint Conference on Artificial Intelligence* (Vol. 22, No. 1, p. 1729).

Shah, R., Krishnaswamy, S., & Gaber, M. (2005, October). Resource-aware very fast k-means for ubiquitous data stream mining. In *Proceedings of the Second International Workshop on Knowledge Discovery in Data Streams, held in conjunction with ECML PKDD 2005.*

Soliman, H., Sudan, K., & Mishra, A. (2010, November). A smart forest-fire early detection sensory system: Another approach of utilizing wireless sensor and neural networks. In *Sensors, 2010 IEEE* (pp. 1900–1904). IEEE.

Stahl, F., Gaber, M. M., Aldridge, P., May, D., Liu, H., Bramer, M., & Philip, S. Y. (2012). Homogeneous and heterogeneous distributed classification for pocket data mining. In *Transactions on Large-Scale Data-And Knowledge-Centered Systems V* (pp. 183–205). Berlin, Germany: Springer.

Vermesan, O., & Friess, P. (Eds.). (2015). *Building the hyperconnected society: Internet of things research and innovation value chains, ecosystems and markets* (Vol. 43). River Publishers.

Wang, L., Gu, T., Tao, X., & Lu, J. (2012). A hierarchical approach to real-time activity recognition in body sensor networks. *Pervasive and Mobile Computing, 8*(1), 115–130.

21

IoT toward Efficient Analysis of Aging, Cardiometabolic, and Neurodegenerative Diseases—An eHealth Perspective

Leandro Cymberknop, Parag Chatterjee, Diego Dujovne,
Luis Romero, and Ricardo Armentano

CONTENTS

21.1 Introduction

Internet of Things (IoT) has been a path-breaking technology backed up by many handshaking research areas to establish a high-end connectivity and communication between several mutually related devices to share information and interact, toward a better user experience. In the field of health care the task of IoT is not only to proffer a truly efficient and personalized health care to the users but also to redefine the health care system by connecting all the stakeholders and the state-of-the-art technologies making the most of the information shared across the closely communicating devices using the IoT platform.

Cardiovascular diseases (CVD), cerebrovascular accidents, and cancer have always been prevalent largely in the elderly population. This includes an increasing incidence of chronic conditions, such as osteoarthritis, chronic airways disease, and diabetes combined with sedentary lifestyles (e.g., obesity). Thus, major health care challenges are posed, primarily focusing on prevention, early detection, and minimally invasive management of such diseases. As a result, new technologies are applied to assist in patient monitoring and care. Accordingly, wearable systems offer users the ability to interact with other tools and physical objects around them, being capable of continuously monitoring vital signs and electrical signals generated by heart and brain, including posture and physical activities (Andreu-Perez et al., 2015; Chan et al., 2012; Krohn et al., 2016). Here, IoT constitutes an emerging paradigm in which everyday objects, devices, and sensors exchange data with little to no human intervention, exploiting the advanced ways of connectivity and computing ability.

This chapter would be providing an approach of IoT in some sectors of health care such as CVD and Parkinson's disease (PD), where this technology can redefine the entire health care experience.

21.2 IoT for Cardiovascular Diseases and Aging

Heart diseases stand as one of the leading causes of human fatalities, in which myocardial infarction constitutes the primary killer worldwide. Risk of several diseases such as coronary heart disease, ischemic stroke, and diabetes mellitus increases because of alterations of blood pressure (BP) and cholesterol. In this perspective, when a chronic disease is diagnosed, continuous and long-term monitoring is required (Kakria et al., 2015; Krohn et al., 2016). Therefore, cardiovascular-focused mobile health technologies count helpful in terms of noninvasive (or implantable) devices that could provide essential physiological information such as continuous BP measurements, continuous multilead electrocardiogram (ECG) monitoring (by means of patches or shirts), transthoracic impedance and cardiac output, arterial oxygen saturation, among others (Chan et al., 2012; Steinhubl and Topol, 2015). Some approaches can address to some diseases as shown in Figure 21.1.

21.2.1 Arterial Indicators: Blood Pressure and Stiffness

BP constitutes a powerful CVD risk factor, which is responsible in part for various cardiovascular events. Traditionally, two specific points of the BP curve, peak systolic BP and end-diastolic BP, are used to define the cardiovascular (CV) risk factor (Nichols et al., 2011). There is also evidence that nocturnal BP is one of the stronger predictors, even when controlling for clinical BP (Steinhubl and Topol, 2015). The determinants of the BP waveform are the cushioning capacity of arteries (influenced by arterial stiffness) and the timing and intensity of wave reflections. It can be seen as the result of the combination of a forward-pressure wave (coming from the heart) and a backward wave (returning from particular sites characterized by specific reflection coefficients) that propagates (in the opposite directions) at a given speed, defined as *Pulse Wave Velocity* (Nichols et al., 2011). Here, wearables are used for self-monitoring and preventing health conditions such as hypertension (systolic BP become persistently elevated) and stress. Wireless upper-arm (or even finger's) BP monitors can be used for extraction and transmission of BP information, toward the IoT approach (Kakria et al., 2015). However, the periodic inflation of the cuff is still annoying and painful in long-term use (Noh et al., 2014). Furthermore, conventional sphygmomanometry underestimates the effect of arterial stiffening on central BP (CBP, which is measured at the output of the left ventricle) and provides less information concerning

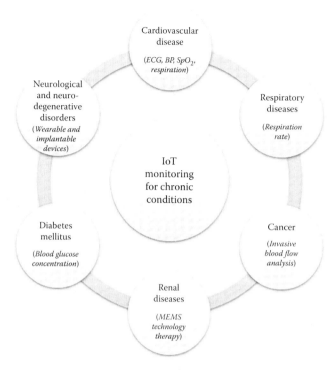

FIGURE 21.1
Chronic conditions monitoring and IoT approach.

systemic arterial stiffness. Several studies have demonstrated that CBP is a better predictor of carotid intima-media thickness, restenosis after coronary angioplasty, coronary artery disease severity, and mortality in end-stage renal disease (Nichols et al., 2011). As this methodology is still challenging, with the help of nonobtrusive wearable devices, monitoring beat-to-beat BP (particularly CBP) 24 h per day could become simple and routine (Steinhubl and Topol, 2015). In this sense, pulse transit time (PTT, the time delay for the pressure wave to travel between two arterial sites) can provide the basis cuff-less BP measurement. Generally, aortic PTT is determined from BP waveforms for large artery stiffness quantification and improved cardiovascular risk stratification. Physical models based on the Moens-Kortweg and Bramwell-Hill equations are usually applied to estimate BP from PTT, parameters of which are adjusted by means of traditional BP cuff measurements (Laurent et al., 2006; Mukkamala et al., 2015).

A simplification of the process has been proposed (named pulse arrival time, PAT), in which PTT is assessed by the time delay between the R-wave of an ECG waveform and the distal arterial waveform foot (Walsh et al., 2014). However, PAT is equal to the sum of PTT and the preejection period, being thus dependent on the ventricular electromechanical delay and isovolumic contraction phase (which can vary with contractility and afterload of the heart). As a result, PAT may not be an adequate surrogate for PTT as a marker of BP, exceeding the food and drug administration (FDA) BP bias and precision error (Zhang et al., 2011). Another approach, focused specifically on CBP assessment, is based on ECG combined with ballistocardiogram recordings (a measurement of the body response caused by the blood ejected during the cardiac systole) to provide an estimation of PTT within the aortic domain. The use of ballistocardiogram recordings to monitor heart rhythm is advantageous in terms of user compliance as very lightweight sensors are utilized (e.g., accelerometers) (Fierro et al., 2016; Noh et al., 2014). It has to be noted that next to the accuracy of readings, the user's compliance with continuous usage is what guarantees the success of a ubiquitous-health system (Noh et al., 2014). In general terms, accurate estimation of PTT (in terms of wearable systems design) is the main determinant in BP, CBP, and stiffness assessment. A perfect complement to this technology could be the pocket mobile echocardiography, a small device capable to provide high-resolution two-dimensional ultrasound imaging, color Doppler, and measurement capabilities, thus revolutionizing the bedside and outpatient management of cardiac patients (Walsh et al., 2014).

21.2.2 Cardiovascular Disease: Monitoring Aging

A considerable share of elderly people has reduced mobility due to age-related diseases leading to a greater need for care and assistance. High age related incidence of CVD arises in part due to broader aging processes, including accumulating morbidities, diminishing homeostasis, and prolonged injurious effects of CV risk factors (Forman et al., 2001). Vascular stiffening predisposes to functional decrements and ultimately to ischemia, heart failure, arrhythmia, and other CV disorders (Forman et al., 2001). An example of CV disease remote monitoring for the elderly is proposed by Armentano and Kun (2014) in which complimentary, gold-standard, noninvasive, and low-cost techniques are applied to obtain parameters used to characterize the structural and functional vascular state. Sphygmomanometry, applanation tonometry, and ultrasound techniques are used to measure the presence of atherosclerotic plaques, intima media thickness variations, instantaneous BP and vessel diameter waveforms, and local and global vessel distensibility (Figure 21.2). It becomes clear that pocket mobile echocardiography devices should be combined with continuous BP waveforms estimations and powerful signal and image-processing systems to achieve this proposal.

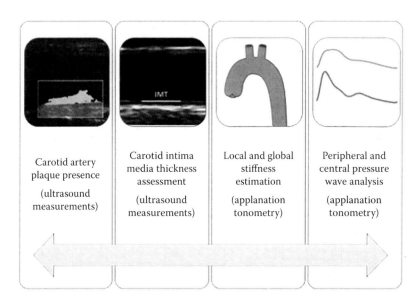

FIGURE 21.2
Integrated approach to characterize the structural and functional vascular state for the elderly.

21.3 IoT in Neuro-Cognitive Systems

After Alzheimer's disease, PD is the second most common neurodegenerative disorders of the central nervous system and affects motor skills and speech (Jankovic, 2008). The decrease in quality of life associated with this disease affects adversely both the patients and their families.

Obtaining an accurate diagnosis is critical, as it is now strongly based on clinical observation and information provided by the patient through his personal impressions and records. Although the clinical symptoms of PD are quite typical, in many cases it is necessary to ensure that these characteristics do not respond to disorders associated with other neurodegenerative diseases.

The primary biochemical abnormality in PD is the deficiency of dopamine due to degeneration of neurons in the substantia nigra (Figure 21.3) pars compact (DeLong and Wichmann, 2007). The four characteristics of motor features associated with PD are bradykinesia, rest tremor, rigidity, postural, and gait impairment. It should be noted that these manifestations do not always occur together in each patient.

Bradykinesia refers to slowness of movements with a progressive loss of amplitude or speed during attempted rapid alternating movements of body segments (Jankovic, 2008; Marsden, 1994). It is crucial to distinguish true bradykinesia from simple slowness. In fact, failing to acknowledge this is a major source of misdiagnosis.

Rest tremor is a rhythmic oscillatory involuntary movement that arises when the affected body part is relaxed and supported by a surface, thus removing the action of gravitational forces (Bain, 2007). It vanishes with active movement. In PD, rest tremor frequency is usually in the low-to-mid range (3–6 Hz), whereas the amplitude is quite variable, from less than 1–10 cm wide.

Rigidity refers to an increased muscle tone felt during examination by passive movement of the affected segment (limbs or neck), involving both flexor and extensor muscle groups (Jankovic, 2008). This resistance is felt throughout the full range of movement and does not increase with higher mobilization speed.

With regards to postural and gait impairment, Parkinsonian patients tend to adopt a stooped posture, owing to the loss of postural reflexes, a major contributor to falls (Jankovic, 2008; Sethi, 2008). The *pull test* is performed to assess postural stability; the examiner stands behind the supine patient who is previously warned of the *pull* applied to her shoulders, then allowing her to step back to regain balance—some patients will fall without any sort of postural response.

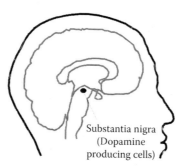

Substantia nigra
(Dopamine
producing cells)

FIGURE 21.3
Location of the substantia nigra in the brain.

21.3.1 Focal Issues toward Treating Parkinson's Disease

In the current medical practice that is widely used, assessment of PD motor disabilities is based on neurological examination during patient's visits to the clinic and home diaries that the patient or the caregiver keeps. However, the short-time examination may not reveal important information to the neurologist, whereas data from the daily diaries are highly subjective as they rely on the patient's memory and perception of his own symptoms. Moreover, the pattern and severity of PD symptoms may vary considerably during the day. So, measurements of motor functions made in the clinic may not reflect in an accurate way the actual motor disabilities experienced by the patients in their daily life.

The first step in the evaluation of a patient with tremor to determine if it has a Parkinsonian origin or not is to categorize the tremor based on its activation condition, topographic distribution, and frequency (Budzianowska and Honczarenko, 2008). Establishing the underlying cause is very important because prognosis and specific treatment plans vary considerably.

The most regularly used instrument for the assessment of PD is the revised version of the unified PD rating scale (UPDRS) (International Parkinson and Movement Disorder Society-UPDRS). The Movement Disorder Society-UPDRS stands for numerous clinical scales and questionnaires that have been regularly used in clinical routine and in studies to evaluate the presence, severity, and progression of PD symptoms. However, the disadvantages of these methods, such as high expenditure of time, investigator and location dependency, limitations in repeating the assessments regularly, and the lack of quantitative outcomes, motivate to search additional, complementary or even supplementary quantitative, and objective assessment strategies to manage PD.

21.3.2 Use of Wearable Systems for Parkinson's Disease

Quantitative assessment and management of PD using new technology-based tools, worn or operated by patients preferentially in the domestic environment, have attracted considerable attention during recent years.

Currently, assessment of motor abnormalities in PD is mainly clinical, based primarily on the UPDRS (part III). Unfortunately, this scale presents intra- and interobserver differences, and its use is limited to the patient's visit to the hospital. To overcome these limitations and difficulties, ambulatory monitoring of PD motor and nonmotor symptoms method or procedure is required.

It is expected that the development of an effective technical solutions to measure specific PD features for clinical management in either a controlled or an uncontrolled environment would be possible with current technology.

Using various techniques, several groups have proposed objective methods to detect and quantify tremor, bradykinesia, and postural instability (Rajaraman et al., 2000; Salarian et al., 2007). Recently, there has been a growing interest in applications of body-fixed sensors or very close to the body over clothes, that is, wearable sensors (WSs), for long-term monitoring of these patients. These sensor units can perform not only motion and physiological data sensing but also some low-level processing.

Although having in mind the possibility of some mean of integration or net of WSs, there has been detailed several aspects of such sensors that need to be taken into consideration (Pastorino et al., 2013). These aspects are resumed in Table 21.1.

Pasluosta et al. (2015) have proposed a new methodology to estimate the outcomes of the Pull Test using inertial sensors attached to patient's shoes. Other research groups have

TABLE 21.1

Aspects to be Taken into Consideration at the Time of Integrate a Network of Wearable Sensors

Aspect	Desired Conditions to Meet
Size	The size of a wearable sensor needs to be considerably small, which is even in the range of micrometer, predominantly in the MEMS-systems. However, small size restricts enough room for power sources and storage
Storage	To compromise on size, small sized sensors have extremely limited storage capacity, which makes it necessary to transmit data to larger repositories or to the cloud
Processing	Due to size and power constraint, processing tried to be minimal in a sensor. However, newer technologies intend to reduce the turnaround time and hence recent trends focus on edge computing for processing the data close to its source (i.e., the sensor)

been used WSs placed at the patient's truck and lower back to measure postural sway as a mean for estimating postural instability (Foerster et al., 1999; Patel et al., 2008).

A group of researchers from Harvard University has developed a wearable, wireless-sensor platform for motion analysis of patients (Lorincz et al., 2009). The system uses eight wireless nodes equipped with sensors for monitoring movement and physiological conditions. In contrast to previous systems intended for short-term use in a laboratory, this system was designed to support long-term, longitudinal data collection on patients in hospital and home settings. Although some of the processings are done at each individual node, a base station is needed to collect data from sensors. This point is a limitation if you want to cover a wide area of circulation of the patient, as it will be limited to the near vicinity of the base station.

Pastorino et al. (2013) have proposed a WS network, called perform system, that uses four triaxial wearable accelerometers (one in each patient's limb) and one accelerometer and gyroscope wear on a belt used to record body-movement accelerations and angular rate. All these information are received and stored on an SD card. The data collected then need to be downloaded to a PC to identify and quantify the patients' symptoms.

Although several technologies and systems have been proposed for the monitoring of some motor symptoms associated with PD, still there is a lack in having a system that allows recording information in the patient remotely using WSs, including the ability to perform this task in an outdoor environment, as well as providing objective parameters near real time to take a certain action (e.g., adjust the dosage of medication delivered to the patient; Figure 21.4). This leads to the development of a WS network that allows, with the

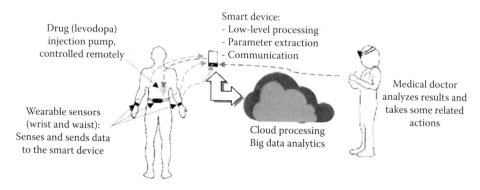

FIGURE 21.4

A possible configuration of an ambulatory monitoring system, with a feedback loop for drug delivery adjustment.

minimum number of sensors, the registration of parameters to achieve a precise diagnosis and following of patients with PD (Romero et al., 2016). A system with such characteristics should include a mobile device (e.g., smartphone or tablet) that would be in charge of making the data collection from the sensors and the parameter extraction and would also upload this information to the cloud, in which additional processing should be done. At the other end of the link, a medical doctor would analyze the results from cloud processing (big data analytics) and could, for example, adjust the dosage of medication needed in a remote way.

21.4 Networking for the Medical IoT Devices

Engineering experience says that one of the key elements in IoT (Da Xu et al., 2014) is the ability to connect, create, and maintain networks autonomously, with minimal user intervention. Connected things are meant to be alone, managing to fulfill their task and to last as long as possible, many of them in harsh environments. Even though IoT networks can be wired, wireless is the most common transmission media, because of the deployment and relocation flexibility it provides and the cost reduction in cables. However, the wireless channel are not free from problems, such as interference and propagation issues, and bandwidth restrictions from local regulations. Following the geographical distribution of the nodes according to the specific measurement requirements from the application, the generated topology has not only to deal with communication impairments but also with range and limited energy challenges. IoT is a very different ecosystem than the original Internet; moreover, given the exponential growth of the number of connected things, they will soon outnumber the amount of people connected to the Internet, raising scalability problems.

From an evolutionary perspective (Mainetti et al., 2011), the ability to create networks from number-constrained devices has progressed during the last 25 years from research proposals at the beginning to commercially available solutions nowadays. Brainstorming in the research community resulted in hundreds of ideas spanning from physical to application network layers; but only the selection of the simplest and efficient ones was done. Those whose design was thought with the whole networking stack interaction in mind and engineering criteria are the current winners. The future is an open opportunity to create new ways to improve both communications and networks together in an effort to enable universal interoperability among all constrained and mainstream devices. It is the task of standardization boards (and the Internet Engineering Task Force in particular) to ease integration and adapt each technology to a single protocol: IPv6 (Ishaq et al., 2013).

21.4.1 Use of IoT to Connect Heterogeneous Devices

Connecting all devices on the IoT is ideal that is pushing industry today to overcome physical limitations, legacy requirements, scalability considerations, and standardization efforts. IoT devices use different type of media to interconnect, betting either on wired solutions on standard cables, or to wireless links, based on traditional wireless local area network (WLAN) infrastructure, Cellular Data (Currently 3G, 4G, and future 5G), and satellite links such as Iridium for remote access.

21.4.1.1 Paradigms

Three main paradigms in IoT for health applications in terms of connectivity are as follows: IoT devices for personal mobile use, such as portable cardiac monitors; devices for home use, such as home activity sensors for the elderly; and integrated systems for medical facilities, such as hospital-wide network deployments. (Baig et al., 2015; Chen et al., 2013).

For the first case, these devices are aimed to be off-the shelf items for massive applications, commonly used during daily activities, either with a body-area network (e.g., to monitor movement using accelerometers) or for cardiac or BP ambulatory instrumentations. These devices are meant to transmit the information via low-bandwidth wireless links, mostly based on cellular data where available. If worldwide connectivity is needed, links based on satellite connections such as Iridium are also available. A special subclass of this case covers temporary deployments to serve special events in which monitoring is required, but a standard infrastructure cannot be installed in a short time. For example, in catastrophe scenarios, such as earthquakes or flooding situations, a portable system must be available and completely functional in a very short time, so as to allow remote access to patient's information and telemedicine links.

The second case implies a system with a local range for multiple IoT interconnected devices at home and a wide-area network access using a third-party Internet service provider. Nowadays, the typical path of the information flow is to send the data to a cloud-based service, which is generally a part of the offer with available analytics both for the user and for the doctors. The main difference with the first case is the concept of having multiple sensors in the same area (namely, home or office where the patient interacts) that corresponds to an integrated view of the behavior of the user, which may be complemented with a body-area network. What is expected in this kind of context is synchronization of the generated data against a single-time source, relative high bandwidth availability to fulfill online video analysis, cardiac data transmission, and environmental sensing data. This type of traffic is highly variable; there is a baseline traffic of constant and periodic data transmitted and a variable component from a different kind of events. If a remote monitoring system is established to react, the balance between local (home or office) processing, called Fog computing, and Cloud-based access must be balanced to provide more reliability in case the domestic Internet service fails. Failure should lead to the use of backup redundancy to a lower bandwidth network, for example, cellular data, which may lead to data loss and reaction delay. It is also relevant to notice that home-based networks do not provide industrial reliability and real-time capabilities, so the design and operational policies must agree with this disadvantage.

The third case includes systems built for hospitals or clinics for remote (in-building) monitoring purposes. These are both mobile devices that are worn by the patients using body-area networks and devices installed on the infrastructure, using both wired and wireless links. Although the devices can be classified as providing critical (e.g., intensive care patients) and noncritical (e.g., temperature monitoring in hallways) information, both share a common infrastructure that belongs to the institution. The most common topology is built around a wired backbone and a mixed wireless/wired capillarity. To guarantee a minimal quality of service to each data flow, data sources must be classified and prioritized accordingly, considering the maximum allowed bandwidth leaving part of the links available to hold peak bandwidth for event-based traffic. Regarding user mobility to locate patients and professionals inside the building, for example, wireless infrastructure and indoor-location services shall be also considered, taking into account emergency situations.

Nowadays, the IoT is a mature technology being applied globally, and in this kind of real applications environments, optimization is based on standardized communication algorithms that can interoperate among nodes. The industry claims for efficient solutions but not at the expense of changing every single deployed device. As a consequence, evolutionary solutions should arise which should take into account legacy device compatibility until the remaining nodes are replaced. So, applied optimization can be done on top of current standard systems, which provide freedom to the implementer in different ways, for example, with the use of objective functions for RPL (Thubert, 2012) routing or by using scheduling functions (Palattella and Accettura, 2015) on 6P resource allocation protocols. Another advantage of standardization is to simplify performance comparison between protocols and design improvements, as they run on the same reference platform (Thubert, 2012).

Finally, another approach for practical performance evaluation is the concept of Lean Sensing (Martinez et al., 2015; Martínez et al., 2015). This idea describes IoT design taking into account energy consumption and network life from the point of view of energy harvesting and establishes a number of limits based on real-field measurements. Another relevant topic of lean sensing is data-compression performance evaluation, showing the ranges in which data compression is efficient for transmission and raw-data transmission is still a valid solution.

The joint effort from both academic and development communities for more than 25 years of evolution since the inception of the first ideas on WSNs has created a solid base where the open problems have also changed; security risks, scalability, management, applications, connectivity, enhanced networks, and energy constraints have become the most relevant drivers of the new wave.

21.5 Data Management and Intelligence

Whatever level of efficiency we use to connect all the devices for sharing data and mutual communication, the vital part of this aspect is to extract the requisite information from these data. The dimensions of the data generated being humongous, big data mechanisms count inevitable for managing these data efficiently and making sense of it by extracting significant information out of these data. This implies, computer and data scientists who write such computational algorithms have to work closely with clinicians to accurately quantify various health conditions and risk factors (Piwek et al., 2016).

Hospital information systems are an integral part of hospitals these days. However, this system primarily focuses on managing the patient-related records, but the analysis of these data is not profusely used directly in medical treatment. IoT in health care primarily intends to bring the data analysis directly beneficial in terms of health care and predictive treatment as well (Figure 21.5).

The data generated in the hospital are multifarious. The variety of the data linked to a patient makes it difficult to manage using any single traditional-data management scheme. However, broadly classifying the data based on its source, the entire data linked to a patient can be classified as personal data (supplied by the patient), machine-generated data (automated medical reports, etc.), and data from the doctor or someone related to the treatment. In the common scenario, personal data of a patient along with other administrative information of the hospital linked to the patient (e.g., billing,

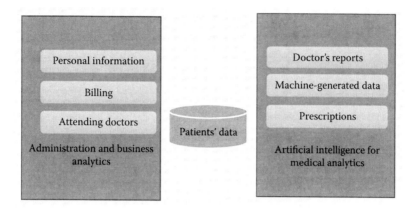

FIGURE 21.5
Medical intelligence versus business intelligence in a typical hospital data-management scenario.

attending doctors) are mostly in structured formats and are easily handled using traditional database-management systems. But medical data (mostly generated from the devices) are more complex and cannot be structured or generalized easily because of its heterogeneous nature and immense volume (containing images, videos, and other reports). IoT puts thrust on analysis of this highly heterogeneous and voluminous medical data because the key of comprehensive and predictive health care lies in these data. Hence, the role of big data analytics becomes inevitable to analyze and make sense of this medical data.

21.5.1 Intelligent Wearables and Machine Learning

Wearables could provide a platform for at-home management of long-term chronic conditions (Piwek et al., 2016). Though analysis of the medical data seems useful in obtaining interesting information regarding a patient, it covers only the first half of the system. To make the most of this information, this needs to be fed to the machine so that it can act accordingly. The task of making the system intelligent by learning from the humongous data generated invokes the need of machine learning and artificial intelligence. Intelligent machines change the efficiency of entire health care system dramatically as the machine-generated reports could be analyzed easily by the intelligent machines with a minimum human intervention. For example, deploying an advanced intelligent image-processing system to analyze videos for PD-affected patients could enhance the efficiency to a large extent. Some phases of treatment varies a lot depending on patients, and analysis of lifestyle data stands very vital in this aspect. Apart from learning from the data generated at the hospital, the system needs to learn from the lifestyle data, remotely generated at the patient's end by the IoT-based wearable devices.

The rapid development of hardware such as biosensors has overshadowed the slower development of the software needed to manage the enormous amount of data that these biosensors can generate. Mobile technologies will go well beyond providing data; physicians already know how to treat (e.g., BP at an office visit) and instead will provide brand new-data streams (e.g., continuous BP during daily activities) that will require tremendous bioinformatics capabilities to eventually understand. The capacity for technological advances in the way we are able to process large amounts of data using predictive analytics and network biology will usher in the era of personalized medicine and the ability to predict clinically significant events well before they occur (Walsh et al., 2014).

A lot of global players have already started advancing in this field of using artificial intelligence in health care. Google has launched its Google DeepMind Health project to mine the data of medical records for providing better and faster health services. IBM Watson launched its special program for oncologists to provide clinicians evidence-based treatment options. Watson for oncology has an advanced ability to analyze the meaning and context of structured and unstructured data in clinical notes and reports that may be critical to selecting a treatment pathway. Then, by combining attributes from the patient's file with clinical expertise, external research, and data, the program identifies potential treatment plans for a patient. Moreover, advancements are ongoing in developing cognitive assistants to doctors in simplifying their task of analysis and decision making. IBM's medical sieve is an ambitious long-term exploratory project to build a next generation *cognitive assistant* with analytical, reasoning capabilities, and a wide range of clinical knowledge. Medical sieve is qualified to assist in clinical decision making in radiology and cardiology. To spot and detect problems faster and in a more reliable way, the cognitive health assistant can analyze radiology images. This leads to a future where radiologists can only look at the most complicated cases in which human supervision is useful (Mesco, 2016).

In this perspective, IoT enables the seamless platform of interaction between the medical devices. However, in the quest of making the system intelligent, the approach can be three fold—mining the medical records, designing the treatment plans, and cognitive assistance.

21.5.2 In-Device-Automated Reporting and Analysis

To extend the reach of health care, it is essential to make sense of the lifestyle data of a person. It helps in either ways, for predictive measures and for supporting a specific treatment. The concept of telehealth—particularly the application of smartphones in health care—is extremely broad and covers all specialties of medicine, from the family doctor taking care of chronic conditions such as diabetes and heart disease, to the cardiologist, ophthalmologist, dermatologist, and psychiatrist (Krohn et al., 2016). Wearable systems are used in areas ranging from *telehealth care*, *telemedicine*, *telecare*, *telehomecare*, *eHealth*, *p-health*, *m-Health*, *assistive technology*, or *gerontechnology* (Chan et al., 2012). Data transmission via wireless-body communication networks enables patient monitoring by health care providers that can be alerted as soon as a dangerous event occurs which can be taken care of by the medical personnel. If cardiac monitoring is not regular, arrhythmias can occur unpredictably without warning, causing fatal damage. Critical abnormalities of cardiac rhythm such as atrial fibrillation can be detected, recorded, and rapidly treated (Andreu-Perez et al., 2015).

Considering the amount of data generated by the IoT-based wearable devices, it is practically beyond a human scope to single-handedly monitor and analyze the lifestyle data. To facilitate this process, minimal human intervention is obvious, in which the IoT-devices would be self-sufficient in recording the lifestyle data and forwarding for analysis in the cloud or at the edge. This counts a special case of intelligence that is based on a specified logic; the lifestyle data can be analyzed to generate summarized reports at intervals. These reports can be analyzed by the machine itself or by a doctor. However, it is possible to embed the logic into the machine that can analyze the report and will report an alarm, only when the data seem beyond the safe range. In practical cases of health care, which is largely dependent on lifestyle data, automated analysis of lifestyle data counts very efficient. Due to its continuous analysis, many diseases at its very inception get detected because of its significant change in this lifestyle data.

21.6 Conclusion

IoT in health care has made possible to extend the reaches of health care beyond the domains of a hospital. The digitization of health care can eventually help us build a markedly improved physician–patient relationship, allowing greater time for interaction when a patient requires the care of a physician (Steinhubl and Topol, 2015). As the era of big data per individual comes into play, with terabytes of biological, anatomic, physiological, and environmental data becoming fully integrated, humans will no longer be capable of processing the information. This requires synergistic interaction between man and machine, which ultimately will transform medicine into a digitized data science and, in the process, markedly improve health care (Steinhubl and Topol, 2015).

The rapid development of the hardware and software involved in the new generation of wireless cardiac-monitoring devices has outpaced the real-world validation; moreover, large-scale, pragmatic studies are needed to validate the enormous amounts of data generated from these monitors. Ongoing clinical trials will be critical to determine the safety, efficacy, and cost effectiveness of this new technology relative to conventional methods of monitoring patients. Moreover, a much greater understanding of individual variability in the acceptance, engagement, and sustainability of these technologies and the most appropriate balance of patient and provider involvement are critically important areas of study (Walsh et al., 2014).

Security (Zhao and Ge, 2013) is a devil in wild, and IoT is no stranger to this world. As in any other connected device, there is a need to attack this problem in an integral way, protecting data from the source to the destination, and protecting devices from unauthorized access and tampering. This is not only a networking but also an IoT architectural problem, with many current solutions, but with more open problems to deal with.

However, the future direction of deployment of IoT in health care opens up several challenges, and aspects such as security risks, scalability, management, applications, interoperability, connectivity, enhanced networks, and energy constraints have become the most relevant drivers of the new wave.

References

Andreu-Perez, J., Leff, D. R., Ip, H. M. D., and Yang, G. Z. (2015). From wearable sensors to smart implants-toward pervasive and personalized healthcare. *IEEE Transactions on Bio-Medical Engineering*, 62(12), 2750–2762.

Armentano, R. and Kun, L. (2014). Multidisciplinary, holistic and patient specific approach to follow up elderly adults. *Health and Technology*, 4(2), 95–100.

Baig, M. M., Gholam Hosseini, H., and Connolly, M. J. (2015). Mobile healthcare applications: System design review, critical issues and challenges. *Australasian Physical & Engineering Sciences in Medicine*, 38(1), 23–38.

Bain, P. G. (2007). Tremor. *Parkinsonism & Related Disorders*, 13, 369–374.

Budzianowska, A. and Honczarenko, K. (2008). Assessment of rest tremor in Parkinson's disease. *Neurologia i Neurochirurgia Polska—Journal*, 42(1), 12–21.

Chan, M., Estéve, D., Fourniols, J. Y., Escriba, C., and Campo, E. (2012). Smart wearable systems: Current status and future challenges. *Artificial Intelligence in Medicine*, 56(3), 137–156.

Chen, C., Knoll, A., Wichmann, H. E., and Horsch, A. (2013). A review of three-layer wireless body sensor network systems in healthcare for continuous monitoring. *Journal of Modern Internet of Things*, 2(3), 24–34.

Da Xu, L., He, W., and Li, S. (2014). Internet of things in industries: A survey. *IEEE Transactions on Industrial Informatics*, 10(4), 2233–2243.

DeLong, M. R. and Wichmann, T. (2007). Circuit and circuit disorders of the basal ganglia. *Archives of Neurology*, 64(1), 20–24.

Fierro, G., Silveira, F., and Armentano, R. (2016). Central blood pressure monitoring method oriented to wearable devices. *Health and Technology 6(3)*. Retrieved from https://www.researchgate.net/publication/309417698_Central_blood_pressure_monitoring_method_oriented_to_wearable_devices

Foerster, F., Smeja, M., and Fahrenberg, J. (1999). Detection of posture and motion by accelerometry: A validation study in ambulatory monitoring. *Computers in Human Behavior*, 15(5), 571–586.

Forman, D. E., Rich, M. W., Alexander, K. P., Zieman, S., Maurer, M. S., Najjar, S. S., Cleveland, J. C. et al. (2011). Cardiac care for older adults time for a new paradigm. *Journal of the American College of Cardiology*, 57(18), 1801–1810.

Ishaq, I., Carels, D., Teklemariam, G. K., Hoebeke, J., Abeele, F. V. D., Poorter, E. D., Moerman, I. et al. (2013). IETF standardization in the field of the Internet of things (IoT): A survey. *Journal of Sensor and Actuator Networks*, 2(2), 235–287.

Jankovic, J. (2008). Parkinson's disease: Clinical features and diagnosis. *Journal of Neurology, Neurosurgery, and Psychiatry*, 79(4), 368–376.

Kakria, P., Tripathi, N. K., and Kitipawang, P. (2015). A real-time health monitoring system for remote cardiac patients using smartphone and wearable sensors. *International Journal of Telemedicine and Applications*, 2015, e373474.

Krohn, R., Metcalf, D., and Salber, P. (Eds.) (2016). *Health-e Everything: Wearables and the Internet of Things for Health*. Merritt Island, FL: DM2 Research and Design.

Laurent, S., Cockcroft, J., Van Bortel, L., Boutouyrie, P., Giannattasio, C., Hayoz, D., Struijker-Boudier, H. (2006). Expert consensus document on arterial stiffness: Methodological issues and clinical applications. *European Heart Journal*, 27(21), 2588–2605.

Lorincz, K., Chen, B. R., Challen, G. W., Chowdhury, A. R., Patel, S., Bonato, P., and Welsh, M. (2009). Mercury: A wearable sensor network platform for high-fidelity motion analysis. *Proceedings of the 7th ACM Conference on Embedded Networked Sensor Systems-SenSys'09*, 183–196.

Mainetti, L., Patrono, L., and Vilei, A. (2011). Evolution of wireless sensor networks towards the Internet of things: A survey. In *Software, Telecommunications and Computer Networks (SoftCOM), 2011 19th International Conference* (pp. 1–6). IEEE.

Marsden, C. D. (1994). Parkinson's disease. *Journal of Neurology, Neurosurgery and Pstchiatry*, 57(6), 672–781.

Martínez, B., Vilajosana, I., and Montón, M. (2015). *Exploiting Spatio-Temporal Correlations for Energy Management Policies*. Universitat Autònoma de Barcelona. Departament de Microelectrònica i Sistemes Electrònics. Barcelona, Spain.

Martinez, B., Vilajosana, X., Vilajosana, I., and Dohler, M. (2015). Lean sensing: Exploiting contextual information for most energy-efficient sensing. *IEEE Transactions on Industrial Informatics*, 11(5), 1156–1165.

Mesco, B. (2016). *Artificial Intelligence will Redesign Healthcare*. Retrieved from http://medicalfuturist.com/artificial-intelligence-will-redesign-healthcare/

Mukkamala, R., Hahn, J.-O., Inan, O. T., Mestha, L. K., Kim, C.-S., Töreyin, H., and Kyal, S. (2015). Towards ubiquitous blood pressure monitoring via pulse transit time: Theory and practice. *IEEE Transactions on Bio-Medical Engineering*, 62(8), 1879–1901.

Nichols, W., O'Rourke, M., and Vlachopoulos, C. (Eds.) (2011). *McDonald's Blood Flow in Arteries, Sixth Edition: Theoretical, Experimental and Clinical Principles* (6th ed.). Boca Raton, FL: CRC Press.

Noh, S., Park, K. S., Chung, T. J., Yoon, H. N., Yoon, C., Hyun, E., and Kim, H. C. (2014). Ferroelectret film-based patch-type sensor for continuous blood pressure monitoring. *The Institution of Engineering and Technology*, 50(3), 143–144.

Palattella, M. R. and Accettura, N. (2015). 6TiSCH D. Dujovne, Ed. Internet-Draft Universidad Diego Portales Intended status: Informational LA. Grieco Expires: September 8, 2015 Politecnico di Bari. University of California Berkeley.

Pasluosta, C. F., Barth, J., Gassner, H., Klucken, J., and Eskoifer, B. M. (2015). Pull test estimation in Parkinson's disease patients using wearable sensor technology. *Conference Proceedings 37th Annual International Conference of the IEEE Engineering in Medicine and Biology Society (EMBC), 2015*, pp. 3109–3112.

Pastorino, M., Arredondo, M. T., Cancela, J., and Guillen, S. (2013). Wearable sensor network for health monitoring: The case of Parkinson's disease. *Journal of Physics: Conference Series*, 450(1), 1–6.

Patel, S., Hughes, R., Huggins, N., Standaert, D., Growdon, J., Dy, J., and Bonato, P. (2008). Using wearable sensors to predict the severity of symptoms and motor complications in late stage Parkinson's disease. *Conference Proceedings 30th Annual International IEEE Engineering in Medicine and Biology Society 2008*, pp. 3686–3689.

Piwek, L., Ellis, D. A., Andrews, S., and Joinson, A. (2016). The rise of consumer health wearables: Promises and barriers. *PLOS Medicine*, 13(2), e1001953.

Rajaraman, V., Jack, D., Adamovich, S. V., Hening, W., Sage, J., and Poizner, H. (2000). A novel quantitative method for 3D measurement of Parkinsonian tremor. *Clinical Neurophysiology*, 111, 338–343.

Romero, L. E., Chatterjee, P., and Armentano, R. L. (2016). An IoT approach for integration of computational intelligence and wearable sensors for Parkinson's disease diagnosis and monitoring. *Health and Technology*, 6(3), 167–172.

Salarian, A., Russmann H., Wider, C., Burkhard, P. R., Vingerhoets, F. J., and Aminian, K. (2007). Quantification of tremor and bradykinesia in Parkinson's disease using a novel ambulatory monitoring system. *IEEE Transactions on Biomedical Engineering*, 54(2), 313–322.

Sethi, K. (2008). Levodopa unresponsive symptoms in Parkinson disease. *Movement Disorders*, 23(3), 521–533.

Steinhubl, S. R. and Topol, E. J. (2015). Moving from digitalization to digitization in cardiovascular care. Why is it important, and what could it mean for patients and providers? *Journal of the American College of Cardiology*, 66(13), 1489–1496.

Thubert, P. (2012). *Objective Function Zero for the Routing Protocol for Low-Power and Lossy Networks* (RPL), RFC 6552. Retrieved from http://www.ietf.org/rfc/rfc6552.txt

Walsh, J. A., Topol, E. J., and Steinhubl, S. R. (2014). Novel wireless devices for cardiac monitoring. *Circulation*, 130(7), 573–581.

Zhang, G., Gao, M., Xu, D., Olivier, N. B., and Mukkamala, R. (2011). Pulse arrival time is not an adequate surrogate for pulse transit time as a marker of blood pressure. *Journal of Applied Physiology*, 111(6), 1681–1686.

Zhao, K. and Ge, L. (2013). A survey on the Internet of things security. In *Computational Intelligence and Security (CIS), 2013 9th International Conference*, pp. 663–667. IEEE.

Index

Note: Page numbers followed by f and t refer to figures and tables, respectively.